TABLES OF STANDARD ELECTRODE POTENTIALS

Project of the Electrochemistry Commission of
The International Union of Pure and Applied Chemistry

TABLES OF STANDARD ELECTRODE POTENTIALS

Guilio Milazzo
and
Sergio Caroli
Istituto Superiore di Sanitá, Roma, Italy

with the cooperation of
V. K. Sharma
Department of Chemistry,
University of Rajasthan, Jaipur, India

Project of the IUPAC Electrochemistry Commission

A Wiley–Interscience Publication

JOHN WILEY & SONS
Chichester · New York · Brisbane · Toronto

Library of Congress Cataloging in Publication Data

Milazzo, Guilio.
Tables of standard electrode potentials.

'A Wiley–Interscience publication.'
Includes bibliographical references and index.
1. Electrodes – Tables. 2. Electrochemistry – Tables, etc. I. Caroli, Sergio, joint author. II. Sharma, V. K. III. Title.
QD571.M47 541′.3724′0212 77-8111

ISBN 0 471 99534 7

Printed in the United States of America.

Contents

Introduction vii
How to Read the Tables xi
Keys and Explanatory Notes for Tables xv

GROUP I
Lithium 2
Sodium 6
Potassium 10
Rubidium 14
Cesium 18
Copper 22
Silver 36
Gold 54

GROUP II
Beryllium 62
Magnesium 66
Calcium 70
Strontium 80
Barium 84
Radium 90
Zinc 94
Cadmium 100
Mercury 106

GROUP III
Boron 114
Aluminum 120
Gallium 124
Indium 128
Thallium 132
Scandium 138
Yttrium 142

GROUP IV
Carbon 146
Silicon 152
Germanium 156
Tin 160
Lead 164
Titanium 174
Zirconium 180
Hafnium 184

GROUP V
Nitrogen 188
Phosphorus 194
Arsenic 202
Antimony 206
Bismuth 210
Vanadium 214
Niobium 220
Tantalum 224

GROUP VI
Oxygen 228
Sulfur 232
Selenium 244
Tellurium 248
Polonium 254
Chromium 258
Molybdenum 264
Tungsten 268

GROUP VII

Fluorine 274
Chlorine 278
Bromine 284
Iodine 290
Astatine 298
Manganese 302
Technetium 310
Rhenium 314

GROUP VIII

Iron 320
Cobalt 336
Nickel 344
Ruthenium 350
Rhodium 356
Palladium 360
Osmium 366
Iridium 372
Platinum 376

LANTHANIDES

Lanthanum 386
Cerium 386
Praseodymium 388
Neodymium 388
Promethium 390
Samarium 390
Europium 390
Gadolinium 390
Terbium 390
Dysprosium 390
Holmium 392
Erbium 392
Thulium 392
Ytterbium 392
Lutetium 392

ACTINIDES

Actinium 398
Thorium 398
Protoactinium 398
Uranium 398
Neptunium 400
Plutonium 402
Americium 402
Curium 404
Berkelium 404
Californium 404
Fermium 404
Nobelium 404

APPENDIX 409

INDEX OF TABLES 421

Introduction

The idea to collect all existing data on electrode potentials came to the mind of one of the authors (G.M.) when, needing the numerical value of a certain redox couple, he was unable to find it in any of the many tables published so far, and was therefore compelled to spend much time in searching for the value he needed in *Chemical Abstracts*. At the same time he realized once more that all available tables always reported the values of more or less the same electrode systems; for example Ag | AgCl | Cl^- or the calomel electrode, known practically by heart by every chemist.

All tables consulted constantly carried titles such as *Selected Tables* or *Critical Tables* or some other similar heading.

The reactions of this author were then:

1. Why collect always the values of the same electrode systems differing from one another by perhaps only a few tenths of millivolt (a difference which is negligible in common practice) and ignore a large number of investigated couples which can be of interest in many other cases?
2. What can be the unambiguous *logical* basis for the decision to accept in tables of so called 'critically selected' values the value given by a certain author for a particular electrode system in preference to all others published? For example, the Ag | AgCl | Cl^- electrode has been thoroughly investigated by Bates, Ives, Janz, and many others. Each paper of each one of these authors is scientifically an excellent paper giving all possible experimental details and a very good logical analysis of the errors. Nevertheless the results each author arrives at are somewhat different. On what logical basis should the value found, for example, by Bates be preferred to the one found by Ives, or vice versa?
3. On what *logical* basis should the value found by a less known investigator for a less familiar electrode system, such as lanthanide or actinide couple, be ignored? Even if a paper containing numerical data for less familiar systems investigated by less known scientists, were not, as a whole, as good as those quoted in point 2; these data, because of their usefulness (at least in first approximation) in specific cases still deserve inclusion in the collections of numerical data. Nevertheless they are practically never found in the collections of so-called 'selected values'.

Then, in view of the practical difficulty, if not of the impossibility, of giving really 'critically selected values' (i.e. *apparently* more reliable values), and in consideration also of the enormous amount of work and time required even for superficial criticism of all the papers read in order to compile these tables (more than 10,000), the decision was taken not to try any critical evaluation of the papers read and of the values given therein, unless they were evidently unreliable or they were found in a very incomplete abstract, the original paper being unavailable. In other words the present authors followed the philosophy that 'it is better to know something than nothing' and adopted the practice of including in the tables all electrode systems and values found, accompanied by any useful information given in the original paper, and by the bibliographic reference, leaving it up to the reader to judge how far the value is reliable and how it answers his particular scientific needs.

At the time this work was started, one of the authors (G.M.) was a member of the Commission on Electrochemistry in the Division of Physical Chemistry of the International Union of Pure and Applied Chemistry (IUPAC). The idea of preparing this kind of table was brought up for discussion with the Colleagues on the Commission and it was accepted as one of its working projects.

Collecting all papers related to the subject and picking up the information required from all papers read, as well as the efforts to arrange such information into a single suitable scheme, as compact and clear as possible, took many years of painstaking work and many discussions at each meeting of the Commission. Criticism and suggestions brought forward at the end of each meeting, not only by colleagues in the Commission itself, but also by authoritative electrochemists outside the Commission, were taken into most serious consideration; their contributions led to consistent improvements in the usefulness of the tables. These tables were then presented in their near final form at the Brighton Meeting of the Commission held on September 20th and 21st, 1974. At the end of the meeting the statement, reproduced herewith in facsimile, was approved.

These tables collect all data published and abstracted in *Chemical Abstracts* from 1945 up to the end of 1973. An appendix contains some new data abstracted from the end of 1973 to the end of 1975 in *Chemical Abstracts* and in other relevant abstracting periodicals (*Analytical Abstracts* and *Electroanalytical Abstracts*).

The aim in presenting these tables is hence to offer a compilation of standard electrode potentials as complete as possible for all the elements, their compounds, and complexes.

By far the most important sources were original papers. A rather limited number of values, however, was taken from existing special compilations such as *The Standard Electrode Potentials and Temperature Coefficients at 25 °C* by A. J. de Bethune and N. A. Swendeman-Loud and the *Atlas d'Equilibres Electrochimiques* by M. Pourbaix, in which practically only standard potential values for some given systems are reported, without any indication of experimental conditions. A certain number of data taken from these special compilations are even calculated data. Finally, a very small number of data were calculated for the first time, following standard procedures, on the basis of other thermodynamically significant data (as in the case of equilibrium constants of complexes).

Not only data obtained using potentiometric procedures were collected, but also those derived using other electrochemical techniques (for example voltammetry), and even those obtained by non-electrochemical ones (for example spectrophotometry).

Some electrode and system compositions may appear strange or unusual, or even impossible. In such cases attention is drawn to the quoted method used to obtain the figure reported for the standard potential, which is in fact calculated theoretically (or obtained using very special procedures) concerning the hypothetical system considered. For instance, the standard potential given for the system $Na \mid (NaH_2P_2O_7)^-$ does not

International Union of Pure and Applied Chemistry

Division of Physical Chemistry
Commission on Electrochemistry

STATEMENT

Since 1967 the Commission on Electrochemistry of the IUPAC has followed with interest the important compilation on Standard Electrode Potentials carried out by Professor G. Milazzo within the frame of the activities of the Commission.

The author has taken into consideration the suggestions and the criticism which were presented on the occasion of the various meetings: Prague (1967), Cortina d'Ampezzo (1969), Paris (1970), Washington (1971), Munich (1973) and Brighton (1974).

The following set of extensive tables will certainly prove very useful for specialists in this field, because these data are widely scattered in the literature. In order to make these tables as comprehensive as possible Professor Milazzo has made no attempt to make a critical selection, but users can readily search in the numerous original contributions referenced for further information.

In view of this, the Commission encourages the publication of this work.

Brighton, Sept. 21st, 1974 The Chairman of the Commission

(Prof. R. Haase)

mean that it was really obtained by dipping a barrel of metallic sodium in a partially ionized acid pyrophosphate solution; it means, on the contrary, that this value was calculated theoretically for the given electrode reaction.

The term *potential* was adopted instead of *electromotive force*, since the latter (as usually employed by a number of electrochemists) is a thermodynamic quantity, not directly measurable, whereas the former is actually the quantity experimentally determined (which should be called much more correctly *electric tension*).[1]

The electrode reaction is always written in the sense of a reduction, according to the general scheme

$$\mathrm{Ox} + e^- \rightleftharpoons \mathrm{Red}$$

where Ox symbolizes the whole system in the oxidized state, while Red stands for the same system in the reduced state.

The convention used to obtain the sign of the potential is that of the *series of phases* in the galvanic half-element. This convention finally gives the same sign as results from the so-called *European Convention*, which in turn coincides also with the sign arising from the *Stockholm Convention*, but is more logical and of easier practical use.[2]

We hope that this set of extensive Tables will prove useful not only for specialists, but also for non-specialists in the field. The exhaustive character of this compilation together with the high number of references cited will give the reader a tool of practical use, which may encourage him and facilitate his task of retrieving in literature any couple of interest to him.

The achievement of such a goal will be for the authors of this compilation a more than satisfactory result.

Acknowledgements

The financial support granted by the (Italian) Consiglio Nazionale delle Ricerche for this compilation is gratefully acknowledged.

1. G. Milazzo, *Anal. Quim.*, **70** (1974) 1214.
2. For more theoretical details on this point and on the practical advantages, see G. Milazzo, *Lehrbuch der Elektrochemie*, Vol. 1. Chap. III in course of publication, Birkhäuser Verlag, Basel.

How to Read the Tables

The Tables are arranged by element in the order of the Mendeleev periodic table. As for Group I, for example, the elements Li, Na, K, Rb, Cs, Fr are followed by Cu, Ag, Au. Elements belonging to the lanthanides and actinides series (from La to Lu and from Ac to No, respectively) are collected in two special tables reported at the end of the elements of Group VIII. For each element an extended table, a reduced table, and tables with notes and bibliography are given. In the extended tables the different electrode-systems are arranged alphabetically according to the right-hand term of the electrode reaction. As a rule, inorganic systems are listed first, followed by systems involving at least one organic component. When a certain electrode-system has been investigated in different media, the standard potential value in water is given first and values in non-aqueous solvents follow. On the other hand, the reduced tables contain for each element only the standard potential values in order of increasing values, together with a reference number for the retrieval of the same electrode system in the extended tables. In these reduced tables, standard potential values for aqueous systems are listed first, followed by those in pure non-aqueous solvents.

The extended Tables contain the following information:

1. **Electrochemically conducting phase**: i.e. what is generally called the *electrode* in a restricted sense. Sometimes it is a composite object, for instance a *glass electrode*. The term Ind. stands for *indifferent* electrode, i.e. not participating chemically in the electrode reaction.
2. **Intermediate species**: The non-ionized or possibly not solute species are given under this heading. For example, AgCl in the Ag | AgCl | Cl^- electrode.
3. **Composition of the solution**: It is obviously reported only when given in the original paper. It is missing for values calculated on the basis of the thermodynamic quantities. The symbols M, m, and N. stand for moles/kg solvent, moles/liter solution, and equivalents/liter solution, respectively.
4. **Solvent**: This is indicated by an Arabic numeral. The complete list of the solvents with the corresponding numeral is given at the end of this introductory section.
5. **Temperature**: This is given in degrees Celsius. The uncertainty is reported whenever quoted in the original paper.
6. **Pressure**: This is given in Torr. When not otherwise specified, the pressure is the uncorrected atmospheric one.
7. **Measuring method**: This is given as a Roman numeral. The explanatory list is also given at the end of the introduction.

8. **Comparison electrode**: This is indicated by an Arabic numeral; the key is given at the end of the introduction*.
9. **Liquid junction**: This is reported whenever present and explicitly mentioned by the author. Otherwise, no indication is given.
10. **Electrode reaction**: This is written as an equilibrium reaction in the sense of a reduction. In the presence of a complex, the ligand, L, is always written in rouna brackets.
11. **Standard value**: Both for water and pure non-aqueous solvent, the standard value of the potential (in V) is referred to the standard hydrogen electrode in the same solvent. When more than a single value was available in the literature for a certain electrode, only that originating from the most complete set of experimental data is given. If the other values are in the same range of uncertainty their average is reported; if not, notice of all of them is given.
12. **Uncertainty**: This is given (in mV) whenever quoted in the original paper for the standard potential reported.
13. **Temperature coefficient**: This is obtained using isothermal cells and includes therefore the individual temperature coefficient of the potential of the standard hydrogen electrode. The temperature coefficients are reported in $\mu V/^{\circ}C$ and are considered positive in sign if an increase in temperature causes an increase in the standard potential value. Considering that the individual temperature coefficient of the potential of the standard hydrogen electrode (i.e. the so-called *non--isothermal* temperature coefficient of the electrode potential) is +859 $\mu V/^{\circ}C$, it is easy to obtain the individual (non-isothermal) temperature coefficient of the standard potential simply by adding this value to that quoted in this volume.[1]
14. **Notes**: Arabic numerals in this column refer to the list given at the end of the tables for each element. Therein useful details on the experimental conditions are given under which the standard potential values were determined, further information on the method employed, etc.
15. **References**: This column gives the bibliographic reference of the papers from which the quoted values were taken. The Arabic numerals refer to the bibliography list given at the end of each group of tables for each element.
16. **Electrode reference number**: This last column contains progressive numbers by which each electrode system is designated.

Each alphabetical table for a single element is followed, as already mentioned, by another condensed table for the same element. In the reduced tables the standard

*Attention is drawn to the fact that whenever concentrations in the composition of the comparison electrode quoted are missing, they were missing in the original paper and the numerical value of the standard potential of the electrode under consideration (vs. the hydrogen standard electrode) were given directly by the author(s) of the paper.

1. For a deeper understanding of this very important point, see G. Milazzo, Ref. 2 of Introduction, or for more details the series of papers by G. Milazzo *et al.*, *J. Electroanal. Chem.*, **2** (1961) 419; *Z. Phys. Chem. N.F.*, **52** (1967) 293; **54** (1967) 1, 13, 27; *J. Res. Inst. Catalysis Hokkaido Univ.*, **16** (1968) 387; *Z. Phys. Chem. N.F.*, **62** (1968) 47; **68** (1969) 250; **76** (1971) 127; **79** (1972) 41; *Anal. Quim.*, **71** (1975) 1033; *Electrochim. Acta*, **21** (1976) 349.

potentials of the electrode systems of each element considered are arranged according to the increasing values, listing aqueous systems first and then those in non-aqueous pure solvents. These condensed tables consist of three columns, with the headings 'Standard Values (V)', 'Solvent', and 'Electrode Reference Number'. This last indication leads the reader, who selected a system on the basis of the value of its standard potential in a given solvent, to locate it immediately in the alphabetical tables and to find all other information requested for the same system.

As already mentioned, at the end of this double series of tables for each element, explanatory notes as well as the related bibliography are given.

Keys and Explanatory Notes for Tables

(a) Ind = indifferent electrode, not participating chemically in the electrode reaction; composite terminal electrodes are given by name.
(b) Species not ionized, or not in solution.
(c) m = mol/kg; M = mol/liter.
(d) See list.
(e) When the value of the pressure is not specifically given, the pressure is the uncorrected atmospheric one.
(f) When more than one reference is given, the standard potential is the mean value of the potentials given by each author, all in the same range of uncertainty.
(g) The values include the temperature coefficient of the potential of the standard hydrogen electrode.
(h) Notes at the end of each group of tables.

Numerical Key for Solvents

1. Water
2. Ammonia
3. Hydrazine
4. Ethane bromide
5. Nitromethane
6. Ethylenediamine
7. Methyl alcohol (methanol)
8. Ethyl alcohol (ethanol)
9. Isopropyl alcohol (isopropanol)
10. n-Propyl alcohol
11. n-Butyl alcohol
12. n-Pentyl alcohol (n-Amyl alcohol)
13. Ethylene glycol
14. Propylene glycol
15. Acetone
16. Formic acid
17. Acetic acid
18. Formamide
19. *N*-Methylformamide
20. *N,N*-Dimethylformamide (DMF)
21. *N*-Methylacetamide
22. *N,N*-Dimethylacetamide
23. Acetonitrile
24. Dimethylsulfoxide
25. Toluene
26. Nitrobenzene
27. Pyridine
28. Hydrofluoric acid
29. Propylene carbonate
30. 1,2-Dimethoxyethane
31. Tetrahydrofurane
32. Glycerol
33. Trifluoroacetic acid
34. Acetamide
35. Deuterium oxide (heavy water)
36. *N*-Methylpyrrolidone
37. n-Hexanol
38. Phosphorous oxychloride
39. Benzonitrile
40. Phenylphosphoxydifluoride
41. Ethylacetate
42. Ethylpropionate
43. Phenylphosphoxydichloride
44. Diethylether
45. Trimethylphosphate
46. Hexamethylphosphoramide

Measuring Methods

(I) Potentiometric.
(II) Polarographic.
(III) Other: voltammetric, conductometric, rotating disk, etc.
(IV) Calculated.
(V) Non-electrochemical method.

Numerical Key for Electrodes*

(1) $Pt-H_2$
(2) N.C.E. (Normal Calomel Electrode)
(3) S.C.E. (Saturated Calomel Electrode)
(4) Ag, AgCl
(5) Ag, AgBr
(6) $Ag \mid Ag^+$ (0.1 M)
(7) $Ag \mid Ag^+$ (0.01 M)
(8) $Ag-AgCl_2(s) \mid LiCl$
(9) $Ag \mid Ag_2O$
(10) $Ag \mid Ag_4[Fe(CN)_6]$
(11) $Ag, Ag_2MoO_4 \left| \begin{matrix} AgNO_3 \\ NaNO_3 \end{matrix} \right.$
(12) $Ag, Ag_2WO_4 \left| \begin{matrix} AgNO_3 \\ NaNO_3 \end{matrix} \right.$
(13) $Ag \mid Ag_2$ (oxalate)
(14) Ag | Ag picrate 0.01 N
(15) $Cu-H_2$
(16) Cu | CuCl
(17) $Cu \mid CuSO_4$ (0.01 M)
(18) Hg, Hg_2Br_2
(19) $Hg \mid Hg_2SO_4, H_2SO_4$ (1 M)
(20) $Hg \mid Hg_x^{2x+}$ (0.01 M)
(21) $(HCOO)_2$ Hg (checked vs. S.C.E.)
(22) Hg | HgO
(23) $Hg \mid (Hg_2)_3[Co(CN)_6]_2$
(24) $Zn(Hg) \mid ZnCl_2$
(25) $Cd \mid CdCl_2$, KCl (satd)
(26) $Cd(Hg) \mid Cd^{2+}$
(27) Tl(Hg)(l)–TlCl(s)
(28) Pb
(29) $Pb \mid PbNO_3$ (0.1 N in NH_3)
(30) Pb(5% wt in Hg) | $Pb_2[Fe(CN)_6]$
(31) $Pb(Hg), PbSO_4 \left| \begin{matrix} K_2SO_4 \\ KNO_3 \end{matrix} \right.$
(32) Na(Hg) | NaF
(33) Pt | Ferricinium / Ferrocene
(34) Glass electrode
(35) Ag, AgCl | $AgClO_4$ (0.01 m)
(36) Fe
(37) Quinhydrone
(38) $Pt-D_2$
(39) K_2SO_4 (satd), Hg_2SO_4 (satd)
(40) Hg pool

*See footnote on page xii.

Group I

Electronically conducting phase	Intermediate species	Composition of the solution	Solvent	Temperature °C	Pressure Torr	Measuring method
a	b	c	d		e	d
Li(Hg)		LiOH: 0.026 up to 2.669 m	1	25 ± 0.02		I
Li			7	25		I
Li(Hg)		Li: 0.035 % wt in Hg; LiHCOO: 0.01 N	16	25 ± 0.01		I
Li		LiCl: 0.01 N	23	25		I
Li		LiCl: 0.01 N	24	25		I
Li		LiCl or $LiClO_4$	29	25		I
Li		$LiNO_3$	8	25		V
Li	LiH		1	25		IV
Ind.	LiH		1	25		IV
Pt, H_2		Li (glycolate)$^-$: 0.003765 up to 0.06515 m; HCl: 0.0001958 up to 0.03388 m	1	25		I
Pt, H_2		Li (lactate)$^-$: 0.004724 up to 0.09660 m; HCl: 0.002367 up to 0.04827 m	1	25		I

Comparison electrode	Liquid junction	Electrode reaction	Standard value V	Uncertainty mV	Temperature Coefficient μV/°C	Notes	Reference	Electrode Reference Number
d			f		g	h		
1	no	$Li^{+} + e^{-} \rightleftharpoons Li$	-3.0401	±1.6	-534		1,11	1
		$Li^{+} + e^{-} \rightleftharpoons Li$	-3.095				2	2
14	no	$Li^{+} + e^{-} \rightleftharpoons Li$	-3.48	±2	850		3	3
7	no	$Li^{+} + e^{-} \rightleftharpoons Li$	-3.23		750		4	4
27	no	$Li^{+} + e^{-} \rightleftharpoons Li$	-2.4212			3	5	5
1		$Li^{+} + e^{-} \rightleftharpoons Li$	-2.8874		877		6	6
1		$Li\,(H_2O)^{+} + e^{-} \rightleftharpoons Li + (H_2O)$	-3.022			2	7	7
		$Li + H^{+} + e^{-} \rightleftharpoons LiH$	0.726				8	8
		$Li^{+} + H^{+} + 2e^{-} \rightleftharpoons LiH$	-1.161				8	9
4		$Li\,(\text{glycolate})^{-} + e^{-} \rightleftharpoons Li + (\text{glycolate})^{2-}$	-3.033			1	9	10
4		$Li\,(\text{lactate})^{-} + e^{-} \rightleftharpoons Li + (\text{lactate})^{2-}$	-3.052			1	10	11

Standard value V	Solvent	Electrode Reference Number	Standard value V	Solvent	Electrode Reference Number	Standard value V	Solvent	Electrode Reference Number
-3.052	1	11						
-3.0401	1	1						
-3.033	1	10						
-1.161	1	9						
0.726	1	8						
-3.095	7	2						
-3.022	8	7						
-3.48	16	3						
-3.23	23	4						
-2.4212	24	5						
-2.8874	29	6						

NOTES : Lithium

1) Potentiometric pH measurements

2) Spectrophotometry data

3) T^{o} ranges from 25 to 35 ^{o}C

BIBLIOGRAPHY : Lithium

1) R. Huston, J.N. Butler, J. Phys. Chem., 72 (1968), 4263 - 4264

2) B. Jakuszewski, S. Taniewska-Osinka, Acta Chim. Soc. Sci. Łodz, 4 (1959), 17

3) V.A. Pleskov, Acta Physicochim. U.S.S.R., 21 (1946), 41 - 54; J. Phys. Chem. U.S.S.R., 20 (1946), 153 - 162

4) V.A. Pleskov, Acta Physicochim. U.S.S.R., 22 (1948), 351 - 364

5) W.H. Smyrl, Dissertation Abstracts, 28 B (1967), 637, Univ. Microfilms, Order n. 67-8655

6) D.P. Boden, Dissertation Abstracts, 31 B (1970), 3223

7) J. Bjerrum, C.K. Jörgensen, Acta Chem. Scand., 7 (1953), 951

8) Data reported by M. Pourbaix "Atlas d'Equilibres Electrochimiques à 25 ^{o}C" (Gauthier-Villars Editeurs, Paris, 1963), pp. 122 - 133

9) P.B. Davies, C.B. Monk, Trans. Faraday Soc., 50 (1954), 128 - 132

10) P.B. Davies, C.B. Monk, Trans. Faraday Soc., 50 (1954), 132 - 136

11) Data reported by A.J. De Bethune, N.A. Swendeman-Loud "Standard Aqueous Electrode Potentials and Temperature Coefficients at 25 ^{o}C" (Clifford A. Hampel, Ill.)

Electronically conducting phase	Intermediate species	Composition of the solution	Solvent	Temperature °C	Pressure Torr	Measuring method
a	b	c	d		e	d
Na(Hg)			1	25		I
Na		NaCl: 0.19 M; $AlBr_3$: 0.24 M	4	25		I
Na			7	25		I
Na(Hg)		Na: 0.205 % wt in Hg; NaHCOO: 0.01 N	16	25 ± 0.01		I
Na		NaI: 0.01 N	23	25		I
Na		NaCl: 0.19 N; $AlBr_3$: 0.24 M	26	25		I
Na	Na (BrO_3)		1	25		III
Na			1	25		III
Pt, H_2	Na (OH)	1) NaOH: 0.01 N; NaBr 2) NaOH: 0.01 N; NaCl	1	25		I
Na		$Na_4P_2O_7$: (0.523 up to 2.340) · 10^{-3} M; HCl: (0.256 up to 1.170) · 10^{-3} N	1	25 ± 0.01		III
Na			1	25		III
Na		$Na_5P_3O_{10}$: (0.1941 up to 2.749) · 10^{-3} M	1	25 ± 0.01		III
Na			1	25		III
Na		1) $Na_2S_4O_3$: 1.014 · 10^{-3} M; NaCl: 0.01 up to 0.2 M 2) $Na_2S_2O_3$: (5.22 up to 26.12) · 10^{-3} M; HCl: (0.245 up to 0.461) M	1	25		V
Na	NaH		1	25	760	IV
Ind.	NaH		1	25	760	IV

Comparison electrode	Liquid junction	Electrode reaction	Standard value	Uncertainty	Temperature Coefficient	Notes	Reference	Electrode Reference Number
			V	mV	μV/°C			
d			f		g	h		
		$Na^+ + e^- \rightleftharpoons Na$	-2.71		-772		1,18	1
3	AgBr(s)	$Na^+ + e^- \rightleftharpoons Na$	-2.45			1	2	2
		$Na^+ + e^- \rightleftharpoons Na$	-2.728				3	3
14	no	$Na^+ + e^- \rightleftharpoons Na$	-3.42	± 10	1000		4	4
7	no	$Na^+ + e^- \rightleftharpoons Na$	-2.87		-700		5	5
3	AgBr(s)	$Na^+ + e^- \rightleftharpoons Na$	-2.545			1	6	6
1		$Na(BrO_3) + e^- \rightleftharpoons Na + (BrO_3)^-$	-2.692			2	7	7
1		$Na(HP_2O_7)^{2-} \rightleftharpoons Na + (HP_2O_7)^{3-}$	-3.610			2	8	8
1) 5		$Na(OH) + e^- \rightleftharpoons Na + (OH)^-$	-2.670		750	5	9	9
2) 4						4	10	
1		$Na(P_2O_7)^{3-} + e^- \rightleftharpoons Na + (P_2O_7)^{4-}$	-2.849			2	11	10
1		$Na(P_3O_9)^{2-} + e^- \rightleftharpoons Na + (P_3O_9)^{3-}$	-2.779			2	12	11
1		$Na(P_3O_{10})^{4-} + e^- \rightleftharpoons Na + (P_3O_{10})^{5-}$	-2.862			2	13	12
1		$Na(SO_4)^- + e^- \rightleftharpoons Na + (SO_4)^{2-}$	-2.752			2	14	13
1		$Na(S_2O_3)^- + e^- \rightleftharpoons Na + (S_2O_3)^{2-}$	-2.747		-1000	3	15	14
						5	16	
		$Na + H^+ + e^- \rightleftharpoons NaH$	0.390				17	15
		$Na^+ + H^+ + 2e^- \rightleftharpoons NaH$	-1.162				17	16

Standard value V	Solvent	Electrode Reference Number
-3.610	1	8
-2.862	1	12
-2.849	1	10
-2.779	1	11
-2.752	1	13
-2.747	1	14
-2.71	1	1
-2.692	1	7
-2.670	1	9
-1.162	1	16
0.390	1	15
-2.45	4	2
-2.728	7	3
-3.42	16	4
-2.87	23	5
-2.545	26	6

NOTES : Sodium

1) Solid junction as reported by the authors

2) Conductometric data

3) Solubility and spectrophotometry data

4) Potentiometric pH measurements

5) The value is an average of those given by several authors when they all are in the same range of uncertainty.

BIBLIOGRAPHY : Sodium

1) V.A. Pleskov, Acta Physicochim. U.S.S.R., 21 (1946), 235 - 238; J. Phys. Chem. U.S.S.R., 20 (1946), 153 - 162

2) Ya. F. Mezhennyi, J. Phys. Chem. U.S.S.R., 21 (1947), 839 - 841

3) B. Jakuszewski, S. Taniewska-Osinka, Acta Chim. Soc. Sci.Łodz, 4 (1959), 17

4) V.A. Pleskov, Acta Physicochim. U.S.S.R., 21 (1946), 41 - 54; J. Phys. Chem. U.S.S.R., 20 (1946), 153 - 162

5) V.A. Pleskov, J. Phys. Chem. U.S.S.R., 22 (1948), 351 - 361

6) Ya. F. Mezhennyi, J. Phys. Chem. U.S.S.R., 20 (1946), 493 - 502

7) R.W. Martel, C.A. Kraus, Proc. Nat. Acad. Sci.USA, 41 (1955), 9

8) J.A. Davis, Thesis, Indiana Univ., 1955, Univ. Microfilms, 14650

9) H.S. Harned, W.T. Hamer, J. Am. Chem. Soc., 55 (1933), 4496 - 4507

10) H.S. Harned, G.E. Mannweiler, J. Am. Chem. Soc., 57 (1935), 1873 - 1876

11) C.B. Monk, J. Chem. Soc., 1949, 423 - 427

12) C.W. Davies, C.B. Monk, J. Am. Chem. Soc., 1949, 413

13) C.B. Monk, J. Chem. Soc., 1949, 427 - 429

14) I.L. Jenkins, C.B. Monk, J. Am. Chem. Soc., 72 (1950), 2695

15) F.G.R. Gimblett, C.B. Monk, Trans. Faraday Soc., 47 (1951), 793 - 802

16) T.O. Denney, C.B. Monk, Trans. Faraday Soc., 47 (1951), 992 - 998

17) Data reported by M. Pourbaix "Atlas d'Equilibres Electrochimiques à 25 °C " (Gauthier-Villars Editeurs, Paris, 1963), pp. 122 - 144

18) Data reported by A.J. De Bethune, N.A. Swendeman-Loud : "Standard Aqueous Electrode Potentials and Temperature Coefficients at 25 °C" (Clifford A. Hampel, Ill.)

Electronically conducting phase	Intermediate species	Composition of the solution	Solvent	Temperature °C	Pressure Torr	Measuring method
a	b	c	d		e	d
K(Hg)		K: 0.02 % wt in Hg; KCl	1	25 ± 0.01		I
K		KCl: 0.19 M; $AlBr_3$: 0.24 M	4	25		I
K(Hg)		KCl: 0.01 up to 0.1 M	7	25		I
K(Hg)		K: 0.02 % wt in Hg; KCl: 0.6266 up to 3.1390 M	8	25 ± 0.01		I
K(Hg)		K: 0.2216 % wt in Hg; KCOOH: 0.01 N	16	25 ± 0.01		I
K(Hg)		K: 0.212 % wt in Hg; KCl: 0.00265 up to 0.444 M	18	25 ± 0.1		I
K		KI: 0.01 N	23	25		I
K		KCl	26	25		I
K		KI or $KClO_4$	29	25		I
K	K $(BrO)_3$		1	25		III
K	K (ClO_3)		1	25		III
K	K (ClO_4)	$KClO_4$: 0.00051512 up to 0.10596 M	1	25		III
K		$K_3[Co(CN)_6]$: (0.07534 up to 1.0514) · 10^{-3} N	1	25		III
K		$K_3[Fe(CN)_6]$: (0.010333 up to 0.1747) · 10^{-3} M	1	25		III
K			1	25		III
K	K (IO_3)	KIO_3 : (3.91 up to 0.182) · 10^{-3} M	1	25		III
K	K (NO_3)		1	25		III
K	K (ReO_4)	$KReO_4$: (0.38123 up to 30.705) · 10^{-3} M	1	25 ± 0.001		III
K			1	25		III
K		$K_2S_2O_3$: 1.014 · 10^{-3} M; KCl: 0.01 up to 2.0 M	1	25		V
K	KH		1	25	760	IV
Ind.	KH		1	25	760	IV

Comparison electrode	Liquid junction	Electrode reaction	Standard value	Uncertainty	Temperature Coefficient	Notes	Reference	Electrode Reference Number
			V	mV	μV/°C			
d			f		g	h		
4	no	$K^+ + e^- \rightleftharpoons K$	-2.931	±0.5	-1080		1,25	1
3	AgBr(s)	$K^+ + e^- \rightleftharpoons K$	-2.516			6	2	2
4	no	$K^+ + e^- \rightleftharpoons K$	-2.921			4	3	3
4	no	$K^+ + e^- \rightleftharpoons K$	-2.843				1	4
14	no	$K^+ + e^- \rightleftharpoons K$	-3.36	±1	1200		4	5
25	no	$K^+ + e^- \rightleftharpoons K$	-2.872	±5	-570		5	6
7	no	$K^+ + e^- \rightleftharpoons K$	-2.91		-300		6	7
3	AgBr(s)	$K^+ + e^- \rightleftharpoons K$	-2.804			6	2	8
1		$K^+ + e^- \rightleftharpoons K$	-2.9504		688		7	9
1		$K(BrO_3) + e^- \rightleftharpoons K + BrO_3^-$	-2.895			1	8	10
1		$K(ClO_3) + e^- \rightleftharpoons K + ClO_3^-$	-2.928			1	9	11
1		$K(ClO_4) + e^- \rightleftharpoons K + ClO_4^-$	-2.902			1	10	12
1		$K[Co(CN)_6]^{2-} + e^- \rightleftharpoons K + [Co(CN)_6]^{3-}$	-3.010			1,2	11,12	13
1		$K[Fe(CN)_6]^{2-} + e^- \rightleftharpoons K + [Fe(CN)_6]^{3-}$	-3.008			1,2	12,13	14
1		$K[Fe(CN)_6]^{3-} + e^- \rightleftharpoons K + [Fe(CN)_6]^{4-}$	-3.067			1,2	11,14,15 16	15
1		$K(IO_3) + e^- \rightleftharpoons K + IO_3^-$	-2.917			1	17,18	16
1		$K(NO_3) + e^- \rightleftharpoons K + NO_3^-$	-2.917			1,2	9,19	17
1		$K(ReO_4) + e^- \rightleftharpoons K + ReO_4^-$	-2.974			1	20	18
1		$K(SO_4)^- + e^- \rightleftharpoons K + SO_4^{2-}$	-2.988			1	21	19
1		$K(S_2O_3)^- + e^- \rightleftharpoons K + S_2O_3^{2-}$	-2.987			2,3	22,23	20
		$K + H^+ + e^- \rightleftharpoons KH$	0.386				24	21
		$K^+ + H^+ + 2e^- \rightleftharpoons KH$	-1.270				24	22

Standard value V	Solvent	Electrode Reference Number	Standard value V	Solvent	Electrode Reference Number	Standard value V	Solvent	Electrode Reference Number
-3.067	1	15						
-3.010	1	13						
-3.008	1	14						
-2.988	1	19						
-2.987	1	20						
-2.974	1	18						
-2.931	1	1						
-2.928	1	11						
-2.917	1	17						
-2.917	1	16						
-2.902	1	12						
-2.895	1	10						
-1.270	1	22						
0.386	1	21						
-2.516	4	2						
-2.921	7	3						
-2.843	8	4						
-3.36	16	5						
-2.872	18	6						
-2.91	23	7						
-2.804	26	8						
-2.9504	29	7						

NOTES : Potassium

1) Conductometric data

2) The value is an average of those given by several authors when they all are in the same range of uncertainty

3) Solubility and spectrophotometric data

4) Activity coefficients in the range of 0.650 - 0.873 in the solution range of 0.1 - 0.01 M

5) Potentiometric pH measurements

6) Solid junction as reported by the authors

BIBLIOGRAPHY : Potassium

1) A.J. Dill, L.M. Itzkovits, V. Popovych, J. Phys. Chem., 72 (1968), 4580 - 4586

2) Ya. F. Mezhennyi, J. Phys. Chem. U.S.S.R., 20 (1946), 493 - 502

3) K. Bräuer, H. Strehlow, Z . Physik. Chem., 17 (1958), 346 - 349

4) V.A. Pleskov, J. Phys. Chem. U.S.S.R., 20 (1946), 153 - 162; Acta Physicochim. U.S.S.R., 21 (1946), 41 - 54

5) T. Pavlopoulos, H. Strehlow, Z. Physik. Chem. (N.F.), 2 (1954), 89 - 103

6) V.A. Pleskov, J. Phys. Chem. U.S.S.R., 22 (1948), 351 - 361

7) D.P. Boden, Diss. Abstr., 31 B (1970), 3223

8) J.H. Jones, J. Am. Chem. Soc., 66 (1944), 1115

9) W.H. Banks, E.C. Righellato, C W. Davies, Trans. Faraday Soc., 27 (1931), 621

10) J.H. Jones, J. Am. Chem. Soc., 67 (1945) 855 - 857

11) J.C. James, Thesis, London, 1947

12) J.C. James, C B. Monk, Trans. Faraday Soc., 46 (1950), 1041 - 1050

13) G.S. Hartley, G.W. Donaldson, Trans. Faraday Soc., 33 (1937), 457

14) G. Jones, F.C. Jelen, J. Am. Chem. Soc., 58 (1936), 2561

15) J.C. James, Trans. Faraday Soc., 45 (1949), 855

16) S.R. Cohen, Thesis, Cornell Univ., Ithaca, N.Y., 1956, Univ. Microfilms, 20, 406

17) K.A. Krieger, M. Kilpatrick, J. Am. Chem. Soc., 64 (1942), 7

18) C.B. Monk, J. Am. Chem. Soc., 70 (1948), 3281 - 3283

19) T. Shedlovsky, J. Am Chem Soc., 23 (1927), 351

20) J.H. Jones, J. Am. Chem. Soc., 68 (1946), 240

21) I.L. Jenkins, C.B. Monk, J. Am. Chem. Soc., 72 (1950), 2695

22) T.O. Denney, C.B. Monk, Trans. Faraday Soc., 47 (1951), 992 - 998

23) F.G.R. Gimblett, C.B. Monk, Trans. Faraday Soc., 51 (1955), 793 - 802

24) Data reported by M. Pourbaix "Atlas d'Equilibres Electrochimiques à 25 °C" (Gauthier-Villars Editeur, Paris, 1963), pp. 122 - 133

25) Data reported by A.J. De Bethune, N.A. Swendeman-Loud "Standard Aqueous Electrode Potentials and Temperature Coefficients at 25 °C" (Clifford A. Hampel, Ill.)

Electronically conducting phase	Intermediate species	Composition of the solution	Solvent	Temperature °C	Pressure Torr	Measuring method
a	b	c	d		e	d
Rb (Hg)			1	25		I
Rb (Hg)		Rb: 0.00045 % wt in Hg; RbHCOO : 0.01 N	16	25 ± 0.01		I
Rb (Hg)		Rb: 0.231 % wt in Hg; RbCl: 0.00759 up to 0.3331 N	18	25		I
Rb (Hg)		RbI: 0.01 N	23	25		I
Rb	RbH		1	25	760	IV
Ind.	RbH		1	25	760	IV

Comparison electrode	Liquid junction	Electrode reaction	Standard value V	Uncertainty mV	Temperature Coefficient μV/°C	Notes	Reference	Electrode Reference Number
d			f		g	h		
1		$Rb^+ + e^- \rightleftharpoons Rb$	-2.98		-374		1,6	1
14	no	$Rb^+ + e^- \rightleftharpoons Rb$	-3.45	±0.8	700		2	2
25	no	$Rb^+ + e^- \rightleftharpoons Rb$	-2.99	± 5	-570		3	3
7	no	$Rb^+ + e^- \rightleftharpoons Rb$	-3.17		-400		4	4
		$Rb + H^+ + e^- \rightleftharpoons Rb$	0.317				5	5
		$Rb^+ + H^+ + 2e^- \rightleftharpoons RbH$	-1.304				5	6

Standard value V	Solvent	Electrode Reference Number	Standard value V	Solvent	Electrode Reference Number	Standard value V	Solvent	Electrode Reference Number
-2.98	1	1						
-1.304	1	6						
0.317	1	5						
-3.45	16	2						
-2.99	18	3						
-3.17	23	4						

BIBLIOGRAPHY : Rubidium

1) J. Bockris, J. Haringshow, Uspekhi Khim., 20 (1951), 216

2) V.A. Pleskov, Acta Physicochim. U.S.S.R., 21 (1946), 41 - 54; J. Phys. Chem. U.S.S.R., 20 (1946), 153 - 162

3) T. Pavlopoulos, H. Strehlow, Z. Physik. Chem. (N.F.), 2 (1954), 89 - 103

4) V.A. Pleskov, J. Phys. Chem. U.S.S.R., 22 (1948), 351 - 369

5) Data reported by M. Pourbaix "Atlas d'Equilibres Electrochimiques à 25 °C" (Gauthier-Villars Editeur, Paris 1963), pp. 122 - 133

6) Data reported by A.J. De Bethune, N.A. Swendeman-Loud "Standard Aqueous Electrode Potentials and Temperature Coefficients at 25 °C" (Clifford A. Hampel, Ill.)

Electronically conducting phase	Intermediate species	Composition of the solution	Solvent	Temperature °C	Pressure Torr	Measuring method
a	b	c	d		e	d
Cs			1	25	760	IV
Cs (Hg)		Cs : 0.2827 % wt. in Hg; $CsNO_3$	2	-35 ± 0.2		I
Cs (Hg)		Cs : 0.02827 % wt. in Hg; CsHCOO : 0.01 N	16	25 ± 0.01		I
Cs		CsCl : 0.01 N	23	25		I
Cs		CsCl : 0.0166 up to 0.0912 m	13	25		I
Cs	CsH		1	25	760	IV
Cs	CsH		1	25	760	IV

Comparison electrode	Liquid junction	Electrode reaction	Standard value	Uncertainty	Temperature Coefficient	Notes	Reference	Electrode Reference Number
			V	mV	μV/°C			
d			f		g	h		
		$Cs^+ + e^- \rightleftharpoons Cs$	-2.92		-326		1,5	1
29	KNO_3 (std)	$Cs^+ + e^- \rightleftharpoons Cs$	-3.93	± 0.1			2	2
14	no	$Cs^+ + e^- \rightleftharpoons Cs$	-3.44	± 0.8	+500		3	3
7	no	$Cs^+ + e^- \rightleftharpoons Cs$	-3.16		-300		4	4
4		$Cs^+ + e^- \rightleftharpoons Cs$	-2.994		500		6	5
		$Cs + H^+ + e^- \rightleftharpoons CsH$	0.317				1	6
		$Cs^+ + H^+ + 2e^- \rightleftharpoons CsH$	-1.304				1	7

Standard value V	Solvent	Electrode Reference Number	Standard value V	Solvent	Electrode Reference Number	Standard value V	Solvent	Electrode Reference Number
-2.92	1	1						
-1.304	1	7						
0.317	1	6						
-3.93	2	2						
-3.44	16	3						
-3.16	23	4						
-2.994	13	5						

BIBLIOGRAPHY : Cesium

1) Data reported by M. Pourbaix "Atlas d'Equilibres Electrochimiques à 25 °C" (Gauthier-Villars Editeur, Paris, 1963), pp. 122 - 133

2) V.A. Pleskov, J. Phys. Chem. U.S.S.R., 20 (1946), 151 - 162; Acta Physicochim. U.S.S.R., 21 (1946), 235 - 238

3) V.A. Pleskov, Acta Physicochim. U.S.S.R., 21 (1946); 41 - 54; J. Phys. Chem. U.S.S.R., 20 (1946), 153 - 162

4) V.A. Pleskov, J. Phys. Chem. U.S.S.R., 22 (1948), 351 - 361

5) Data reported by A.J. De Bethune, N.A. Swendeman-Loud "Standard Aqueous Electrode Potentials and Temperature Coefficients at 25 °C" (Clifford A. Hampel, Ill.)

6) K.K. Kundu, A.K. Rakshit, Ind. J.Chem., 9 (1971), 439 - 443

Electronically conducting phase	Intermediate species	Composition of the solution	Solvent	Temperature °C	Pressure Torr	Measuring method
a	b	c	d		e	d
Cu			1	25	760	IV
Cu (Hg)		$Cu(ClO_4)_2$: (0.08 up to 2.06) · 10^{-3} M	7	25 ± 0.2		I
Cu (Hg)		$CuClO_4$: 0 up to 0.01 M; $AgClO_4$: 0 up to 0.01 M; $(C_2H_5)_4NClO_4$: 0.1 M	23	25		I
Cu			1	25	760	IV
Cu (Hg)		$Cu(ClO_4)_2$: (1.54 up to 9.50) · 10^{-3} M	7	25 ± 0.2		I
Ind.			1	25	760	IV
Cu (Hg)		$CuClO_4$: (0.08 up to 2.06) · 10^{-3} M; $Cu(ClO_4)_2$: (1.54 up to 9.50) · 10^{-3} M	7	25 ± 0.2		I
Pt		$Cu(ClO_4)_2(CH_3CN)_4$; $CuClO_4(CH_3CN)_4$	23	25		I
Cu	CuBr		1	25	760	IV
Cu		$CuBr_2$, $HClO_4$ or $LiClO_4$ or $NaClO_4$: not reported	1	25		V
Cu	CuBr	KBr : 0.1 N; KCl : 0.1 N	1	25		I
Cu	$CuBr_2$	$CuBr_2$: 0.00242; 0.00682; 0.0158 M	18	25		I
Cu	CuCl		1	25	760	IV
Cu	CuCN		1	25		I
Cu			1	25	760	IV
Cu			1	25		V
Cu			1	25		V
Cu	$Cu_2[Fe(CN)_6]$		1	25		V
Cu	$CuHPO_4$		1	25	760	IV
Cu	$Cu(H_2PO_4)_2$		1	25	760	IV
Cu	CuI		1	25	760	IV
Cu	$Cu(IO_3)_2$	$Cu(IO_3)_2$: (3.245 up to 4.694) · 10^{-3} M; KCl : 0.1005 M	1	25		V
Cu	CuN_3		1	25		I
Glass	NH_3 (aq)		1	25		I
Cu	NH_3 (aq)		1	25	760	IV
Glass	NH_3 (aq)		1	25		I
Cu		$NaNO_3$, $Cu(NO_3)_2$, $NaNO_2$: 0 up to 0.7 M	1	25		V
Cu	CuO	H_2O_2 : 10^{-4} up to 1 M; pH : 1 up to 9	1	25 ± 0.05		I
Ind.	CuO		1	25	760	IV
Cu			1	25	760	IV
Ind.			1	25	760	IV

Comparison electrode	Liquid junction	Electrode reaction	Standard value V	Uncertainty mV	Temperature Coefficient µV/°C	Notes	Reference	Electrode Reference Number
d			f		g	h		
		$Cu^+ + e^- \rightleftharpoons Cu$	0.521		-58		1	1
3	KCl	$Cu^+ + e^- \rightleftharpoons Cu$	0.64	±10			2	2
7		$Cu^+ + e^- \rightleftharpoons Cu$	-0.344	±7			3	3
		$Cu^{2+} + 2e^- \rightleftharpoons Cu$	0.3419		+8		1	4
3	KCl	$Cu^{2+} + 2e^- \rightleftharpoons Cu$	0.513	±50			2	5
		$Cu^{2+} + e^- \rightleftharpoons Cu^+$	0.153		73		1	6
3	KCl	$Cu^{2+} + e^- \rightleftharpoons Cu^+$	0.40	±10			2	7
7		$Cu^{2+} + e^- \rightleftharpoons Cu^+$	0.133	±5			3	8
		$CuBr + e^- \rightleftharpoons Cu + Br^-$	0.033		-445	9	1	9
1		$Cu(Br)^+ + 2e^- \rightleftharpoons Cu + Br^-$	0.343				4	10
16		$Cu(Br)_2^- + e^- \rightleftharpoons Cu + 2\,Br^-$	0.170				5	11
25		$CuBr_2 + 2e^- \rightleftharpoons Cu + 2\,Br^-$	0.28	±12	-140		6	12
		$CuCl + e^- \rightleftharpoons Cu + Cl^-$	0.137		+635		1	13
1		$CuCN + e^- \rightleftharpoons Cu + CN^-$	-0.639				7	14
		$Cu(CN)_2^- + e^- \rightleftharpoons Cu + 2\,CN^-$	-0.429				1	15
1		$Cu(CN)_3^{2-} + e^- \rightleftharpoons Cu + 3\,CN^-$	-1.170			1	7,8	16
1		$Cu(CN)_4^{3-} + e^- \rightleftharpoons Cu + 4\,CN^-$	-1.281			1	7,8	17
1		$Cu_2[Fe(CN)_6] + 4e^- \rightleftharpoons 2\,Cu + [Fe(CN)_6]^{4-}$	-0.130			5	9	18
		$CuHPO_4 + 2e^- \rightleftharpoons Cu + HPO_4^{2-}$	0.172				10	19
		$Cu(H_2PO_4)_2 + 2e^- \rightleftharpoons Cu + 2H_2PO_4^-$	0.204				10	20
		$CuI + e^- \rightleftharpoons Cu + I$	-0.1852		-200		1	21
1		$Cu(IO_3)_2 + 2e^- \rightleftharpoons Cu + 2\,IO_3^-$	-0.079			2,5	11,12	22
1		$CuN_3 + e^- \rightleftharpoons Cu + N_3^-$	0.03				13	23
3		$Cu(NH_3)^{2+} + 2e^- \rightleftharpoons Cu + NH_3$	-0.223				14	24
		$Cu(NH_3)_2^+ + e^- \rightleftharpoons Cu + 2\,NH_3$	-0.12		-780		1	25
		$Cu(NH_3)_3^{2+} + 2e^- \rightleftharpoons Cu + 3\,NH_3$	0.025			4	15	26
1		$Cu(NO_2)^+ + 2e^- \rightleftharpoons Cu + NO_2^-$	0.305			9	16	27
3		$CuO + H_2O + 2e^- \rightleftharpoons Cu + 2OH^-$	-0.262			3	17	28
		$CuO + 2H^+ + e^- \rightleftharpoons Cu^+ + H_2O$	0.620				18	29
		$CuO_2^{2-} + 4H^+ + 2e^- \rightleftharpoons Cu + 2H_2O$	1.515				18	30
		$CuO_2^{2-} + 4H^+ + e^- \rightleftharpoons Cu^+ + 2H_2O$	2.510				18	31

Electronically conducting phase	Intermediate species	Composition of the solution	Solvent	Temperature °C	Pressure Torr	Measuring method
a	b	c	d		e	d
Cu	Cu_2O	H_2O_2 : 10^{-4} up to 1 M; pH : 1 up to 9	1	25 ± 0.05		I
Glass			1	25		I
Cu	$Cu(OH)_2$	H_2O_2 : 10^{-4} up to 1 M; pH : 1 up to 9	1	25 ± 0.05		I
Cu	$Cu_3(PO_4)_2$		1	25	760	I
Cu	CuS		1	25		IV
Ind.	Cu_2S		1	25	760	IV
Cu	CuSCN		1	25	760	IV
Cu (Hg) (DME)			1	25		II
Cu	$Cu(SO_4)$	$Cu(ClO_4)_2$: 0.0600 and 0.0863 M; Li_2SO_4 : 2.13 M	1	25		V
Cu			1	25		I
Cu			1	25	760	IV
Ind.			1	25	760	IV
Ind.	CuBr		1	25	760	IV
Ind.	CuCl		1	25	760	IV
Ind.			1	25	760	IV
Cu	CuH		1	25	760	IV
Ind.	CuI		1	25	760	IV
Ind.	Cu_2O		1	25	760	IV
Ind.	CuO (α), Cu_2O		1	25	760	IV
Ind.	Cu_2O, CuO (β)		1	25	760	IV
Ind.	Cu_2		1	25	760	IV
Ind.	$Cu(OH)_2$		1	25	760	IV
Ind.	Cu_2O		1	25	760	IV
Cu		Acetic acid: (18.50 up to 153.3) · 10^{-3} M; $Cu(IO_3)_2$ satd.; NaOH : (11.19 up to 44.52) · 10^{-3} M	1	25		V
Glass	Cu (adipate)	NaH adipate : 5 · 10^{-3} M; $CuCl_2$: 5 · 10^{-3} and 15 · 10^{-3} M	1	25		I
Glass	Alanine	1) Alanine: (37.52 up to 42.25) · 10^{-3} M; $CuCl_2$: (3.81 up to 4.28) · 10^{-3} M 2) Alanine: 0.01252 up to 0.2027 M; $Cu(IO_3)_2$: (3.98 up to 12.30) · 10^{-3} M	1	25 ± 0.02		I, V
Glass	Alanine	1) Alanine: (37.52 up to 42.25) · 10^{-3} M; $CuCl_2$: (3.81 up to 4.28) · 10^{-3} M 2) Alanine: 0.01252 up to 0.2027 M; $Cu(IO_3)_2$: (3.98 up to 12.30) · 10^{-3} M	1	25 ± 0.02		I, V

Comparison electrode	Liquid junction	Electrode reaction	Standard value	Uncertainty	Temperature Coefficient	Notes	Reference	Electrode Reference Number
			V	mV	μV/°C			
d			f		g	h		
3		$Cu_2O + H_2O + 2e^- \rightleftharpoons 2\,Cu + 2OH^-$	-0.360			3	17	32
1		$Cu(OH)^+ + 2e^- \rightleftharpoons Cu + OH^-$	0.153				20	33
3		$Cu(OH)_2 + 2e^- \rightleftharpoons Cu + 2OH^-$	-0.222			3	17	34
15		$Cu_3(PO_4)_2 + 6e^- \rightleftharpoons 3\,Cu + 2\,PO_4^{3-}$	0.325				10	35
		$CuS + 2e^- \rightleftharpoons Cu + S^{2-}$	-0.698				21	36
		$Cu_2S + 2e^- \rightleftharpoons 2\,Cu + S^{2-}$	-0.89		-1040		1	37
		$CuSCN + e^- \rightleftharpoons Cu + SCN^-$	-0.27				1	38
		$Cu(SCN)_4^{3-} + e^- \rightleftharpoons Cu + 4\,SCN^-$	-0.020				22	39
1		$Cu(SO_4) + 2e^- \rightleftharpoons Cu + SO_4^{2-}$	0.274			2,8	19,23	40
		$Cu(S_2O_3)_2^{2-} + 2e^- \rightleftharpoons Cu + 2\,S_2O_3^{2-}$	-0.022				24	41
		$HCuO_2^- + 3H^+ + 2e^- \rightleftharpoons Cu + 2H_2O$	1.127				18	42
		$HCuO_2^- + 3H^+ + e^- \rightleftharpoons Cu^+ + 2H_2O$	1.733				18	43
		$Cu^{2+} + Br^- + e^- \rightleftharpoons CuBr$	0.640		+460		1	44
		$Cu^{2+} + Cl^- + e^- \rightleftharpoons CuCl$	0.538		+650		1	45
		$Cu^{2+} + 2\,CN^- + e^- \rightleftharpoons [Cu(CN)_2]^-$	1.103				1	46
		$Cu + H^+ + e^- \rightleftharpoons CuH$	-2.775				18	47
		$Cu^{2+} + I^- + e^- \rightleftharpoons CuI$	0.86		+215		1	48
		$2\,Cu^{2+} + H_2O + 2e^- \rightleftharpoons Cu_2O + 2H^+$	0.203				18	49
		$2\,CuO(\alpha) + 2H^+ + 2e^- \rightleftharpoons Cu_2O + H_2O$	0.669				18	50
		$2\,CuO(\beta) + 2H^+ + 2e^- \rightleftharpoons Cu_2O + H_2O$	0.747				18	51
		$2\,CuO_2^{2-} + 6H^+ + 2e^- \rightleftharpoons Cu_2O + 3H_2O$	2.560				18	52
		$2\,Cu(OH)_2 + 2e^- \rightleftharpoons Cu_2O + 2OH^- + H_2O$	-0.080		-725		1	53
		$2\,HCuO_2^- + 4H^+ + 2e^- \rightleftharpoons Cu_2O + 3H_2O$	1.783				18	54
1		$Cu(acetate) + e^- \rightleftharpoons Cu + (acetate)^-$	0.276			5	12	55
3		$Cu(adipate) + 2e^- \rightleftharpoons Cu + (adipate)^{2-}$	0.243			4	25	56
1		$Cu(alanine)^{2+} + 2e^- \rightleftharpoons Cu + (alanine)$	0.092			2,6	26 11	57
1		$Cu(alanine)^{2+} + 2e^- \rightleftharpoons Cu + 2\,(alanine)$	-0.118			2,6	26 11	58

Electronically conducting phase	Intermediate species	Composition of the solution	Solvent	Temperature °C	Pressure Torr	Measuring method
a	b	c	d		e	d
Cu		Bromoacetic acid (54.75 up to 89.74) · 10^{-3} M; $Cu(IO_3)_2$;satd. NaOH (52.01 up to 87.28)· 10^{-3} M	1	25		V
Cu		α-bromobutyric acid:(20.06 and 25.64)· 10^{-3} M; $Cu(IO_3)_2$ satd. NAOH;(18.94 and 21.73) · 10^{-3} M	1	25		V
Cu		iso-butyric acid: (31.33 up to 93.99)· 10^{-3} M; $Cu(IO_3)_2$ satd.; NaOH: (10.06 and 25.80)·10^{-3} M	1	25		V
Cu		n-butyric acid: (760 up to 191.0) · 10^{-3} M; $Cu(IO_3)_2$ satd.; NaOH:(19.90 up to 49.74) ·10^{-3} M	1	25		V
Cu		Chloroacetic acid: (39.77 up to 88.72) · 10^{-3} M; NaOH: (37.67 up to 87.28) · 10^{-3} M; $Cu(IO_3)_2$ satd.	1	25		V
Glass	1,2-diaminopropane		1	25		V
Glass	1,2-diaminopropane		1	25		I
Glass	1,3-diaminopropane	$Cu(NO_3)_2$: 0.1 M; HNO_3: 0.1 M; KNO_3: 1 M; 1,3-diaminopropane: 0.05 up to 0.5 M	1	25		I
Glass	1,3-diaminopropane	$Cu(NO_3)_2$: 0.1 M; HNO_3 : 1 M; 1,3-diaminopropane: 0.05 up to 0.5 M	1	25		I
Cu	Cu (dimethylmalona-te)		1	25		III
Cu			1	25		I
Cu	Ethylenediamine	$Cu(NO_3)_2$: 0.00972 M; HNO_3 : 0.0499 M; KNO_3 : 1 M; ethylenediamine : 0.1698 up to 0.7094 M	1	25		I
Cu(Hg) (DME)	EDTA	$CuSO_4$: 10^{-5} up to 0.0145 M; EDTA : 0.25 M	1	25 ± 0.1		II
Cu	Cu (ethylmalonate)		1	25		III
Cu		Formic acid : (30.40 up to 43.43) · 10^{-3} M; $Cu(IO_3)_2$ satd.; NaOH: (12.37 up to 31.65)· 10^{-3} M	1	25		V
Glass	Cu (fumarate)	NaH fumarate : 5 · 10^{-3} M; $CuCl_2$: 5 · 10^{-3} M	1	25		I
Cu(Hg) (DME)	Cu (gluconate)	Na_2 gluconate : 0.1 M; NaOH : 0.34 M; $CuSO_4$	1	25.0		II
Cu (Hg) (DME)		Na_2 gluconate : 0.1 M; NaOH : 0.34 M; $CuSO_4$	1	25.0		II
Glass	Glycine	1) Glycine : (61.36 up to 76.72) · 10^{-3} M 2) Glycine : 0.01253 up to 0.2025 M; $Cu(IO_3)_2$: (4.096 up to 13.52) · 10^{-3} M	1	25 ± 0.02		I
Glass	Glycine	1) Glycine : (61.36 up to 76.72) · 10^{-3} M; 2) Glycine : 0.01253 up to 0.2025 M; $Cu(IO_3)_2$: (4.96 up to 13.52) · 10^{-3} M	1	25 ± 0.02		I

Comparison electrode	Liquid junction	Electrode reaction	Standard value V	Uncertainty mV	Temperature Coefficient μV/°C	Notes	Reference	Electrode Reference Number
d			f		g	h		
1		$Cu(bromoacetate)^+ + 2e^- \rightleftharpoons Cu + (bromoacetate)^-$	0.294			5	12	59
1		$Cu(\alpha\text{-bromobutyrate})^+ + 2e^- \rightleftharpoons Cu + (\alpha\text{-bromobutyrate})^-$	0.299			5	12	60
1		$Cu(iso\text{-butyrate})^+ + 2e^- \rightleftharpoons Cu + (iso\text{-butyrate})^-$	0.278			5	12	61
1		$Cu(n\text{-butyrate})^+ + 2e^- \rightleftharpoons Cu + (n\text{-butyrate})^-$	0.216			5	12	62
1		$Cu(chloroacetate)^+ + 2e^- \rightleftharpoons Cu + (chloroacetate)^-$	0.294			5	12	63
3		$Cu(1,2\text{-diaminopropane})^{2+} + 2e^- \rightleftharpoons Cu + (1,2\text{-diaminopropane})$	0.024			4	27	64
3		$Cu(1,2\text{-diaminopropane})_2^{2+} + 2e^- \rightleftharpoons Cu + 2(1,2\text{-diaminopropane})$	-0.268			4	27	65
3		$Cu(1,3\text{-diaminopropane})^{2+} + 2e^- \rightleftharpoons Cu + (1,3\text{-diaminopropane})$	0.051			2 4	28 29	66
3		$Cu(1,3\text{-diaminopropane})_2^{2+} + 2e^- \rightleftharpoons Cu + 2(1,3\text{-diaminopropane})$	-0.163			2 4	28 29	67
1		$Cu(dimethylmalonate) + 2e^- \rightleftharpoons Cu + (dimethylmalonate)^{2-}$	0.199			7	30	68
17		$Cu(dimethylmalonate) + 2e^- \rightleftharpoons Cu + 2(dimethylmalonate)^{2-}$	0.153				31	69
1		$Cu(ethylenediamine)_2^+ + e^- \rightleftharpoons Cu + 2(ethylenediamine)$	-0.119				32	70
3		$Cu(EDTA)^{2-} + 2e^- \rightleftharpoons Cu + (EDTA)^{4-}$	-0.216	± 2			33	71
1		$Cu(ethylmalonate) + 2e^- \rightleftharpoons Cu + (ethylmalonate)^{2-}$	0.190			7	30	72
1		$Cu(formate)^+ + 2e^- \rightleftharpoons Cu + (formate)^-$	0.2835			5	12	73
3		$Cu(fumarate) + 2e^- \rightleftharpoons Cu + (fumarate)^{2-}$	0.268			4	25	74
3		$Cu(gluconate) + 2e^- \rightleftharpoons Cu + (gluconate)^{2-}$	-0.023				34	75
3		$Cu(gluconate)_2^{2-} + 2e^- \rightleftharpoons Cu + 2(gluconate)^{2-}$	-0.233				33	76
1		$Cu(glycine)_2^{2+} + 2e^- \rightleftharpoons Cu + (glycine)$	0.092			2	26 11	77
1		$Cu(glycine)_2^{2+} + 2e^- \rightleftharpoons Cu + 2(glycine)$	-0.123			2	26 11	78

Electronically conducting phase	Intermediate species	Composition of the solution	Solvent	Temperature °C	Pressure Torr	Measuring method
a	b	c	d		e	d
Cu		Glycolic acid: (13.67 up to 22.78) · 10^{-3} M; $Cu(IO_3)_2$ satd. NaOH: (13.11 up to 22.16) · 10^{-3} M	1	25		V
Glass	Glycylglycine	Glycylglycine: (27.42 up to 30.59) · 10^{-3} M; $CuCl_2$: (4.73 up to 5.38) · 10^{-3} M	1	25 ± 0.02		I
Glass	Glycylglycine	Glycylglycine: (27.42 up to 30.59) · 10^{-3} M; $CuCl_2$: (4.73 up to 5.38) · 10^{-3} M	1	25 ± 0.02		I
Glass	Glycylglycylglycine	Glycylglycylglycine : (16.74 up to 20.26) · 10^{-3} M; $CuCl_2$: (2.93 up to 3.547) · 10^{-3} M	1	25		I
Glass	Glycylglycylglycine	Glycylglycylglycine : (16.74 up to 20.26) · 10^{-3} M; $CuCl_2$: (2.93 up to 3.547) · 10^{-3} M	1	25 ± 0.03		I
Cu		n-hexanoic acid: (35.0 up to 40.0) · 10^{-3} M; $Cu(IO_3)_2$ satd. NaOH: (7.95 up to 10.33) · 10^{-3} M	1	25		V
Glass	Histidine		1	25		I
Cu		β-hydroxypropionic acid: (53.20 up to 85.12) · 10^{-3} M; $Cu(IO_3)_2$ satd. NaOH: (9.88 up to 47.82) · 10^{-3} M	1	25		V
Glass		8-hydroxyquinoline 5-sulfonic acid : 2.954 · 10^{-3} M; $Cu(ClO_4)_2$	1	25	743	I
Glass	Cu (8-hydroxyquinoline 5-sulfonate)$_2$	8-hydroxyquinoline 5-sulfonic acid: 2.954 · 10^{-3} M; $Cu(ClO_4)_3$	1	25	743	I
Glass	Leucine		1	25		I
Glass	Leucine		1	25		I
Glass	Cu (maleate)	NaH maleate : 5 · 10^{-3} M; $CuCl_2$: 5 · 10^{-3} M	1	25		I
Ind.	Cu (malonate)		1	25	760	IV
Cu	1-methylallylalcohol	CuCl	1	25.0		V
Cu	2-methylallylalcohol	CuCl	1	25.0		V
Glass		6-methylpicolinic acid: 10^{-3} M; $Cu(ClO_4)_2$: 5 · 10^{-4} M	1	25 ± 0.1		I
Glass	Cu (6-methylpicolinate)$_2$	6-methylpicolinic acid: 10^{-3} M; $Cu(ClO_4)_2$: 5 · 10^{-4} M	1	25 ± 0.1		I
Glass	Norvaline		1	25		I
Glass	Norvaline		1	25		I
Glass	Ornithine		1	25		I
Glass	Ornithine		1	25		I
Cu		Phenylacetic acid: (23.50 up to 44.06) · 10^{-3} M; $Cu(IO_3)_2$: satd.; NaOH: (10.39 up to 19.48) · 10^{-3} M	1	25		V

Comparison electrode	Liquid junction	Electrode reaction	Standard value V	Uncertainty mV	Temperature Coefficient μV/°C	Notes	Reference	Electrode Reference Number
d			f		g	h		
1		Cu (glycolate) + $2e^- \rightleftharpoons$ Cu + (glycolate)$^{2-}$	0.257			5	12	79
4		Cu (glycylglycine)$^{2+}$ + $2e^- \rightleftharpoons$ Cu + (glycylglycine)	0.163				26	80
4		Cu (glycylglycine)$^{2+}$ + $2e^- \rightleftharpoons$ Cu + 2(glycylglycine)	-0.004			4	26	81
4		Cu (glycylglycylglycine)$^{2+}$ + $2e^- \rightleftharpoons$ Cu + (glycylglycylglycine)	0.182			4	35	82
4		Cu (glycylglycylglycine)$_2^{2+}$ + $2e^- \rightleftharpoons$ Cu + 2 (glycylglycylglycine)	0.030			4	35	83
1		Cu (n-hexanoate)$^+$ + $2e^- \rightleftharpoons$ Cu + (n-hexanoate)$^-$	0.282			5	12	84
		Cu (histidine)$_2^{2+}$ + $2e^- \rightleftharpoons$ Cu + 2 (histidine)	0.023			4	36	85
1		Cu (β-hydroxypropionate)$^+$ + $2e^- \rightleftharpoons$ Cu + (β-hydroxypropionate)$^-$	0.255			5	12	86
3	KCl (satd.)	Cu (8-hydroxyquinoline 5-sulfonate)$^+$ + $2e^- \rightleftharpoons$ Cu + (8-hydroxyquinoline 5-sulfonate)$^-$	-0.100			2 4	37 38	87
3	KCl (satd.)	Cu (8-hydroxyquinoline 5-sulfonate)$_2$ + $2e^- \rightleftharpoons$ Cu + 2 (8-hydroxyquinoline 5-sulfonate)$^-$	-0.308			4	37 38	88
		Cu (leucine)$^{2+}$ + $2e^- \rightleftharpoons$ Cu + (leucine)	0.109			4	37	89
		Cu (leucine)$_2^{2+}$ + $2e^- \rightleftharpoons$ Cu + 2 (leucine)	-0.083			4	37	90
3		Cu (maleate) + $2e^- \rightleftharpoons$ Cu + (maleate)$^{2-}$	0.226			4	25	91
		Cu (malonate) + $2e^- \rightleftharpoons$ Cu + (malonate)$^{2-}$	0.203				39	92
1		Cu (1-methylallylalcohol)$^+$ + $e^- \rightleftharpoons$ Cu + (1-methylallylalcohol)	0.247			5	40	93
1		Cu (2-methylallylalcohol)$^+$ + $e^- \rightleftharpoons$ Cu + (2-methylallylalcohol)	0.296			5	40	94
3		Cu (6-methylpicolinate)$^+$ + $2e^- \rightleftharpoons$ Cu + (6-methylpicolinate)$^-$	0.155			4	41	95
3		Cu (6-methylpicolinate)$_2$ + $2e^- \rightleftharpoons$ Cu + 2 (6-methylpicolinate)$^-$	-0.010			4	41	96
		Cu (norvaline)$^{2+}$ + $2e^- \rightleftharpoons$ Cu + (norvaline)	0.085				42	97
		Cu (norvaline)$_2^{2+}$ + $2e^- \rightleftharpoons$ Cu + 2 (norvaline)	-0.124			4	42	98
		Cu (ornithine)$^{2+}$ + $2e^- \rightleftharpoons$ Cu + (ornithine)	0.138			4	42	99
		Cu (ornithine)$_2^{2+}$ + $2e^- \rightleftharpoons$ Cu + 2 (ornithine)	-0.028			4	42	100
1		Cu (phenylacetate)$^+$ + $2e^- \rightleftharpoons$ Cu + (phenylacetate)$^-$	0.284			5	12	101

Electronically conducting phase	Intermediate species	Composition of the solution	Solvent	Temperature °C	Pressure Torr	Measuring method
a	b	c	d		e	d
Cu		Pivalic acid : (48.02 up to 123.0) · 10^{-3} M; $Cu(IO_3)_2$ satd.; NaOH: (10.06 up to 25.80)· 10^{-3} M	1	25		V
Glass		Na (salicylaldehyde-5-sulfonate) : 0.01, 0.1 M; $CuCl_2$: 0.167 M; NaOH: 0.33 M	1	25 ± 0.01		I
Glass	Cu (salicylaldehyde-5-sulfonate)$_2$	Na (salicylaldehyde-5-sulfonate): 0.01, O.1 M; $CuCl_2$: 0.167 M; NaOH: 0.33 M	1	25 ± 0.01		I
Glass	Cu (succinate)	NaH succinate : 5 · 10^{-3} M ; $CuCl_2$: 5 · 10^{-3} , 15 · 10^{-3} M	1	25		I
Cu		iso-valeric acid: (34.68 up to 93.76) · 10^{-3} M; Cu $(IO_3)_2$ satd.; NaOH: (11.98 up to 29.86)·10^{-3} M	1	25		V
Cu		n-valeric acid : (61.60 up to 123.2) · 10^{-3} M; Cu $(IO_3)_2$ satd.; NaOH : (16.72 up to 33.44)·10^{-3} M	1	25		V
Glass	Valine		1	25		I
Glass	Valine		1	25		I

Comparison electrode	Liquid junction	Electrode reaction	Standard value V	Uncertainty mV	Temperature Coefficient μV/°C	Notes	Reference	Electrode Reference Number
d			f		g	h		
1		$Cu(pivalate)^+ + 2e^- \rightleftharpoons Cu + (pivalate)^-$	0.278			5	12	102
		$Cu(salicylaldehyde\text{-}5\text{-}sulfonate)^+ + 2e^- \rightleftharpoons$ $Cu + (salicylaldehyde\text{-}5\text{-}sulfonate)^-$	0.184			4	43	103
		$Cu(salicylaldehyde\text{-}5\text{-}sulfonate)_2 + 2e^- \rightleftharpoons$ $Cu + 2\,(salicylaldehyde\text{-}5\text{-}sulfonate)^-$	0.067				43	104
3		$Cu(succinate) + 2e^- \rightleftharpoons Cu + (succinate)^{2-}$	0.243			4	25	105
1		$Cu(iso\text{-}valerate)^+ + 2e^- \rightleftharpoons Cu + (iso\text{-}valerate)^-$	0.281			5	12	106
1		$Cu(n\text{-}valerate)^+ + 2e^- \rightleftharpoons Cu + (n\text{-}valerate)^-$	0.276			5	12	107
		$Cu(valine)^{2+} + 2e^- \rightleftharpoons Cu + (valine)$	0.095			4	42	108
		$Cu(valine)_2^{2+} + 2e^- \rightleftharpoons Cu + 2\,(valine)$	-0.114			4	42	109

Standard value V	Solvent	Electrode Reference Number
-2.775	1	47
-1.281	1	17
-1.170	1	16
-0.89	1	37
-0.698	1	36
-0.639	1	14
-0.429	1	15
-0.360	1	32
-0.308	1	88
-0.27	1	38
-0.268	1	65
-0.262	1	28
-0.233	1	76
-0.223	1	24
-0.222	1	34
-0.216	1	71
-0.1852	1	21
-0.163	1	67
-0.130	1	18
-0.124	1	98
-0.123	1	78
-0.12	1	25
-0.119	1	70
-0.118	1	58
-0.114	1	109
-0.100	1	87
-0.083	1	90
-0.080	1	53
-0.079	1	22
-0.028	1	100
-0.023	1	75
-0.022	1	41
-0.020	1	39
-0.010	1	96
-0.004	1	81
0.023	1	85
0.024	1	64
0.025	1	26
0.03	1	23
0.030	1	83
0.033	1	9
0.051	1	66
0.067	1	104
0.085	1	97
0.092	1	57
0.092	1	77
0.095	1	108
0.109	1	89
0.137	1	13
0.138	1	99
0.153	1	33
0.153	1	6
0.153	1	69
0.155	1	95
0.163	1	80
0.170	1	11
0.172	1	19
0.182	1	82
0.184	1	103
0.190	1	72
0.199	1	68
0.203	1	49
0.203	1	92
0.204	1	20
0.216	1	62
0.226	1	91
0.243	1	56
0.243	1	105
0.247	1	93
0.255	1	86
0.257	1	79
0.268	1	74
0.274	1	40
0.276	1	107
0.276	1	55
0.278	1	102
0.278	1	61
0.281	1	106
0.282	1	84
0.284	1	101
0.2835	1	73
0.294	1	63
0.294	1	59
0.296	1	94
0.299	1	60
0.305	1	27
0.325	1	35
0.3419	1	4
0.343	1	10
0.521	1	1
0.538	1	45
0.620	1	29
0.640	1	44
0.669	1	50
0.747	1	51
0.86	1	48
1.103	1	46
1.127	1	42
1.515	1	30
1.733	1	43
1.783	1	54
2.510	1	31
2.560	1	52
0.40	7	7
0.513	7	5

Copper 2

Standard value V	Solvent	Electrode Reference Number
0.64	7	2
0.28	18	12
-0.344	23	3
1.33	23	8

Standard value V	Solvent	Electrode Reference Number

Standard value V	Solvent	Electrode Reference Number

NOTES : Copper

1) I.R. Data
2) The value is an average of those given by several authors; they all are in the same range of uncertainty
3) T^{o} ranges from 25 to 50 ^{o}C
4) Potentiometric pH measurements
5) Solubility data
6) Potentiometry and solubility data
7) Conductometric data
8) Solubility and spectrophotometric data
9) Spectrophotometric data

BIBLIOGRAPHY : Copper

1) Data reported by A.J. De Bethune, N.A. Swendeman-Loud "Standard Aqueous Electrode Potentials and Temperature Coefficients at 25 °C" (Clifford A. Hampel, Ill.)

2) J.P. Desmarquet, C. Trinh-Dinh, O. Bolck, J. Electroanal. Chem., 27 (1970), 101 - 108

3) J.K. Senne, B. Kratochvil, Anal. Chem., 43 (1971), 79 - 82

4) R. Näsänen, Acta Chem. Scand., 4 (1950), 816 - 820

5) G. Bödlander, O. Starbeck, Z. Anorg. Chem., 31 (1902), 458 - 476

6) T. Pavlopoulos, H. Strehlow, Z. Physik. Chem. (N.F.), 2 (1954), 89 - 103

7) M.G. Vladimirova, I.A. Kakovskii, Zh. Priklad. Khim., 23 (1950), 580

8) R.A. Penneman, L.H. Jones, J. Chem. Phys., 24 (1956), 293 - 296

9) I.V. Tananaev, M.A. Glukova, G.B. Seifer, Zh. Neorgan. Khim., 1 (1956), 66

10) K.P. Batashev, E.N. Nikitin, Zh. Priklad. Khim., 23 (1950), 263 - 270

11) R.M. Keefer, J. Am. Chem. Soc., 70 (1948), 476 - 479

12) N. Lloyd, V. Wycherley, C.B. Monk, J. Chem. Soc., 1951, 1786 - 1789

13) S. Suzuki, J. Chem. Soc. Japan, Pure Chem. Sect., 73 (1952), 278 - 280

14) R. Näsänen, Suomen Kem., 17 B (1944), 31

15) R. Lloyd, Thesis, Temple Univ. Philadelphia, Pa., 1954, Univ. Microfilms, 12401

16) A. Kossiskoff, D.V. Sickman, J. Am. Chem. Soc., 68 (1946), 442 - 444

17) E.M. Khairy, N.A. Darwish, Z. Phys. Chem., 230 (1965), 327 - 346

18) Data reported by M. Pourbaix "Atlas d'Equilibres Electrochimiques à 25 °C" (Gauthier-Villars Editeur, Paris, 1963) pp. 385 - 386

19) B.B. Owen, R.W. Gurry, J. Am. Chem. Soc., 60 (1938), 3074

20) C.W. Davies, J. Chem. Soc., 1951, 1256 - 1258

21) F.D. Rossini, Nat. Bureau Standards, Circular 500, 1952

22) I.A. Korshunov, N.I. Nalyugina, Zh. Obsh. Khim., 20 (1950), 1309

BIBLIOGRAPHY: Copper

23) R. Näsänen, Acta Chem. Scand., 3 (1949), 176 - 189; 959 - 966

24) A.I. Levin, G.A. Ukshe, Sb. Stat. Obsh. Khim., 2 (1953), 798

25) J.M. Peacock, J.C. James, J. Chem. Soc., 1951, 2233 - 2239

26) C.B. Monk, Trans. Faraday Soc., 47 (1951), 285 - 291

27) F. Basolo, Y.T. Chen, J. Am. Chem. Soc., 76 (1954), 956 - 959

28) J. Poulsen, J. Bjerrum, Acta Chem. Scand., 9 (1955), 1407 - 1420

29) H. Irving, J. Chem. Soc., 1954, 3494 - 3504

30) D.J. Ives, H.L. Riley, J. Chem. Soc., 1931, 1938

31) H. Riley, J. Chem. Soc., 1930, 1642

32) J.Bjerrum, E.J. Nielson, Acta Chem. Scand., 2 (1948), 297 - 318

33) R. Pecsok, Anal. Chem., 25 (1953), 561 - 564

34) R. Pecsok, R.S. Juvet, J. Am. Chem. Soc., 77 (1955), 20

35) J.I. Evans, C.B. Monk, Trans. Faraday Soc., 51 (1955), 1244 - 1250

36) I.E. Maley, D.S. Mellor, Nature, 165 (1950), 453

37) L.E. Maley, D.P. Mellor, Australian J. Sci. Res., 2 A (1949), 579

38) R. Näsänen, E. Uisitalo, Acta Chem. Scand., 8 (1954), 112 - 118

39) N. Bonnet, V. Mihalova, Dokl. Bolg. Acad. Nauk, 21 (1968), 1295 - 1298

40) R.N. Keefer, L.J. Andrews, R.E. Kepner, J. Am. Chem. Soc., 71 (1949), 3906 - 3909

41) F. Holmes, W.R.C. Crimmin, J. Chem. Soc., 1955, 1175 - 1180

42) R.L. Rebertus, Diss. Urbana, 1952

43) N. Calvin, N.C. Melchier, J. Am. Chem. Soc., 70 (1948), 3270 - 3273

44) D.A. Zatko, Diss. Abstr. 28 B (1967), 524, Univ. Microfilms, Order n. 67 - 3422

Electronically conducting phase	Intermediate species	Composition of the solution	Solvent	Temperature °C	Pressure Torr	Measuring method
a	b	c	d		e	d
Ag		$AgClO_4$, $NaClO_4$	1	25		I
Ag		AgCl: 0.0035 up to 0.1192 m	6	25 ± 0.1		I
Ag		HCl: 0.0056 up to 0.149 m	15	25	760	IV
Ag		$AgNO_3$	17	22		I
Ag(Hg) (D.M.E.)		1) Na trifluoroacetate: 1 M; ferricinium trifluoroacetate; Ag trifluoroacetate; 2) $(ethyl)_4$ $NClO_4$: 0.5 M; (ferricinium) ClO_4 ; $AgClO_4$	33	25		II
Ag			18	25	760	IV
Ag		KNO_3 , $NaNO_3$ and $AgNO_3$: all up to 0.05 m	20	25		I
Ag		$AgClO_4$	23	22 ± 3		I
Ag		AgCl as well as AgCN, AgCNS, $AgNO_3$, Ag picrate	27	25± 0.5		I
Ind.			1	25	760	IV
Ag	AgBr	HBr: 0.005 up to 1.0 m	1	25		I
Ag	AgBr	HBr: 0.005 up to 0.100 N	1	25± 0.05		I
Ag	Ag (Br)		1	25		V
Ag	AgBr	NaBr, succinic acid, Li succinate: all in the range of 0.0003906 - 0.02050 m	7	25± 0.5		I
Ag	AgBr	HBr: 0.00229 up to 0.0692 M	8	25± 0.05		I
Ag	AgBr	HBr: 0.00440 up to 0.09319 m	13	25		I
Ag	AgBr	NaBr: 0.00518 up to 0.0502 m; Na acetate: 0.00512 up to 0.0497 m; acetic acid: 0.00496 up to 0.0505 m	14	25		I
Ag	AgBr	HBr	18	25		I
Ag		$AgClO_4$ + NaBr: 1.50 up to 5.0 m; $NaClO_4$: up to 3.50 m	1	25.0		I, V
Ag		$AgClO_4$ + NaBr: 1.50 up to 5.0 m; $NaClO_4$: up to 3.50 m	1	25.0		I, V
Ag		$AgClO_4$ + NaBr: 1.50 up to 5.0 m; $NaClO_4$: 0 up to 3.50 m	1	25.0		I, V
Ag		AgBr satd.;KCl: 0.05 up to 4 M; KBr : 0.05 up to 4 M	1	25 ± 0.1		I

Comparison electrode	Liquid junction	Electrode reaction	Standard value V	Uncertainty mV	Temperature Coefficient μV/°C	Notes	Reference	Electrode Reference Number
d			f		g	h		
1	no	$Ag^{+} + e^{-} \rightleftharpoons Ag$	0.800		-1000		1,2	1
3	Ethylenediamine with Ag^{+} and Hg^{2+} (satd.)	$Ag^{+} + e^{-} \rightleftharpoons Ag$	0.757	±3			3	2
	no	$Ag^{+} + e^{-} \rightleftharpoons Ag$	0.49				4	3
3		$Ag^{+} + e^{-} \rightleftharpoons Ag$	0.87	±3			5	4
33		$Ag^{+} + e^{-} \rightleftharpoons Ag$	0.31				6	5
		$Ag^{+} + e^{-} \rightleftharpoons Ag$	0.6902				7	6
1	$NaNO_3$ (aq.)	$Ag^{+} + e^{-} \rightleftharpoons Ag$	0.812	±3		1	8	7
33		$Ag^{+} + e^{-} \rightleftharpoons Ag$	0.240	±7			9	8
8,24		$Ag^{+} + e^{-} \rightleftharpoons Ag$	0.551	±2		2	10	9
		$Ag^{2+} + e^{-} \rightleftharpoons Ag^{+}$	1.980				2	10
1	no	$AgBr + e^{-} \rightleftharpoons Ag + Br^{-}$	0.07133	±7	-410		11,12	11
37	HBr	$AgBr + e^{-} \rightleftharpoons Ag + Br^{-}$	0.0714		-500		13	12
1		$Ag(Br) + e^{-} \rightleftharpoons Ag + Br^{-}$	0.654			3	14	13
1	no	$AgBr + e^{-} \rightleftharpoons Ag + Br^{-}$	-0.139	±1			15	14
1	no	$AgBr + e^{-} \rightleftharpoons Ag + Br^{-}$	-0.1939				16	15
1	no	$AgBr + e^{-} \rightleftharpoons Ag + Br^{-}$	-0.1007		-1100		17	16
1	no	$AgBr + e^{-} \rightleftharpoons Ag + Br^{-}$	-0.1633		-1480		18	17
1	no	$AgBr + e^{-} \rightleftharpoons Ag + Br^{-}$	0.0967	±7			19	18
18		$Ag(Br)_2^{-} + e^{-} \rightleftharpoons Ag + 2\,Br^{-}$	0.379			4	14	19
18		$Ag(Br)_3^{-} + e^{-} \rightleftharpoons Ag + 3\,Br^{-}$	0.328			4	14	20
18		$Ag(Br)_4^{-} + e^{-} \rightleftharpoons Ag + 4\,Br^{-}$	0.274			4	14	21
3	KNO_3 2 M	$Ag(Br)(Cl)_3^{3-} + e^{-} \rightleftharpoons Ag + Br^{-} + 3\,Cl^{-}$	0.330				20	22

Electronically conducting phase	Intermediate species	Composition of the solution	Solvent	Temperature °C	Pressure Torr	Measuring method
a	b	c	d		e	d
Ag		AgBr satd.; KCl: 0.05 up to 4 M; KBr: 0.05 up to 4 M	1	25± 0.1		I
Ag	$AgBrO_3$		1	25	760	IV
Ag	AgCl	HCl : 0.001 up to 1 m	1	25		I
Ag	AgCl	HCl : 0.001 up to 3.0 m	1	25± 0.02		I
Ag	AgCl	HCl : 0.0001 up to 0.1 M	7	25	760	IV
Ag	AgCl	HCl : 0.00123 up to 0.0933 m	8	25		I
Ag	AgCl	HCl : 0.00123 up to 0.0933 m	10	25		I
Ag	AgCl	HCl : 0.00123 up to 0.0933 m	9	25		I
Ag	AgCl	HCl : 10^{-4} up to 4.15 m	11	25± 0.01		I
Ag	AgCl	HCl : 0.005 up to 0.0933 m	13	25± 0.1		I
Ag	AgCl	HCl : 0.00188 up to 0.09087 m	14	25		I
Ag	AgCl	HCl : 0.003 up to 0.07 m	32	25		I
Ag	AgCl	HCl : 0.005 up to 0.149 m	15	25		I
Ag	AgCl	KCl : 0.005 up to 0.28 m	18	25	760	I
Ag	AgCl	HCl : 0.0100 up to 0.0684 m	19	25± 0.05		I
Ag	AgCl	HCl : 0.004834 up to 0.155698 m	21	35		I
Ag	AgCl	HCl : 0.0052 up to 0.0584 M	22	25± 0.05		I
Ag	$Ag(ClO_4)$	$AgClO_4$: 0.0005 up to 0.1 M	1	25± 0.001		III
Ag	AgCN		1	25	760	IV
Ag			1	25	760	IV
Ag	Ag_2CO_3		1	25	760	IV
Ag	$Ag_3[Co(CN)_6]$	$K_3[Co(CN)_6]$: 0.10, 0.20, 0.50 m	1	25		I
Ag	Ag_2CrO_4	K_2CrO_4, KNO_3, $AgNO_3$: all in the range of 0.02 - 0.05 m	1	25		I
Ag	Ag (F)		1	25	760	IV
Ag	$Ag_4[Fe(CN)_6]$	$K_4[Fe(CN)_6]$: 0.05, 0.1 and 0.25 m	1	25		I
Ag	AgI	HI : 0.005 up to 0.9 m	1	25± 0.1		I
Ag	AgI	NaI, succinic acid, Li succinate: all in the range of 0.0003906 - 0.02050 m	7	25		I
Ag	AgI	NaI : 0.00493 up to 0.0502 m; Na acetate : 0.00465 up to 0.0504 m; Acetic acid: 0.00463 up to 0.0472 m	13	25		I

Comparison electrode	Liquid junction	Electrode reaction	Standard value	Uncertainty	Temperature Coefficient	Notes	Reference	Electrode Reference Number
			V	mV	μV/°C			
d			f		g	h		
3	KNO_3 2 M	$Ag(Cl)(Br)_3^{3-} + e^- \rightleftharpoons Ag + Cl^- + 3\,Br^-$	0.239				20	23
		$AgBrO_3 + e^- \rightleftharpoons Ag + BrO_3^-$	0.546				2	24
1	no	$AgCl + e^- \rightleftharpoons Ag + Cl^-$	0.22230	±0.06	-543		21,22 23,24	25
1	no	$AgCl + e^- \rightleftharpoons Ag + Cl^-$	0.22233	±0.04			25	26
	no	$AgCl + e^- \rightleftharpoons Ag + Cl^-$	-0.0221				26	27
1	no	$AgCl + e^- \rightleftharpoons Ag + Cl^-$	-0.0848		-13000		27	28
1	no	$AgCl + e^- \rightleftharpoons Ag + Cl^-$	-0.1135		-2000		27	29
1	no	$AgCl + e^- \rightleftharpoons Ag + Cl^-$	-0.1345		-4500		27	30
1	no	$AgCl + e^- \rightleftharpoons Ag + Cl^-$	-0.1430	±0.2			28	31
1	no	$AgCl + e^- \rightleftharpoons Ag + Cl^-$	0.0235	±0.3			29	32
1	no	$AgCl + e^- \rightleftharpoons Ag + Cl^-$	-0.0323		-1800		17	33
1		$AgCl + e^- \rightleftharpoons Ag + Cl^-$	0.02077	±0.06			30	34
1	no	$AgCl + e^- \rightleftharpoons Ag + Cl^-$	-0.54				4	35
1	no	$AgCl + e^- \rightleftharpoons Ag + Cl^-$	0.1979	±0.7	-999		7,31	36
1		$AgCl + e^- \rightleftharpoons Ag + Cl^-$	0.208	±3			32	37
1	no	$AgCl + e^- \rightleftharpoons Ag + Cl^-$	0.21187		-1000		33	38
1	no	$AgCl + e^- \rightleftharpoons Ag + Cl^-$	0.1500	±0.6			34	39
1		$Ag(ClO_4) + e^- \rightleftharpoons Ag + ClO_4^-$	0.787			5	35	40
		$AgCN + e^- \rightleftharpoons Ag + CN^-$	-0.017		121		2	41
		$Ag(CN)_2^- + e^- \rightleftharpoons Ag + 2\,CN^-$	-0.31		87		2	42
		$Ag_2CO_3 + 2e^- \rightleftharpoons 2Ag + CO_3^{2-}$	0.47		-1377		2	43
23	0.01% K amalgam	$Ag_3[Co(CN)_6] + 3e^- \rightleftharpoons 3\,Ag + [Co(CN)_6]^{3-}$	0.298	±1			36	44
31	KNO_3	$Ag_2CrO_4 + 2e^- \rightleftharpoons 2\,Ag + CrO_4^{2-}$	0.4470		-1100		37	45
		$Ag(F) + e^- \rightleftharpoons Ag + F^-$	0.779				38	46
30	0.01% K amalgam	$Ag_4[Fe(CN)_6] + 4\,e^- \rightleftharpoons 4\,Ag + [Fe(CN)_6]^{4-}$	0.1478	±0.3			39	47
1	no	$AgI + e^- \rightleftharpoons Ag + I^-$	-0.15224	±0.05	-315		40,41	48
1	no	$AgI + e^- \rightleftharpoons Ag + I^-$	-0.3176	±0.8			14	49
1	no	$AgI + e^- \rightleftharpoons Ag + I^-$	-0.2928		-1030		18	50

Electronically conducting phase	Intermediate species	Composition of the solution	Solvent	Temperature °C	Pressure Torr	Measuring method
a	b	c	d		e	d
Ag	AgI	NaI : 0.00436 up to 0.0492 m; Na acetate: 0.00450 up to 0.0492 m; Acetic acid: 0.00539 up to 0.0484 m	14	25		I
Ag	$AgIO_3$		1	25	760	IV
Ag	$Ag(IO_3)$		1	25		V
Ag			1	25		V
Ag	Ag_2MoO_4	Na_2MoO_4, $NaNO_3$, $AgNO_2$: all in the range of 0.02 - 0.05 m	1	25± 0.02		I
Ag	AgN_3		1	25		I
Ag	NH_3 (aq.)	1) $AgIO_3$ and AgCl satd.; NH_4NO_3 : 0.0099 up to 0.1424 m; NH_3 : 0.01241 up to 0.03628 m; 2) Ag_2SO_4 and $AgIO_3$: 0.00501 up to 0.01020 m; NH_3 : 0.0940 up to 0.823 m	1	25		V
Ag	NH_3 (aq.)		1	25	760	IV
Ag	$AgNO_2$		1	25	760	IV
Ag			1	25	760	IV
Ind.	AgO		1	25	760	IV
Ind.			1	25	760	IV
Ind.			1	25	760	IV
Ag	Ag_2O	NaOH	1	25± 0.1		I
Ind.	Ag_2O_3		1	25	760	IV
Ind.	Ag_2O_3		1	25	760	IV
Ind.	Ag_2O_3		1	25	760	IV
Ind.	AgO		1	25	760	IV
Ind.			1	25	760	IV
Ind.	Ag_2O_3		1	25	760	IV
Ind.	Ag_2O_3		1	25	760	IV
Ind.	Ag_2O, AgO		1	25	760	IV
Ag	Ag_2O_3		1	25	760	IV
Ag	Ag_3PO_4	H_3PO_4 : 0.025 up to 0.1 M	1	25		I
Ag	Ag_2S	NaOH : 0.0120 up to 1.20 m; S^{2-}: 10^{-6} up to 10^{-3} m	1	25		I
Ag	AgSCN	$HClO_4$: 0.00071 up to 0.0162 m; KCNS : 0.00074 up to 0.0183 m	1	25± 0.01		I
Ag		KSCN: up to 2.2 m; AgSCN satd.	1	25		V
Ag		KSCN: up to 2.2 m; AgSCN satd.	1	25		V
Ag		KSCN: up to 2.2 m; AgSCN satd.	1	25		III

Comparison electrode	Liquid junction	Electrode reaction	Standard value V	Uncertainty mV	Temperature Coefficient µV/°C	Notes	Reference	Electrode Reference Number
d			f		g	h		
1	no	$AgI + e^- \rightleftharpoons Ag + I^-$	-0.3468		-1380		18	51
		$AgIO_3 + e^- \rightleftharpoons Ag + IO_3^-$	0.354		-568		2	52
1		$Ag(IO_3) + e^- \rightleftharpoons Ag + IO_3^-$	0.763			3	42	53
1		$Ag(IO_3)_2^- + e^- \rightleftharpoons Ag + 2\,IO_3^-$	0.725			3	42	54
11	KNO_3 (aq.)	$Ag_2MoO_4 + 2e^- \rightleftharpoons 2\,Ag + MoO_4^{2-}$	0.4573		1190		43	55
		$AgN_3 + e^- \rightleftharpoons Ag + N_3^-$	0.29				44	56
1		$Ag(NH_3)^+ + e^- \rightleftharpoons Ag + NH_3$	0.603				45 46 47	57
		$Ag(NH_3)_2^+ + e^- \rightleftharpoons Ag + 2\,NH_3$	0.373		-460		2	58
		$AgNO_2 + e^- \rightleftharpoons Ag + NO_2^-$	0.564		-265		2	59
		$AgO^- + 2H^+ + e^- \rightleftharpoons Ag + H_2O$	2.220				48	60
		$AgO + 2H^+ + e^- \rightleftharpoons Ag^+ + H_2O$	1.772				48	61
		$AgO^+ + 2H^+ + 2e^- \rightleftharpoons Ag^+ + H_2O$	1.998				48	62
		$AgO^+ + 2H^+ + e^- \rightleftharpoons Ag^{2+} + H_2O$	2.016				48	63
22	no	$Ag_2O + H_2O + 2e^- \rightleftharpoons 2\,Ag + 2OH^-$	0.342	± 1	-2000		49	64
		$Ag_2O_3 + 6H^+ + 4e^- \rightleftharpoons 2\,Ag^+ + 3H_2O$	1.670				48	65
		$Ag_2O_3 + 3H_2O + 4e^- \rightleftharpoons 2\,Ag^+ + 6OH^-$	1.757				50	66
		$Ag_2O_3 + 6H^+ + 2e^- \rightleftharpoons 2\,Ag^{2+} + 3\,H_2O$	1.360				48	67
		$AgO + e^- \rightleftharpoons AgO^-$	0.351				48	68
		$AgO^+ + 2e^- \rightleftharpoons AgO^-$	1.288				48	69
		$Ag_2O_3 + H_2O + 2e^- \rightleftharpoons 2\,AgO + 2OH^-$	0.739				2	70
		$Ag_2O_3 + 2H^+ + 4e^- \rightleftharpoons 2\,AgO^- + H_2O$	0.960				48	71
		$2\,AgO + H_2O + 2e^- \rightleftharpoons Ag_2O + 2OH^-$	0.607		-1117		2	72
		$Ag_2O_3 + H_2O + 2e^- \rightleftharpoons Ag_2O_2 + 2\,OH^-$	1.711				50	73
1	no	$Ag_3PO_4 + 3e^- \rightleftharpoons 3\,Ag + PO_4^{3-}$	0.3402	± 0.1	-1520		51	74
1	no	$Ag_2S + 2e^- \rightleftharpoons 2\,Ag + S^{2-}$	-0.691	± 1	-1080		2, 52	75
1	no	$AgSCN + e^- \rightleftharpoons Ag + SCN^-$	0.08951	± 0.01	+110		53	76
1		$Ag(SCN)_2^- + e^- \rightleftharpoons Ag + 2\,SCN^-$	0.304			3	54	77
1		$Ag(SCN)_3^{2-} + e^- \rightleftharpoons Ag + 3\,SCN^-$	0.231			3	54	78
1		$Ag(SCN)_4^{3-} + e^- \rightleftharpoons Ag + 4\,SCN^-$	0.214			3	54	79

Electronically conducting phase	Intermediate species	Composition of the solution	Solvent	Temperature °C	Pressure Torr	Measuring method
a	b	c	d		e	d
Ag	Ag_2SeO_3	$NaHSeO_3$: 0.00923 up to 0.0922 m	1	25± 0.02		I
Ag			1	25	760	IV
Ag			1	25	760	IV
Ag	Ag_2SO_4		1	25	760	IV
Ag	Ag_2WO_4	Na_2WO_4 , $NaNO_3$, $AgNO_3$: all in the range of 0.02 - 0.05 m	1	25± 0.02		I
Ag	Ag (acetate)		1	25	760	IV
1)Ag 2)Glass		1) Ag acetate: 0.01 up to 0.02 M; Na acetate: 0.04 up to 0.7 M; 2) Ag acetate: satd.; Me acetate: 0 up to 1.0987 m (Me = Na, K, Cs, Sr) 3) Ag acetate: 0.6160 m; $NaClO_4$: 2.95 M; 4) CH_3COOH: 0.05418 up to 0.07760 M; NaOH: 0.03180 up to 0.07152 N; $AgBrO_3$ satd.	1	25		I V
Glass	Alanine	Alanine: (24.95 up to 26.30) · 10^{-3} M; $AgNO_3$: (18.20 up to 19.19) · 10^{-3} M	1	25± 0.02		I
Ag	Ag (benzoate)	$AgClO_4$ + $NaClO_4$: 0.2 or 1 M	1	25		I
Ag	Ag (chloroacetate)	Ag chloroacetate : up to 1.0 M	1	25.00± 0.02		I
Ag		Ag chloroacetate : 1.0 M	1	25.00± 0.02		I
Ag	Ag_3 (citrate)	Citric acid: 0.2 up to 0.1 M	1	25	751	I
Ag	3-cyanopyridine		1	25.0± 0.1		I
Ag	4-cyanopyridine		1	25.0± 0.1		I
Pt, H_2	Dimethylamine		1	25		I
Ind.	Dipyridyl	Ag (dipyridyl$)_2^{2+}$, Ag (dipyridyl$)_2^{+}$, KNO_3	1	25± 0.1		I
Glass	Ethylamine		1	25		I
Ag	Ethylamine		1	25		III
Ag	Ethylenediamine	KNO_3 : 0.1 M	1	25		I
Ag	Ag (formate)	Na acetate: 0.005 up to 3 M	1	25± 0.05		I
Glass	Glycine	Glycine: (21.27 up to 21.90)· 10^{-3} M; $AgNO_3$: (18.23 up to 18.78) · 10^{-3} M	1	25		I

Comparison electrode	Liquid junction	Electrode reaction	Standard value V	Uncertainty mV	Temperature Coefficient μV/°C	Notes	Reference	Electrode Reference Number
d			f		g	h		
33	KNO_3 (aq.)	$Ag_2SeO_3 + 2e^- \rightleftharpoons 2\ Ag + SeO_3^{2-}$	0.3629				55	80
		$Ag(SO_3)_2^{3-} + e^- \rightleftharpoons Ag + 2\ SO_3^{2-}$	0.295				2	81
		$Ag(S_2O_3)_2^{3-} + e^- \rightleftharpoons Ag + 2\ S_2O_3^{2-}$	0.017				2	82
		$Ag_2SO_4 + 2e^- \rightleftharpoons 2\ Ag + SO_4^{2-}$	0.654		-1182		2	83
12	KNO_3 (aq.)	$Ag_2WO_4 + 2e^- \rightleftharpoons 2\ Ag + WO_4^{2-}$	0.4660		1200		56	84
		Ag (acetate) + e^- ⇌ Ag + acetate$^-$	0.643		-797		2	85
1) 7 2)3		Ag $(acetate)_2^-$ + e^- ⇌ Ag + 2 acetate$^-$	0.421			4	57 58 59 60	86
3		Ag $(alanine)_2^+$ + e^- ⇌ Ag + 2 alanine	0.375			6	61	87
4	$NaClO_4$ (aq.)	Ag (benzoate) + e^- ⇌ Ag + benzoate$^-$	0.746				62	88
7		Ag (chloroacetate) + e^- ⇌ Ag + chloroacetate$^-$	0.422				57	89
7		Ag $(chloroacetate)_2^-$ + e^- ⇌ Ag + 2 chloroacetate$^-$	0.486				57	90
1	no	Ag_3 (citrate) + $3e^-$ ⇌ 3 Ag + citrate^{3-}	0.5046	± 0.2	-890		63	91
3	KNO_3 satd. KCl satd.	Ag $(3\text{-cyanopyridine})_2^+$ + e^- ⇌ Ag + 2(3-cyano-pyridine)	0.629				64	92
3	KNO_3 satd. KCl satd.	Ag$(4\text{-cyanopyridine})_2^+$ + e^- ⇌ Ag + 2 (4-cyanopyridine)	0.618				64	93
3	KCl satd.	Ag $(dimethylamine)_2^+$ + e^- ⇌ Ag + 2 dimethylamine	0.482			6	65	94
33		Ag $(dipyridyl)_2^{2+}$ + e^- ⇌ Ag $(dipyridyl)_2^+$	-0.758		1200	7	66	95
3		Ag $(ethylamine)^+$ + e^- ⇌ Ag + ethylamine	0.600		1000	6	67	96
1		Ag $(ethylamine)_2^+$ + e^- ⇌ Ag + 2 ethylamine	0.368				65, 67 68, 69	97
		Ag $(ethylenediamine)^+$ + e^- ⇌ Ag + ethylenediamine	0.361				70	98
21		Ag (formate) + e^- ⇌ Ag + formate$^-$	0.502	± 1000		2	71	99
3		Ag $(glycine)_2^+$ + e^- ⇌ Ag + 2 glycine	0.392			6	61	100

Electronically conducting phase	Intermediate species	Composition of the solution	Solvent	Temperature °C	Pressure Torr	Measuring method
a	b	c	d		e	d
Glass	Glycylglycine	Glycylglycine : (20.40 up to 24.05) · 10^{-3} M; $AgNO_3$:(14.64 up to 19.48) · 10^{-3} M	1	25		I
Glass	Glycylglycine	Glycylglycine : (20.40 up to 24.05) · 10^{-3} M; $AgNO_3$: (14.64 up to 19.48) · 10^{-3} M	1	25		I
Glass	3-methoxypyridine		1	25.0± 0.1		I
Glass	4-methoxypyridine		1	25.0± 0.1		I
Glass	3-methoxypyridine		1	25.0± 0.1		I
Glass	4-methoxypyridine		1	25.0± 0.1		I
Glass	Methylamine	$CH_3NH_3^+NO_3^-$: 0.1 M	1	25		I
Pt, H_2	Methylamine		1	25		I
Ag	Methylnicotinamide		1	25± 0.1		I
Glass			1	25.0± 0.1		I
Ag	Nicotinamide		1	25.0± 0.1		I
Ag	Isonicotinamide		1	25.0± 0.1		I
Ag	Ag_2(oxalate)	Oxalic acid: 0.02 up to 0.66 M	1	25		I
Ag			16	25.00± 0.01		I

Comparison electrode	Liquid junction	Electrode reaction	Standard value V	Uncertainty mV	Temperature Coefficient μV/°C	Notes	Reference	Electrode Reference Number
d			f		g	h		
3		$Ag(glycylglycine)^+ + e^- \rightleftharpoons Ag + glycylglycine$	0.639			6	61	101
3		$Ag(glycylglycine)_2^+ + e^- \rightleftharpoons Ag + 2\ glycylglycine$	0.505			6	61	102
3	KNO_3 satd. KCl satd.	$Ag(3\text{-methoxypyridine})^+ + e^- \rightleftharpoons$ $Ag + 3\text{-methoxypyridine}$	0.707			6	64	103
3	KNO_3 satd. KCl satd.	$Ag(4\text{-methoxypyridine})^+ + e^- \rightleftharpoons$ $Ag + 4\text{-methoxypyridine}$	0.665			6	64	104
3	KNO_3 satd. KCl satd.	$Ag(3\text{-methoxypyridine})_2^+ + e^- \rightleftharpoons$ $Ag + 2\ (3\text{-methoxypyridine})$	0.583			6	64	105
3	KNO_3 satd. KCl satd.	$Ag(4\text{-methoxypyridine})_2^+ + e^- \rightleftharpoons$ $Ag + 2\ (4\text{-methoxypyridine})$	0.538			6	64	106
		$Ag(methylamine)^+ + e^- \rightleftharpoons Ag + methylamine$	0.614				70	107
3	KCl satd.	$Ag(methylamine)_2^+ + e^- \rightleftharpoons Ag + 2\ methylamine$	0.399			6	65	108
3	KNO_3 satd. KCl satd.	$Ag(methylnicotinamide)_2 + e^- \rightleftharpoons$ $Ag + 2\ methylnicotinamide$	0.625				64	109
3	KNO_3 satd. KCl satd.	$Ag(methylisonicotinate)_2^- + e^- \rightleftharpoons$ $Ag + 2\ methylisonicotinate^-$	0.655			6	64	110
3	KNO_3 satd. KCl satd.	$Ag(nicotinamide)_2^+ + e^- \rightleftharpoons Ag + 2\ nicotinamide$	0.609				64	111
3	KNO_3 satd. KCl satd.	$Ag(isonicotinamide)_2^+ + e^- \rightleftharpoons Ag + 2\ isonicotinamide$	0.699				64	112
1		$Ag_2(oxalate) + 2e^- \rightleftharpoons 2\ Ag + oxalate^{2-}$	0.4647		971	8	72	113
1		$Ag(picrate) + e^- \rightleftharpoons Ag + picrate^-$	0.17		-340		73	114

Electronically conducting phase	Intermediate species	Composition of the solution	Solvent	Temperature °C	Pressure Torr	Measuring method
a	b	c	d		e	d
Ag	Pyridine	1) not reported 2) $AgIO_3$, $AgBrO_3$ and $AgSO_4$ satd.	1	25.0± 0.1		I
Ag	Pyridine	1) Not reported 2) $AgIO_3$, $AgBrO_3$ and $AgSO_4$ satd.	1	25.0± 0.1		I, V
Ag	Ag_2 (tartrate)	Tartaric acid: 0.010 up to 0.10 M	1	25	760	I
Ag	Thiourea	Thiourea : 0.0584 up to 0.1963M; $AgNO_3$: $1 \cdot 10^{-3}$M	1	25		I

Comparison electrode	Liquid junction	Electrode reaction	Standard value	Uncertainty	Temperature Coefficient	Notes	Reference	Electrode Reference Number
			V	mV	μV/°C			
d			f		g	h		
1	KNO_3 satd. KCl satd.	$Ag(pyridine)^+ + e^- \rightleftharpoons Ag + pyridine$	0.682				64, 74	115
1	KNO_3 satd. KCl satd.	$Ag(pyridine)_2^+ + 2e^- \rightleftharpoons Ag + 2\ pyridine$	0.549				64, 74	116
1	no	$Ag_2(tartrate) + 2e^- \rightleftharpoons 2\ Ag + tartrate^{2+}$	0.5610		-1340		75	117
6	NH_4-NO_3	$Ag(thiourea)_3^+ + e^- \rightleftharpoons Ag + 3\ thiourea$	0.026				76	118

Silver 1

Standard value V	Solvent	Electrode Reference Number
-0.758	1	95
-0.691	1	75
-0.31	1	42
-0.15224	1	48
-0.017	1	41
0.017	1	82
0.026	1	118
0.07133	1	11
0.0714	1	12
0.08951	1	76
0.1478	1	47
0.214	1	79
0.22230	1	25
0.22233	1	26
0.231	1	78
0.239	1	23
0.274	1	21
0.29	1	56
0.295	1	81
0.298	1	44
0.304	1	77
0.328	1	20
0.330	1	22
0.3402	1	74
0.342	1	64
0.351	1	68
0.354	1	52
0.361	1	98
0.3629	1	80
0.368	1	97
0.373	1	58
0.375	1	87
0.379	1	19
0.392	1	100
0.399	1	108

Standard value V	Solvent	Electrode Reference Number
0.421	1	86
0.422	1	89
0.4470	1	45
0.4573	1	55
0.4647	1	113
0.4660	1	84
0.47	1	43
0.482	1	94
0.486	1	90
0.502	1	99
0.5046	1	91
0.505	1	102
0.538	1	106
0.546	1	24
0.549	1	116
0.5610	1	117
0.564	1	59
0.583	1	105
0.600	1	96
0.603	1	57
0.607	1	72
0.609	1	111
0.614	1	107
0.618	1	93
0.625	1	109
0.629	1	92
0.639	1	101
0.643	1	85
0.654	1	13
0.654	1	83
0.655	1	110
0.665	1	104
0.682	1	115
0.699	1	112
0.707	1	103

Standard value V	Solvent	Electrode Reference Number
0.725	1	54
0.739	1	70
0.746	1	88
0.763	1	53
0.779	1	46
0.787	1	40
0.800	1	1
0.960	1	71
1.288	1	69
1.360	1	67
1.670	1	65
1.711	1	73
1.757	1	66
1.772	1	61
1.980	1	10
1.998	1	62
2.016	1	63
2.220	1	60
0.757	6	2
-0.3176	7	49
-0.139	7	14
-0.0221	7	27
-0.1939	8	15
-0.0848	8	28
-0.1135	10	29
-0.1345	9	30
-0.1430	11	31
-0.2928	13	50
-0.1007	13	16
0.0235	13	32
-0.3468	14	51
-0.1633	14	17
-0.0323	14	33
0.02077	32	34
-0.54	15	35

Standard value V	Solvent	Electrode Reference Number
0.49	15	3
0.17	16	114
0.87	17	4
0.31	33	5
0.0967	18	18
0.1979	18	36
0.6902	18	6
0.208	19	37
0.812	20	7
0.21187	21	38
0.1500	22	39
0.240	23	8
0.551	27	9

Standard value V	Solvent	Electrode Reference Number

Standard value V	Solvent	Electrode Reference Number

1) Liquid junction correction was applied
2) Liquid junction correction was neglected
3) Solubility data
4) Potentiometric and solubility data
5) Conductometric data
6) Potentiometric pH measurements
7) Liquid junction electric tension was minimized
8) Only the most recent value is reported. Other two equally reliable values for the same system are : 0.4591 V (72') and 0.4725 V (72")

BIBLIOGRAPHY: Silver

1) C. Papon, J. Jacq, Bull. Soc. Chim. France, 1965, 13 - 17

2) Data reported by A.J. De Bethune, N.A. Swendeman-Loud, "Standard Aqueous Electrode Potentials and Temperature Coefficients at 25 °C" (Clifford A. Hampel, Ill.)

3) L.M. Mukherjee, S. Bruckenstein, F.A.K. Badawi, J. Phys. Chem., 69 (1965), 2544 - 2557

4) D.M. Everett, S.E. Rasmussen, J. Chem. Soc., 1954, 2812 - 2815

5) P.S. Tutundžik, P. Putanov, Glasnik Khim. Drustva, 21 (1956), 19 - 31

6) G. Petit, J. Bessiere, J. Electroanal. Chem., 34 (1972), 489 - 497

7) R.W. Broadbank, S. Dhabanandana, H.W. Marcom, B.L. Muju, Trans. Faraday Soc., 64 (1968), 3311 - 3317

8) M. Chateau, M.C. Moncet, Compt. Rend., 256 (1963), 1504 - 1506

9) R.L. Benoit, Inorg. Nuclear Chem. Letters, 4 (1968), 723 - 729

10) L.M. Mukherjee, J.J. Kelly, D. McRichards, J.M. Luckacs, J. Phys. Chem., 71 (1969), 580 - 586

11) H.B. Hetzer, R.A. Robinson, R.G. Bates, J. Phys. Chem., 66 (1962), 1423 - 1426

12) M.B. Towns, R.S. Greeley, N.H. Lietzke, J. Phys. Chem., 64 (1960), 1861 - 1863

13) H.K. Sinha, B. Prasad, J. Indian Chem. Soc., 47 (1970), 901 - 903

14) B. Berne, I. Leden, Z. Naturforsch., 8 A (1953), 719 - 727

15) M. Alfenaar, C.L. De Ligny, A.G. Remijnze, Rec. Trav. Chim., 84 (1967), 555

16) L.J. Nunez, M.C. Day, J. Phys. Chem., 65 (1961), 164 - 166

17) K.K. Kundu, P.K. Chattopadhyay, D. Jana, M.N. Das, J. Chem. Eng. Data, 15 (1970), 209 - 213

18) K.K. Kundu, D. Jana, M.N. Das, J. Phys. Chem., 74 (1970), 2625 - 2632

19) K.W. Marcom, B.L. Muju, Nature, 217 (1968), 1046 - 1047

20) H. Chateau, B. Hervier, J. Chim. Phys., 54 (1957), 356 - 362

21) B. Sen, D.A. Johnson, R.N. Roy, J. Phys. Chem., 71 (1967), 1523

22) R.G. Bates, V.E. Bowers, J. Research N.B.S., 53 (1954), 283 - 290

23) R.S. Greeley, W.T. Smith, R.W. Stoughton, M.H. Lietzke, J. Phys. Chem., 64 (1960), 652 - 657

24) N.H. Lietzke, R.W. Stoughton, J. Chem. Educ., 39 (1962), 230 - 235

25) N.P. Komar, A.Z. Kaftanov, Zh. Anal. Khim., 26 (1971), 2086 - 2089

26) H.P. Marshall, E. Grünwald, J. Chem. Phys., 21 (1953), 2143 - 2151

27) A. Tézé, R. Schaal, Compt. Rend., 252 (1961), 3995 - 3997

28) E.M. Ryzhov, A.N. Sukhotin, Zh. Fiz. Khim., 34 (1960), 1402 - 1406

29) U. Sen, K. Kundu, M.N. Das, J. Phys. Chem., 71 (1967), 3665

BIBLIOGRAPHY: Silver

30) R.N. Roy, W. Vernon, J.J. Gibbons, A.L.M. Bothwell, J. Electroanal. Chem., 34 (1972), 101 - 107

31) R.K. Agarwal, R. Nayak, J. Phys. Chem., 71 (1967), 2062

32) M.L. Berardelli, G. Pecci, B. Scrosati, J. Electrochem. Soc., 117 (1970), 781 - 783

33) R.L. Dawson, R.C. Sheridan, H.C. Eckstrom, J. Phys. Chem., 69 (1965), 1335

34) B. Scrosati, G. Pecci, G. Pistoia, J. Electrochem. Soc., 115 (1968), 506 - 507

35) J.H. Jones, J. Am. Chem. Soc., 67 (1945), 855 - 857

36) P.A. Rock, Inorg. Chem., 4 (1965), 1667 - 1668

37) K. Pan, J. Chinese Chem. Soc., Ser. II, 2 (1955), 15 - 22

38) A.D. Paul, Univ. California, Berkeley, Calif., 1955 , UCRL - 2526

39) P.A. Rock, Inorg. Chem., 3 (1964), 1593 - 1597

40) H.B. Hetzer, R.A. Robinson, J. Phys. Chem., 68 (1964), 1929 - 1933

41) V.G. Kortüm, W. Häussermann, Ber. Bunsen-Ges.Physik. Chem., 69 (1965), 594 - 604

42) J.J. Ranier, D.S. Martin, USAEC ISC, 668 - 1955

43) K. Pan, J. Chinese Chem. Soc., Ser. II, 1 (1954), 16 - 25

44) S. Suzuki, J. Chem. Soc. Japan, Pure Chem. Sect. 73 (1952), 153 - 155

45) P.D. Dorr, R.M. Stockdale, W.C. Vosburgh, J. Am. Chem. Soc., 63 (1941), 2670 - 2674

46) W.C. Vosburgh, R.S. McClure, J. Am. Chem. Soc., 65 (1943), 1060 - 1063

47) S. Kilpi, R. Näsänen, Suomen Kem., 17 B (1944), 9

48) Data reported by M. Pourbaix, "Atlas d'Equilibres Electrochimiques à 25 ^{o}C" (Gauthier-Villars Editeur, Paris, 1963), pp. 394 - 395

49) L.V. Gregor, K.S. Pitzer, J. Am. Chem. Soc., 84 (1962), 2671 - 2673

50) B. Stehlik, Chem. Zvesti, 17 (1963), 6

51) S.M. Naqvi, P.B. Mathur, Bull. Acad. Polon. Sci., 16(1968), 479 - 484

52) R.M. Golding, J. Chem. Soc., 1959 , 1838 - 1839

53) C.E. Vanderzee, W.E. Smith, J. Am. Chem. Soc., 78 (1956), 721 - 725

54) G.B.C. Cave, D.N. Hume, J. Am. Chem. Soc., 75 (1953), 2893 - 2897

55) C.L. Lin, K. Pan, J. Chinese Chem. Soc., Ser. II , 8 (1961), 14 - 22

56) K. Pan, J. Chinese Chem. Soc., Ser. II, 1 (1954), 26 - 38

57) F.H. MacDougall, I.E. Topol, J. Phys. Chem., 56 (1952), 1090 - 1093

58) F.H. MacDougall, M. Allen, J. Phys. Chem., 46 (1942), 730 - 738

59) F.H. MacDougall, S. Peterson, J. Phys. Chem., 51 (1947), 1346 - 1361

BIBLIOGRAPHY: Silver

60) C.W. Davies, C.B. Monk, J. Chem. Soc., 1951, 2718 - 2723

61) C.B. Monk, Trans. Faraday Soc., 47 (1951), 292 - 297

62) I. Leden, Acta Chem. Scand., 3 (1949), 1318 - 1325

63) P.B. Mathur, S.M.A. Naqvi, Proc. Symp. Electrode Processes, 1966, 39 - 45

64) R.K. Murmann, F. Basolo, J. Am. Chem. Soc., 77 (1955), 3484 - 3486

65) W.S. Fyfe, J. Chem. Soc., 1955, 1347

66) E. Scrocco, G. Marmani, Atti Ac. Lincei, Rendiconti Cl. Sci. Fis. Mat. Nat., 16 (1954), 637 - 640

67) R.J. Bruehlman, F.H. Verhoak, J. Am. Chem. Soc., 70 (1948), 1401

68) H. Euler, Ber., 36 (1903), 2878

69) G. Bädländer, W. Eberlein, Ber., 36 (1903), 2878

70) J. Bjerrum, Chem. Rev., 46 (1950), 381

71) J. Chauchard, J. Gauthier, Bull. Soc. Chim. France, 1966, 2635

72) D.J. Ferrel, I. Blackburn, W. Vosburgh, J. Am. Chem. Soc., 70 (1968), 3812 - 3815

73) P.B. Mathur, S.M.A. Naqvi, Indian J. Chem., 6 (1968), 311 - 313

74) S.C. Sircair, B. Prasad, J. Indian Chem. Soc., 38 (1961), 361 - 363

75) V. Pleskov, Acta Physicochim. U.S.S.R., 21 (1946), 41 - 54

76) W.C. Vosburgh, S.A. Cogswell, J. Am. Chem. Soc., 65 (1943), 2412 - 2413

77) S.M.A. Naqvi, P.B. Mathur, Electrochim. Acta, 13 (1968), 1569 - 1573

78) W.S. Fyfe, J. Chem. Soc., 1955, 1032

Electronically conducting phase	Intermediate species	Composition of the solution	Solvent	Temperature °C	Pressure Torr	Measuring method
a	b	c	d		e	d
Au			1	25	760	IV
Au			1	25	760	IV
Ind.			1	25	760	IV
Au		HBr: 0 1 up to 1 M	1	25.00		I
Au		HBr: 0.1 up to 1 M	1	25.00		I
Au		KBr: 10^{-3} - 1.0 N; pH: 3.1 - 7.3; Au(I): (0.09 - 5.20) · 10^{-4} M; Au(III): (2.0 - 6.8) · 10^{-4} M	1	25 ± 0.5		I
Au		HCl: 0.2 up to 2 M	1	25.00		I
Au		HCl: 0.2 up to 2 M	1	25.00		I
Au		NaCl: 9.6 · 10^{-3} - 7.9 · 10^{-1} N; pH: 3.2 - 6.3; Au (I): (0.33 - 3.92) · 10^{-4} M; Au (III): (2.0 - 20.0) · 10^{-4} M	1	25 ± 0.5		I
Ind.			1	25	760	IV
Au		$H[AuCl_4]$, $H[AuBr_4]$, KI from 0.15 up to 1 M	1	25 ± 0.1		I
Au			1	25		IV
Au	Au_2O_3	H_2SO_4 : 2N; O_2 (satd.)	1	25 ± 1		III
Ind.	AuO_2		1	25	760	IV
Au			1	25 ± 0.5		I
Au	Au $(OH)_3$		1	25	760	IV
Ind.	Au $(OH)_3$		1	25	760	IV
Au			1	25 ± 0.5		I
Au		HCl : 5 · 10^{-2} up to 2 M; KCl : 0 up to 1 M; NaSCN: 4 · 10^{-2} up to 3 · 10^{-1} M; Au $(SCN)_4^-$: 10^{-5} up to 6 · 10^{-3} M	1	25 ± 0.1		I
Au			1	25	760	IV
Au			1	25	760	IV
Ind.			1	25	760	IV
Ind.			1	25	760	IV
Ind.	AuBr		1	25	760	IV
Ind.		HBr : 0.1 up to 1 M	1	25.00		I
Ind.		HCl : 0.2 up to 2 M	1	25.00		I
Ind.	$Au_2O_3(\alpha)$, AuO_2		1	25	760	IV
Ind.	$Au_2O_3(\beta)$, AuO_2		1	25	760	IV
Ind.	AuO_2		1	25	760	IV

Comparison electrode	Liquid junction	Electrode reaction	Standard value	Uncertainty	Temperature Coefficient	Notes	Reference	Electrode Reference Number
			V	mV	μV/°C			
d			f		g	h		
		$Au^{+} + e^{-} \rightleftharpoons Au$	1.692				1	1
		$Au^{3+} + 3e^{-} \rightleftharpoons Au$	1.498				1	2
		$Au^{3+} + 2e^{-} \rightleftharpoons Au^{+}$	1.401				2	3
5	no	$Au(Br)_2^{-} + e^{-} \rightleftharpoons Au + 2\,Br^{-}$	0.959	±1			3	4
5	no	$Au(Br)_4^{-} + 3e^{-} \rightleftharpoons Au + 4\,Br^{-}$	0.854	±2			3	5
3		$[Au(Br)(OH)]^{-} + e^{-} \rightleftharpoons Au + Br^{-} + OH^{-}$	0.63				4	6
4	no	$Au(Cl)_2^{-} + e^{-} \rightleftharpoons Au + 2\,Cl^{-}$	1.1540	± 0.5			5	7
4	no	$Au(Cl)_4^{-} + 3e^{-} \rightleftharpoons Au + 4\,Cl^{-}$	1.002	± 0.5	-630		5, 1	8
3		$[Au(Cl)(OH)]^{-} + e^{-} \rightleftharpoons Au + Cl^{-} + OH^{-}$	0.75				4	9
		$Au(CN)_2^{-} + e^{-} \rightleftharpoons Au + 2\,CN^{-}$	-0.669				6	10
3		$Au(I)_2^{-} + e^{-} \rightleftharpoons Au + 2I^{-}$	0.578	±5			7	11
		$Au(I)_4^{-} + 3e^{-} \rightleftharpoons Au + 4\,I^{-}$	0.56	± 50			8	12
1	no	$Au_2O_3 + 6H^{+} + 6e^{-} \rightleftharpoons 2Au + 3H_2O$	1.360	±1			9	13
		$AuO_2 + 4H^{+} + e^{-} \rightleftharpoons Au^{3+} + 2H_2O$	2.507				2	14
3		$Au(OH)_2^{-} + e^{-} \rightleftharpoons Au + 2\,OH^{-}$	0.4				4	15
		$Au(OH)_3 + 3H^{+} + 3e^{-} \rightleftharpoons Au + 3H_2O$	1.45		-206		1	16
		$Au(OH)_3 + 3H^{+} + 2e^{-} \rightleftharpoons Au^{+} + 3H_2O$	1.502				2	17
3	KNO_3	$Au(SCN)_2^{-} + e^{-} \rightleftharpoons Au + 2\,SCN^{-}$	0.661				10	18
3	KNO_3 0.1M	$Au(SCN)_4^{-} + 3e^{-} \rightleftharpoons Au + 4\,SCN^{-}$	0.636	±5			11	19
		$HAuO_3^{2-} + 5H^{+} + 3e^{-} \rightleftharpoons Au + 3H_2O$	2.059				2	20
		$H_2AuO_3^{-} + 4H^{+} + 3e^{-} \rightleftharpoons Au + 3H_2O$	1.796				2	21
		$HAuO_3^{2-} + 5H^{+} + 2e^{-} \rightleftharpoons Au^{+} + 3H_2O$	2.243				2	22
		$H_2AuO_3^{-} + 4H^{+} + 2e^{-} \rightleftharpoons Au^{+} + 3H_2O$	1.849				2	23
		$Au(Br)_4^{-} + 2e^{-} \rightleftharpoons AuBr + 3\,Br^{-}$	0.820				12	24
5	no	$Au(Br)_4^{-} + 2e^{-} \rightleftharpoons Au(Br)_2^{-} + 2\,Br^{-}$	0.802	±2			3	25
4	no	$Au(Cl)_4^{-} + 2e^{-} \rightleftharpoons Au(Cl)_2^{-} + 2\,Cl^{-}$	0.926	±0.5			5	26
		$2AuO_2 + 2H^{+} + 2e^{-} \rightleftharpoons Au_2O_3\ (\alpha) + H_2O$	2.630				2	27
		$2AuO_2 + 2H^{+} + 2e^{-} \rightleftharpoons Au_2O_3\ (\beta) + H_2O$	2.465				2	28
		$AuO_2 + H_2O + H^{+} + e^{-} \rightleftharpoons Au(OH)_3$	2.305				2	29

Electronically conducting phase	Intermediate species	Composition of the solution	Solvent	Temperature °C	Pressure Torr	Measuring method
a	b	c	d		e	d
Ind.		HCl : 10^{-2} up to 1 M; NaSCN : $2.95 \cdot 10^{-1}$ up to 3 M; $Au(SCN)_2^-$: $6 \cdot 10^{-5}$ up to $8 \cdot 10^{-1}$ M; $Au(SCN)_2^-$: 10^{-5} up to $6 \cdot 10^{-3}$	1	25 ± 0.1		I
Ind.	AuO_2		1	25	760	IV
Ind.	AuO_2		1	25	760	IV

Comparison electrode	Liquid junction	Electrode reaction	Standard value	Uncertainty	Temperature Coefficient	Notes	Reference	Electrode Reference Number
			V	mV	μV/°C			
d			f		g	h		
3	KNO_3 0.1 M	$Au(SCN)_4^- + 2e^- \rightleftharpoons Au(SCN)_2^- + 2\ SCN^-$	0.623	± 5			11	30
		$AuO_2 + H_2O + e^- \rightleftharpoons HAuO_3^{2-} + H^+$	0.822				2	31
		$AuO_2 + H_2O + e^- \rightleftharpoons H_2AuO_3^-$	1.611				2	32

Standard value V	Solvent	Electrode Reference Number
-0.669	1	10
0.4	1	15
0.56	1	12
0.578	1	11
0.623	1	30
0.63	1	6
0.636	1	19
0.661	1	18
0.75	1	9
0.802	1	25
0.820	1	24
0.822	1	31
0.854	1	5
0.926	1	26
0.959	1	4
1.002	1	8
1.1540	1	7
1.360	1	13
1.401	1	3
1.45	1	16
1.498	1	2
1.502	1	17
1.611	1	32
1.692	1	1
1.796	1	21
1.849	1	23
2.059	1	20
2.243	1	22
2.305	1	29
2.465	1	28
2.507	1	14
2.630	1	27

Standard value V	Solvent	Electrode Reference Number

Standard value V	Solvent	Electrode Reference Number

BIBLIOGRAPHY : Gold

1) Data reported by A.J. De Bethune, N.A. Swendeman-Loud, "Standard Aqueous Electrode Potentials and Temperature Coefficients at 25 °C" (Clifford A. Hampel, Ill.)

2) Data reported by M. Pourbaix, "Atlas d'Equilibres Electrochimiques à 25 °C" (Gauthier-Villars Editeur, Paris, 1963), pp. 400 - 401

3) D.H. Evans, J.J. Lingane, J. Electroanal. Chem., 6 (1963), 1 - 10

4) M.C. Gadet, J. Pouradier, Compt. Rend., 275 C (1972), 1061 - 1064

5) J.J. Lingane, J. Electroanal. Chem., 4 (1962), 332 - 342

6) W.M. Latimer, "Oxidation Potentials", 1952, 195 (Prentice - Hall, New York)

7) A.M. Erenburg, B.I. Peshchevitskii, Zh. Neorgan. Khim., 14 (1969), 932 - 935

8) A.M.Erenburg, B.I. Peshchevitskii, Russian J. Inorg. Chem., 14 (1969), 1429 - 1431

9) J.P. Hoare, Electrochim. Acta, 11 (1966), 203 - 210

10) A. De Cugnac-Pailliotet, J. Pouradier, Compt. Rend., 273 C (1971), 1565 - 1568

11) J. Pouradier, M.G. Gadet, J. Chim. Phys., 63 (1966), 1467 - 1473

12) B.I. Peshchevitskii, V.P. Kazakov, A.M. Erenburg, Zh. Neorgan. Khim., 8 (1963), 853

Group II

Electronically conducting phase	Intermediate species	Composition of the solution	Solvent	Temperature °C	Pressure Torr	Measuring method
a	b	c	d		e	d
Be			1	25	760	IV
Be	BeO		1	25	760	IV
Be	BeO		1	25	760	IV
Be			1	25	760	IV
Be			1	25	760	IV
Be			1	25	760	IV

Comparison electrode	Liquid junction	Electrode reaction	Standard value V	Uncertainty mV	Temperature Coefficient μV/°C	Notes	Reference	Electrode Reference Number
d			f		g	h		
		$Be^{2+} + 2e^- \rightleftharpoons Be$	-1.847		+565		1	1
		$BeO + 2H^+ + 2e^- \rightleftharpoons Be + H_2O$	-1.785				2	2
		$BeO + H_2O + 2e^- \rightleftharpoons Be + 2OH^-$	-2.613		-1172		1	3
		$BeO_2^{2-} + 4H^+ + 2e^- \rightleftharpoons Be + 2H_2O$	-0.909				2	4
		$Be_2O_3^{2-} + 6H^+ + 4e^- \rightleftharpoons 2Be + 3H_2O$	-1.387				2	5
		$Be_2O_3^{2-} + 3H_2O + 4e^- \rightleftharpoons 2Be + 6OH^-$	-2.63				1	6

Standard value V	Solvent	Electrode Reference Number	Standard value V	Solvent	Electrode Reference Number	Standard value V	Solvent	Electrode Reference Number
-2.63	1	6						
-2.613	1	3						
-1.847	1	1						
-1.785	1	2						
-1.387	1	5						
-0.909	1	4						

BIBLIOGRAPHY : Beryllium

1) Data reported by A.J. De Bethune, N.A. Swendeman-Loud "Standard Aqueous Electrode Potentials and Temperature Coefficients at 25 °C" (Clifford A. Hampel, Ill.)

2) Data reported by M. Pourbaix "Atlas d'Equilibres Electrochimiques à 25 °C" (Gauthier-Villars Editeur, Paris, 1963), pp. 146 - 154

Electronically conducting phase	Intermediate species	Composition of the solution	Solvent	Temperature °C	Pressure Torr	Measuring method
a	b	c	d		e	d
Mg			1	25	760	IV
Cd	Cd $(OH)_2$, $Mg(OH)_2$	$MgCl_2$	1	25± 0.01		I
Cd	Cd $(OH)_2$, $Mg(OH)_2$	$MgCl_2$	7	25± 0.01		I
Mg			1	25		V
Mg			1	25		V
Glass			1	25		I
Mg	Mg $(OH)_2$		1	25	760	IV
Glass, quinhydrone		$MgCl_2$: (6.67 up to 66.67) · 10^{-3}M; H_3PO_4 : (1.116 up to 11.16) · 10^{-3} M; NaOH: (1.60 up to 2.80) · 10^{-3} M	1	25		I
Mg		$Na_4P_4O_{12}$:(0.76728 up to 0.77536) · 10^{-3} M; $MgCl_2$: (0.16834 up to 5.2934) · 10^{-3} M	1	25		III
Mg	Mg (SO_4)	1) Mg (SO_4): (0.16196 up to 1.7692) · 10^{-3} M; 2) Mg (SO_4): (3.795 up to 6.933) · 10^{-3} M; HCl : (7.666 up to 7.762) · 10^{-3} M	1	25		III, I
Glass	Glycine	$MgCl_2$: (21.40 up to 38.74) · 10^{-3} M; glycine: (65.23 up to 85.67) · 10^{-3} M; NaOH : (1.76 up to 6.33) · 10^{-3} M	1	25		I
Pt, H_2	Mg (glycolate)	Mg(glycolate): (10.576 up to 38.37) · 10^{-3} M; HCl : (6.287 up to 35.33) · 10^{-3} M	1	25		I
Glass	Glycylglycine	$MgCl_2$: (2.693 up to 2.764) · 10^{-3} M; glycylglycine: (9.69 up to 9.94) · 10^{-3} M; NaOH : (1.32 up to 2.66) · 10^{-3} M	1	25± 0.02		I
Glass	Mg (8-hydroxyquinoline-5-sulfonate)	8-hydroxyquinoline-5-sulfonate : 2.954 · 10^{-3} M; NaOH : 1.4 · 10^{-3} M; $MgCl_2$	1	25	743	I
Pt, H_2	Mg (lactate)	Mg (lactate) : (7.141 up to 18.43) · 10^{-3} M; HCl : (9.389 up to 45.418) · 10^{-3} M	1	25		I
Pt, H_2	Mg (malonate)	$MgCl_2$: (1.650 up to 10.333) · 10^{-3} M; malonic acid : (5.688 up to 18.156) · 10^{-3} M; NaOH : (8.469 up to 29.620) · 10^{-3} M	1	25		I
Mg	Mg (oxalate)	Mg (oxalate) : (2.5 up to 3.1) · 10^{-3} M; K_2(oxalate): (0 up to 3.0) · 10^{-3} M	1	25 ± 0.03		V

Comparison electrode	Liquid junction	Electrode reaction	Standard value	Uncertainty	Temperature Coefficient	Notes	Reference	Electrode Reference Number
			V	mV	μV/°C			
d			f		g	h		
		$Mg^+ + e^- \rightleftharpoons Mg$	-2.70				1	1
3		$Cd(OH)_2 + Mg^{2+} + 2e^- \rightleftharpoons Cd + Mg(OH)_2$	-2.372		+103	1	2,8	2
3		$Cd(OH)_2 + Mg^{2+} + 2e^- \rightleftharpoons Cd + Mg(OH)_2$	-2.478			1	2	3
1		$Mg[Fe(CN)_6]^{2-} + 2e^- \rightleftharpoons Mg + [Fe(CN)_6]^{4-}$	-2.478			2	3	4
1		$Mg(IO_3)^+ + 2e^- \rightleftharpoons Mg + IO_3^-$	-2.384			3	4,5,6	5
3		$Mg(OH)^+ + 2e^- \rightleftharpoons Mg + OH^-$	-2.440			4	7	6
		$Mg(OH)_2 + 2e^- \rightleftharpoons Mg + 2OH^-$	-2.690		-945		8	7
3		$Mg(PO_4)^- + 2e^- \rightleftharpoons Mg + PO_4^{3-}$	-2.437			4	9	8
1		$Mg(P_4O_{12})^{2-} + 2e^- \rightleftharpoons Mg + P_4O_{12}^{4-}$	-2.515			5	10	9
1		$Mg(SO_4) + 2e^- \rightleftharpoons Mg + SO_4^{2-}$	-2.430		-200	6	11,12	10
		$Mg(glycine)^{2+} + 2e^- \rightleftharpoons Mg + glycine$	-2.465			4	13	11
3		$Mg(glycolate) + 2e^- \rightleftharpoons Mg + glycolate^{2-}$	-2.402			4	14	12
3		$Mg(glycylglycine)^{2+} + 2e^- \rightleftharpoons Mg + glycylglycine$	-2.394			4	13	13
4		$Mg(8\text{-hydroxyquinoline-5-sulfonate}) + 2e^- \rightleftharpoons Mg + (8\text{-hydroxyquinoline-5-sulfonate})^{2-}$	-2.606			4	15	14
3		$Mg(lactate) + 2e^- \rightleftharpoons Mg + lactate^{2-}$	-2.404			4	16	15
4		$Mg(malonate) + 2e^- \rightleftharpoons Mg + malonate^{2+}$	-2.447			4	17	16
1		$Mg(oxalate) + 2e^- \rightleftharpoons Mg + oxalate^{2-}$	-2.493			3	18	17

Standard value V	Solvent	Electrode Reference Number	Standard value V	Solvent	Electrode Reference Number	Standard value V	Solvent	Electrode Reference Number
-2.70	1	1						
-2.690	1	7						
-2.606	1	14						
-2.515	1	9						
-2.493	1	17						
-2.478	1	4						
-2.465	1	11						
-2.447	1	16						
-2.440	1	6						
-2.437	1	8						
-2.430	1	10						
-2.404	1	15						
-2.402	1	12						
-2.394	1	13						
-2.384	1	5						
-2.372	1	2						
-2.478	7	3						

NOTES: Magnesium

1) Reaction equivalent to: $Mg^{2+} + 2e^- \rightleftharpoons Mg$
2) Spectrophotometric data
3) Solubility data
4) Potentiometric pH measurements
5) Conductometric data
6) Potentiometric and conductometric data

BIBLIOGRAPHY: Magnesium

1) S.M. Pobedinskii, G.A. Krestov, L.L. Kuzmin, Izv. Vysskh. Ucheb. Zavedenii, Khim. i Khim. Tekhnol., 65 (1963), 768 - 773
2) B.J. Jakuszewski, S. Taneewska-Osinska, Zesz. Nauk Univ. Lodz, Ser., 2 (1961), 10, 195
3) S.R. Cohen, Thesis, Cornell Univ. Ithaca, N.Y., 1956, Univ. Microfilms 20,406
4) C.W. Davies, J. Chem. Soc., 1930, 2410
5) G. Kilde, Z. Anorg. Chem., 218 (1934), 113
6) C.W. Wise, C.W. Davies, J. Chem. Soc., 1938, 273
7) D.I. Stock, C.W. Davies, Trans. Faraday Soc., 44 (1948), 856
8) Data reported by A.J. De Bethune, N.A. Swendeman-Loud, "Standard Aqueous Electrode Potentials and Temperature Coefficients at 25 °C" (Clifford A. Hampel, Ill.)
9) I. Greenwald, J. Redish, A.C. Kibrick, J. Biol. Chem., 135 (1940), 65 - 76
10) H.W. Jones, C.B. Monk, J.Chem. Soc., 1950, 3475
11) H.S. Dunsmore, J.C. James, J. Chem. Soc., 1951, 2925
12) H.W. Jones, C.B. Monk, Trans. Faraday Soc., 48 (1952), 929
13) C.B. Monk, Trans. Faraday Soc., 47 (1951), 292 - 297
14) P.B. Davies, C.B. Monk, Trans. Faraday Soc., 50 (1954), 128
15) R. Näsänen, E. Uisitalo, Acta Chem. Scand., 8 (1954), 112 - 118
16) C.W. Davies, C.B. Monk, Trans. Faraday Soc., 50 (1954), 132
17) J.I. Evans, C.B. Monk, Trans. Faraday Soc., 48 (1952), 934
18) J.E. Barney, W.J. Argersiger, Jr., G.A. Reynolds, J. Am. Chem. Soc., 73 (1951), 3785

Electronically conducting phase	Intermediate species	Composition of the solution	Solvent	Temperature °C	Pressure Torr	Measuring method
a	b	c	d		e	d
Ca	CaH_2		1	25	760	IV
Ca			1	25	760	IV
Pb	$CaCO_3$, $PbCO_3$	$CaCl_2$	1	25		I
Ca		1) $Pb(NO_3)_2$: 0.1 N; $Ca(NO_3)_2$: 0.24 N; 2) NH_4SCN : 20 mol %; $Pb(NO_3)_2$: 0.01 N; $Ca(NO_3)_2$: 0.054 up to 0.28 N	2	-34 ± 0.5		I
Ca			3	25	760	IV
Ca (Hg)		$Ca(ClO_4)_2$: $7.17 \cdot 10^{-4}$ M	29	25		I
Pb	$CaCO_3$, $PbCO_3$	$CaCl_2$	7	25		I
Ca			8	25	760	IV
Ca			11	25	760	IV
Ca			12	25	760	IV
Ca			15	25	760	IV
Ca (Hg)		Ca : 0.0253 % wt in Hg; $Ca(HCOO)_2$: 0.01 N	16	25 ± 0.01		I
Ca		$Ca(NO_3)_2$: 0.01 N	23	25		I
Ind.	CaH_2		1	25	760	IV
Ca	$Ca(HPO_4)$	$Ca(IO_3)_2$ satd.; KH_2PO_4 : up to $66.67 \cdot 10^{-3}$ M or K_2HPO_4 : 0.95 M + Na_2HPO_4 : 0.05 M	1	25		V
Ca		$Ca(IO_3)_2$ satd.; KH_2PO_4 : up to $66.67 \cdot 10^{-3}$ M or K_2HPO_4 : 0.95 + Na_2HPO_4 : 0.05 M	1	25		V
Ca	CaO		1	25	760	IV
Ind.	CaH_2, CaO		1	25	760	IV
Ind.	CaO_2		1	25	760	IV
Ind.	CaO, CaO_2		1	25	760	IV
Ind.	$Ca(OH)_2$, CaO_2		1	25	760	IV
Ca		1) Not reported 2) Not reported 3) NaOH : (20.4 up to 26.2) $\cdot 10^{-3}$ M; $CaCl_2$: (19.8 up to 59.3) $\cdot 10^{-3}$ M; 4) $Ca(OH)_2$: (11.80 up to 18.77) $\cdot 10^{-3}$ M; $CaCl_2$: (13.96 up to 57.4) $\cdot 10^{-3}$ M	1	25		I IV IV
Ca	$Ca(OH)_2$		1	25	760	IV
Ca	$Ca(OH)_2$		1	25	760	IV
Ind.	CaH_2, $Ca(OH)_2$		1	25	760	IV
Ca			1	25		V, III
Ca		1) Not reported 2) $Na_4P_4O_{12}$: (0.87892 up to 0.92256) $\cdot 10^{-3}$ N; $CaCl_2$ (0.43470 up to 1.10498) $\cdot 10^{-3}$ N	1	25		V

Comparison electrode	Liquid junction	Electrode reaction	Standard value V	Uncertainty mV	Temperature Coefficient μV/°C	Notes	Reference	Electrode Reference Number
d			f		g	h		
		$Ca + 2H^+ + 2e^- \rightleftharpoons CaH_2$	0.776				1	1
1		$Ca^+ + e^- \rightleftharpoons Ca$	-3.80				2	2
3		$PbCO_3 + Ca^{2+} + 2e^- \rightleftharpoons Pb + CaCO_3$	-2.868		-175	1	3,15	3
29		$Ca^{2+} + 2e^- \rightleftharpoons Ca$	-2.185	± 4			4	4
		$Ca^{2+} + 2e^- \rightleftharpoons Ca$	-1.91				5	5
35		$Ca^{2+} + 2e^- \rightleftharpoons Ca$	-2.65	± 10			6	6
3		$PbCO_3 + Ca^{2+} + 2e^- \rightleftharpoons Pb + CaCO_3$	-2.929			1	3	7
		$Ca^{2+} + 2e^- \rightleftharpoons Ca$	-2.870				7	8
		$Ca^{2+} + 2e^- \rightleftharpoons Ca$	-2.869				7	9
		$Ca^{2+} + 2e^- \rightleftharpoons Ca$	-2.808				7	10
		$Ca^{2+} + 2e^- \rightleftharpoons Ca$	-2.895				7	11
14		$Ca^{2+} + 2e^- \rightleftharpoons Ca$	-3.20		+300		8	12
7		$Ca^{2+} + 2e^- \rightleftharpoons Ca$	-2.75				9	13
		$Ca^{2+} + 2H^+ + 4e^- \rightleftharpoons CaH_2$	-1.045				1	14
1		$Ca(HPO_4) + 2e^- \rightleftharpoons Ca + HPO_4^{2-}$	-2.948			2	10	15
1		$Ca(H_2PO_4)^+ + 2e^- \rightleftharpoons Ca + H_2PO_4^-$	-2.900			2	10	16
		$CaO + 2H^+ + 2e^- \rightleftharpoons Ca + H_2O$	-1.902				1	17
		$CaO + 4H^+ + 4e^- \rightleftharpoons CaH_2 + H_2O$	-0.563				1	18
		$CaO_2 + 4H^+ + 2e^- \rightleftharpoons Ca^{2+} + 2H_2O$	2.224				1	19
		$CaO_2 + 2H^+ + 2e^- \rightleftharpoons CaO + H_2O$	1.260				1	20
		$CaO_2 + 2H^+ + 2e^- \rightleftharpoons Ca(OH)_2$	1.547				1	21
1		$Ca(OH)^+ + 2e^- \rightleftharpoons Ca + OH^-$	-2.906		-150	3	11 12 13 14	22
		$Ca(OH)_2 + 2H^+ + 2e^- \rightleftharpoons Ca + 2H_2O$	-2.189				1	23
		$Ca(OH)_2 + 2e^- \rightleftharpoons Ca + 2OH^-$	-3.02		965		15	24
		$Ca(OH)_2 + 4H^+ + 4e^- \rightleftharpoons CaH_2 + 2H_2O$	-0.706				1	25
1		$Ca(P_3O_9)^- + 2e^- \rightleftharpoons Ca + P_3O_9^{3-}$	-2.971			4	16,17	26
1		$Ca(P_4O_{12})^{2-} + 2e^- \rightleftharpoons Ca + P_4O_{12}^{4-}$	-3.021			4	16,18	27

Electronically conducting phase	Intermediate species	Composition of the solution	Solvent	Temperature °C	Pressure Torr	Measuring method
a	b	c	d		e	d
Ca	Ca (SO_4)	K_2SO_4 : (3.05 up to 27.71) · 10^{-3} M; Ca $(IO_3)_2$ satd.	1	25		V
Ca	Ca (S_2O_3)	1) Not reported 2) $Na_2S_2O_3$: 2.013 · 10^{-3} M; Ca $(ClO_4)_2$(7.3 up to 31.8) · 10^{-3} M	1	25		III
Ca		Na acetate : (37.27 up to 72.13) . 10^{-3} M; Ca (IO_3): satd.	1	25		V
Ca	Ca (adipate)		1	25		V
Ca	Alanylglycine	Na alanylglycine : 61.22 · 10^{-3} M; Ca $(IO_3)_2$: satd.	1	25		V
Ca		1) Na aminoacetate: up to 0.1494 M; Ca $(IO_3)_2$: satd 2) Not reported 3) Na aminoacetate : (35.40 up to 66.34) · 10^{-3} M; Ca $(IO_3)_2$: satd.	1	25		V
Ca		Na α - aminopropionate: 0.01962 up to 0.1096 M; Ca $(IO_3)_2$: satd.	1	25		V
Ca		Na bromoacetate : (50.03 up to 72.99) · 10^{-3} M; Ca $(IO_3)_2$: satd.	1	25		V
Ca		Na n-butyrate:(57.73 up to 78.72) · 10^{-3} M; Ca $(IO_3)_2$: satd.	1	25		V
Ca		Na iso-butyrate : (32.63 up to 74.32) · 10^{-3} M; Ca $(IO_3)_2$: satd.	1	25		V
Ca	di-Iodotyrosine	Na di -iodotyrosine : (18.64 up to 41.89) · 10^{-3} M; Ca $(IO_3)_2$: satd.	1	25		V
Ca		Na formate : (51.10 up to 75.00) · 10^{-3} M; Ca $(IO_3)_2$: satd.	1	25		V
Ca	Ca (fumarate)		1	25		III
Ca	Ca (glutamate)	Na_2 glutamate : (5.62 up to 22.49) · 10^{-3} M; Ca $(IO_3)_2$: satd.	1	25		V
Ca			1	25		
Ca	Glycylglycine	Na glycylglycine : (11.55 up to 48.52) · 10^{-3} M; Ca $(IO_3)_2$: satd.	1	25		V
Glass	Ca (8-hydroxyquinoli-ne-5-sulfonate)	8-hydroxyquinoline-5-sulfonate : 2.954 · 10^{-3} M; NaOH : 1.4 · 10^{-3} M; $CaCl_2$	1	25	743	I
Ca		Na hippurate : (24.15 up to 50.22) · 10^{-3} M; Ca $(IO_3)_2$: satd.	1	25		V
Ca	Leucylglycine	Na leucylglycine : 52.55 · 10^{-3} M; Ca $(IO_3)_2$:satd.	1	25		V
Ca	Ca (malate)	1) Not reported 2) Malic acid : 0.0397 up to 0.1093 N; NaOH : 0.0111 up to 0.0598 N; $CaCl_2$: 0.0141 up to 0.0546 N	1	25 ± 0.01		IV
Ca	Ca (maleate)		1	25		III

Comparison electrode	Liquid junction	Electrode reaction	Standard value V	Uncertainty mV	Temperature Coefficient $\mu V/^{o}C$	Notes	Reference	Electrode Reference Number
d			f		g	h		
1		$Ca(SO_4) + 2e^- \rightleftharpoons Ca + SO_4^{2-}$	-2.936		30	5	11	28
1		$Ca(S_2O_3) + 2e^- \rightleftharpoons Ca + S_2O_3^{2-}$	-2.926			6	19, 20	29
1		$Ca(\text{acetate})^+ + 2e^- \rightleftharpoons Ca + \text{acetate}^-$	-2.891			5	21	30
1		$Ca(\text{adipate}) + 2e^- \rightleftharpoons Ca + \text{adipate}^{2-}$	-3.468			7	22	31
1		$Ca(\text{alanylglycine})^{2+} + 2e^- \rightleftharpoons Ca + \text{alanylglycine}$	-2.887			5	23	32
1		$Ca(\text{aminoacetate})^{2+} + 2e^- \rightleftharpoons Ca + \text{aminoacetate}$	-2.909				23 25 21	33
1		$Ca(\alpha\text{-aminopropionate})^{2+} + 2e^- \rightleftharpoons$ $Ca + \alpha\text{-aminopropionate}$	-2.904			5	23	34
1		$Ca(\text{bromoacetate})^+ + 2e^- \rightleftharpoons Ca + \text{bromoacetate}^-$	-2.884			5	21	35
1		$Ca(\underline{n}\text{-butyrate})^+ + 2e^- \rightleftharpoons Ca + \underline{n}\text{-butyrate}^-$	-2.884			5	21	36
1		$Ca(\underline{iso}\text{-butyrate})^+ + 2e^- \rightleftharpoons Ca + \underline{iso}\text{-butyrate}^-$	-2.883			5	21	37
1		$Ca(\underline{di}\text{-iodotyrosine})^{2+} + 2e^- \rightleftharpoons Ca + \underline{di}\text{-iodotyrosine}$	-2.913			5	23, 24	38
1		$Ca(\text{formate})^+ + 2e^- \rightleftharpoons Ca + \text{formate}^-$	-2.892			5	21	39
1		$Ca(\text{fumarate}) + 2e^- \rightleftharpoons Ca + \text{fumarate}^{2-}$	-2.927			7	22	40
1		$Ca(\text{glutamate}) + 2e^- \rightleftharpoons Ca + \text{glutamate}^{2-}$	-2.928			5	23	41
		$Ca(\text{glycolate})^+ + 2e^- \rightleftharpoons Ca + \text{glycolate}^-$	-2.915				25	42
1		$Ca(\text{glycylglycine})^{2+} + 2e^- \rightleftharpoons Ca + \text{glycylglycine}$	-2.904			5	23	43
4		$Ca(\text{8-hydroxyquinoline-5-sulfonate}) + 2e^- \rightleftharpoons$ $Ca + \text{8-hydroxyquinoline-5-sulfonate}^{2-}$	-2.972			8	26	44
1		$Ca(\text{hippurate})^+ + 2e^- \rightleftharpoons Ca + \text{hippurate}^-$	-2.881			5	23	45
1		$Ca(\text{leucylglycine})^{2+} + 2e^- \rightleftharpoons Ca + \text{leucylglycine}$	-2.889			5	23	46
		$Ca(\text{malate}) + 2e^- \rightleftharpoons Ca + \text{malate}^{2-}$	-2.942				22 27	47
1		$Ca(\text{maleate}) + 2e^- \rightleftharpoons Ca + \text{maleate}^{2-}$	-2.940			7	22	48

Electronically conducting phase	Intermediate species	Composition of the solution	Solvent	Temperature °C	Pressure Torr	Measuring method
a	b	c	d		e	d
Ca	Ca (malonate)		1	25		V
Ca			1	25		V
Ca			1	25		V
Ca	Ca (isophthalate)		1	25		III
Ca		Na propionate : (46.71 up to 75.00) · 10^{-3} M; $Ca(IO_3)_2$: satd.	1	25		V
Ca			1	25		V
Ca	Ca (salicylate)	Salicylic acid : 0.0399 up to 0.0653 N; NaOH: 0.0336 up to 0.0598 N; $CaCl_2$: 0.0425 up to 0.0506 N	1	25		V
Ca	Serine	Na serine : (11.22 up to 56.10) · 10^{-3} N; $Ca(IO_3)_2$: satd.	1	25		V
Ca	Ca (succinate)		1	25		III
Ca	Ca (tartrate)		1	25		III
Ca			1	25		V
Ca	Tyrosine	Na tyrosine : (19.09 up to 35.89) · 10^{-3} M; $Ca(IO_3)_2$: satd.	1	25		V
Ca		Na n-valerate : (33.69 up to 53.34) · 10^{-3} M; $Ca(IO_3)_2$: satd.	1	25		V
Ca		Na iso -valerate : (32.82 up to 62.46) · 10^{-3} M; $Ca(IO_3)_2$: satd.	1	25		V

Comparison electrode	Liquid junction	Electrode reaction	Standard value	Uncertainty	Temperature Coefficient	Notes	Reference	Electrode Reference Number
			V	mV	μV/°C			
d			f		g	h		
1		Ca (malonate) + 2e$^-$ ⇌ Ca + malonate^{2-}	-3.608			9	24	49
1		Ca (mandelate)$^+$ + 2e$^-$ ⇌ Ca + mandelate$^-$	-2.911			5	25	50
1		Ca (methoxyacetate)$^+$ + 2e$^-$ ⇌ Ca + methoxyacetate$^-$	-2.901			5	25	51
1		Ca (isophthalate) + 2e$^-$ ⇌ Ca + isophthalate^{2-}	-2.927			7	22	52
1		Ca (propionate)$^+$ + 2e$^-$ ⇌ Ca + propionate$^-$	-2.888			5	21	53
1		Ca (pyruvate)$^+$ + 2e$^-$ ⇌ Ca + pyruvate$^-$	-2.900			5	23	54
1		Ca (salicylate) + 2e$^-$ ⇌ Ca + salicylate^{2-}	-2.881			5	25, 27	55
1		Ca (serine)$^{2+}$ + 2e$^-$ ⇌ Ca + serine	-2.910			5	23	56
1		Ca (succinate) + 2e$^-$ ⇌ Ca + succinate^{2-}	-2.927			7	22	57
1		Ca (tartrate) + 2e$^-$ ⇌ Ca + tartrate^{2-}	-2.951			7	22	58
1		Ca (trimethylacetate)$^+$ + 2e$^-$ ⇌ Ca + trimethylacetate$^-$	-2.884			5	18	59
1		Ca (tyrosine)$^{2+}$ + 2e$^-$ ⇌ Ca + tyrosine	-2.912			5	23	60
1		Ca (_n_-valerate)$^+$ + 2e$^-$ ⇌ Ca + _n_-valerate$^-$	-2.877			5	21	61
1		Ca (_iso_-valerate)$^+$ + 2e$^-$ ⇌ Ca + _iso_-valerate$^-$	-2.874			5	21	62

Calcium 1

Standard value V	Solvent	Electrode Reference Number	Standard value V	Solvent	Electrode Reference Number	Standard value V	Solvent	Electrode Reference Number
-3.80	1	2	-2.884	1	59			
-3.608	1	49	-2.884	1	35			
-3.468	1	31	-2.883	1	37			
-3.021	1	27	-2.881	1	55			
-3.02	1	24	-2.881	1	45			
-2.972	1	44	-2.877	1	61			
-2.971	1	26	-2.874	1	62			
-2.951	1	58	-2.868	1	3			
-2.948	1	15	-2.189	1	23			
-2.942	1	47	-1.902	1	17			
-2.940	1	48	-1.045	1	14			
-2.936	1	28	-0.706	1	25			
-2.928	1	41	-0.563	1	18			
-2.927	1	57	0.776	1	1			
-2.927	1	52	1.260	1	20			
-2.927	1	40	1.547	1	21			
-2.926	1	29	2.224	1	19			
-2.915	1	42	-2.185	2	4			
-2.913	1	38	-1.91	3	5			
-2.912	1	60	-2.65	29	6			
-2.911	1	50	-2.929	7	7			
-2.910	1	56	-2.870	8	8			
-2.909	1	33	-2.869	11	9			
-2.906	1	22	-2.808	12	10			
-2.904	1	32	-2.895	15	11			
-2.904	1	43	-3.20	16	12			
-2.901	1	51	-2.75	23	13			
-2.900	1	54						
-2.900	1	16						
-2.892	1	39						
-2.891	1	30						
-2.889	1	46						
-2.888	1	53						
-2.887	1	32						
-2.884	1	36						

NOTES: Calcium

1) Reaction equivalent to: $Ca^{2+} + 2e^{-} \rightleftharpoons Ca$

2) Solubility and colorimetry data

3) Solubility, potentiometry and thermodynamics data

4) Solubility and conductometry data

5) Solubility data

6) Solubility and spectrophotometry data

7) Conductometric data

8) Potentiometric pH measurements

9) Colorimetric data

BIBLIOGRAPHY : Calcium

1) Data reported by M. Pourbaix "Atlas d'Equilibres Electrochimiques à 25 ^{o}C" 146 - 154 (Gauthier-Villars Editeur, Paris, 1968)

2) S.N. Pobedinskii, G.A. Krestov, L.L. Kuz'min, Izv. Vyss. Uchebn. Zavedenii, Khim. i Khim. Tekhnol., 73 (1963), 768

3) B. Jakuszewski, S. Taniewska-Osenka, Rocz. Chem., 36 (1962), 329 - 334

4) F.E. Rosztoczy, C.W. Tobias, Electrochim. Acta, 11 (1966), 857 - 861

5) R. Parsons, Handbook of Electrochemical Constants, Butterworths, London, 1959

6) M. L'Her, J. Courtot-Coupez, Bull. Soc. Chim. France, 1972 ,3645 - 3653

7) N.E. Khomutov, Zh. Fiz. Khim., 42 (1968), 2223 - 2229

8) V.A. Pleskov, Acta Physicochim. U.S.S.R., 21 (1946), 41 - 54; J. Phys. Chem.U.S.S.R., 20 (1946), 153 - 162

9) V.A. Pleskov, J. Phys. Chem. U.S.S.R., 22 (1948), 351 - 361

10) C.W. Davies, B.E. Hoyle, J. Chem. Soc., 1953, 4134 - 4136

11) R.P. Bell, J.H.B. George, Trans. Faraday Soc., 49 (1953), 619 - 627

12) C.W. Davies, B.E. Hoyle, J. Chem. Soc., 1938, 277

13) R.P. Bell, M.H. Pauckhurst, J. Chem. Soc., 1956, 2836 - 2840

14) G.R. Gimblett, C.B. Monk, Trans. Faraday Soc., 50 (1954), 965 - 972

15) Data reported by A.J. de Bethune, N.A. Swendeman-Loud "Standard Aqueous Electrode Potentials and Temperature Coefficients at 25 ^{o}C" (Clifford A. Hampel, Ill.)

16) C.W. Davies, C.B. Monk, J. Chem. Soc., 1949, 413

17) H.W. Jones, C.B. Monk, C.W. Davies, J. Chem. Soc., 1949, 2693

18) H.W. Jones, C.B. Monk, J. Chem. Soc., 1950, 3475 - 3478

19) C.W. Davies, P.A.H. Wyatt, Trans. Faraday Soc., 45 (1949), 470

20) F.G.R. Gimblett, C.B. Monk, Trans. Faraday Soc., 51 (1955) 793 - 802

21) C.A. Colman-Porter, C.B. Monk, J. Chem. Soc., 1952, 4363 - 4368

22) N.E. Topp, C.W. Davies, J. Chem. Soc., 1940, 87

23) C.W. Davies, G.M. Waind, J. Chem. Soc., 1950, 301 - 303

24) P.S. Winnek, C.L.A. Schmidt, J. Gen. Physiol., 18 (1935)

25) C.W. Davies, J. Chem. Soc., 1938, 277

26) R. Näsänen, E. Uisitalo, Acta Chem. Scand., 8 (1954), 112 - 118

27) R.P. Bell, G.M. Waind, J. Chem. Socn, 1951, 2357 - 2362

28) D.I. Stock, C.W. Davies, J. Chem. Soc., 1949, 1371

Electronically conducting phase	Intermediate species	Composition of the solution	Solvent	Temperature °C	Pressure Torr	Measuring method
a	b	c	d		e	d
Sr			1	25	760	IV
Pb	$PbCO_3$, $SrCO_3$	$SrCl_2$	1	25		I
Pb	$PbCO_3$, $SrCO_3$	$SrCL_2$	7	25		I
Sr			8	25	760	IV
Sr			11	25	760	IV
Sr			12	25	760	IV
Sr			15	25	760	IV
Sr			1	25		III
Sr			1	25		III, V
Sr	SrO		1	25	760	IV
Ind.	SrO_2		1	25	760	IV
Sr			1	25		I, V
Sr	$Sr(OH)_2$		1	25	760	IV
Sr	$Sr(OH)_2$		1	25	760	IV
Sr			1	25		III, V
Sr			1	25		III, V
Sr	$Sr(S_2O_3)$		1	25		V
Sr	SrH_2		1	25	760	IV
Ind.	SrH_2		1	25	760	IV
Ind.	SrH_2 , SrO		1	25	760	IV
Ind.	SrH_2 $Sr(OH)_2$		1	25	760	IV
Ind.	SrO, SrO_2		1	25	760	IV
Ind.	$Sr(OH)_2$ SrO_2		1	25	760	IV
Sr	Alanine		1	25		V
Sr			1	25		V
Sr			1	25		V
Sr	Glycine		1	25		V
Glass		8-hydroxyquinoline-5-sulfonate: $2.954 \cdot 10^{-3}$ M; NaOH: $1.4 \cdot 10^{-3}$ M ; $SrCl_2$	1	25	743	I
Sr			1	25		I, V
Sr			1	25		V
Sr			1	25		V
Sr			1	25		V

Comparison electrode	Liquid junction	Electrode reaction	Standard value V	Uncertainty mV	Temperature Coefficient μV/°C	Notes	Reference	Electrode Reference Number
d			f		g	h		
		$Sr^+ + e^- \rightleftharpoons Sr$	-4.10				1	1
3		$PbCO_3 + Sr^{2+} + 2e^- \rightleftharpoons Pb + SrCO_3$	-2.886		-191	1	2,6	2
3		$PbCO_3 + Sr^{2+} + 2e^- \rightleftharpoons Pb + SrCO_3$	-2.938			1	2	3
		$Sr^{2+} + 2e^- \rightleftharpoons Sr$	-2.892				3	4
		$Sr^{2+} + 2e^- \rightleftharpoons Sr$	-2.991				3	5
		$Sr^{2+} + 2e^- \rightleftharpoons Sr$	-2.830				3	6
		$Sr^{2+} + 2e^- \rightleftharpoons Sr$	-2.902				3	7
1		$Sr[Fe(CN)_6]^- + 2e^- \rightleftharpoons Sr + [Fe(CN)_6]^{3-}$	-2.972			2	4	8
1		$Sr(IO_3)^+ + 2e^- \rightleftharpoons Sr + IO_3^-$	-2.931			3	5	9
		$SrO + 2H^+ + 2e^- \rightleftharpoons Sr + H_2O$	-1.972				1	10
		$SrO_2 + 4H^+ + 2e^- \rightleftharpoons Sr^{2+} + 2H_2O$	2.333				1	11
1		$Sr(OH)^+ + 2e^- \rightleftharpoons Sr + OH^-$	-2.913		-100	3	5,6	12
		$Sr(OH)_2 + 2H^+ + 2e^- \rightleftharpoons Sr + 2H_2O$	-2.047				1	13
		$Sr(OH)_2 + 2e^- \rightleftharpoons Sr + 2\,OH^-$	-2.88		-960		7	14
1		$Sr(P_3O_9)^- + 2e^- \rightleftharpoons Sr + P_3O_9^{3-}$	-2.987			3	8	15
1		$Sr(P_4O_{12})^{2+} + 2e^- \rightleftharpoons Sr + P_4O_{12}^{4-}$	-3.039			3	8	16
1		$Sr(S_2O_3) + 2e^- \rightleftharpoons Sr + S_2O_3^{2-}$	-2.948			4	9	17
		$Sr + 2H^+ + 2e^- \rightleftharpoons SrH_2$	0.718				1	18
		$Sr^{2+} + 2H^+ + 4e^- \rightleftharpoons SrH_2$	-1.085				10	19
		$SrO + 4H^+ + 4e^- \rightleftharpoons SrH_2 + H_2O$	-0.477				10	20
		$Sr(OH)_2 + 4H^+ + 4e^- \rightleftharpoons SrH_2 + 2H_2O$	-0.665				10	21
		$SrO_2 + 2H^+ + 2e^- \rightleftharpoons SrO + H_2O$	1.116				10	22
		$SrO_2 + 2H^+ + 2e^- \rightleftharpoons Sr(OH)_2$	1.492				10	23
1		$Sr(alanine)^{2+} + 2e^- \rightleftharpoons Sr + alanine$	-2.909			4	11	24
1		$Sr(bromoacetate)^+ + 2e^- \rightleftharpoons Sr + bromoacetate^-$	-2.896			4	11	25
1		$Sr(formate)^+ + 2e^- \rightleftharpoons Sr + formate^-$	-2.907			4	11	26
1		$Sr(glycine)^{2+} + 2e^- \rightleftharpoons Sr + glycine$	-2.915			4	11	27
4		$Sr(8\text{-hydroxyquinoline-5-sulfonate}) + 2e^- \rightleftharpoons$ $Sr + 8\text{-hydroxyquinoline-5-sulfonate}^{2-}$	-2.969			5	12	28
1		$Sr(lactate)^+ + 2e^- \rightleftharpoons Sr + lactate^-$	-2.917			6	11,13	29
1		$Sr(pivalate)^+ + 2e^- \rightleftharpoons Sr + pivalate^-$	-2.892			4	11	30
1		$Sr(propionate)^+ + 2e^- \rightleftharpoons Sr + propionate^-$	-2.895			4	11	31
1		$Sr(\underline{n}\text{-valerate})^+ + 2e^- \rightleftharpoons Sr + \underline{n}\text{-valerate}^-$	-2.894			4	11	32

Standard value V	Solvent	Electrode Reference Number	Standard value V	Solvent	Electrode Reference Number	Standard value V	Solvent	Electrode Reference Number
-4.10	1	1						
-3.039	1	16						
-2.987	1	15						
-2.972	1	8						
-2.969	1	28						
-2.948	1	17						
-2.931	1	9						
-2.917	1	29						
-2.915	1	28						
-2.913	1	12						
-2.909	1	24						
-2.907	1	27						
-2.896	1	25						
-2.895	1	31						
-2.894	1	32						
-2.892	1	30						
-2.886	1	2						
-2.88	1	14						
-2.047	1	13						
-1.972	1	10						
-1.085	1	19						
-0.665	1	21						
-0.477	1	20						
0.718	1	18						
1.116	1	22						
1.492	1	23						
2.333	1	11						
-2.938	7	3						
-2.892	8	4						
-2.991	11	5						
-2.830	12	6						
-2.902	15	7						

NOTES: Strontium

1) Reaction equivalent to: $Sr^{2+} + 2e^- \rightleftharpoons Sr$
2) Conductometric data
3) Solubility and potentiometric data
4) Solubility data
5) Potentiometric pH measurements
6) Stability and potentiometric data

BIBLIOGRAPHY: Strontium

1) S.N. Pobedinski, G.A. Krestov, L.L. Kuzmin, Izv. Vyss. Ucheb. Zavedenii, Khim. i Khim. Teknol., 6 (1963), 5, 768 - 773
2) B. Jakuszewski, S. Taniewska-Osenka, Rocz. Chem., 36 (1962), 329 - 334
3) N.E. Khomutov, Zh. Fiz. Khim., 42 (1968), 2223 - 2229
4) C.W. Gibby, C.B. Monk, Trans. Faraday Soc., 48 (1952), 632
5) C.A. Colman-Porter, C.B. Monk, J. Chem. Soc., 1952, 1312
6) H.S. Harned, T.R. Paxton, J. Phys. Chem., 57 (1953), 531
7) Data reported by A.J. De Bethune, N.A. Swendeman-Loud, "Standard Aqueous Electrode Potentials and Temperature Coefficients at 25 °C" (Clifford A. Hampel, Ill.)
8) C.B. Monk, J. Chem. Soc., 1952, 1314
9) T.O. Denney, C.B. Monk, Trans. Faraday Soc., 47 (1951), 992
10) Data reported by M. Pourbaix, "Atlas d'Equilibres Electrochimiques à 25 °C" (Gauthier-Villars Editeur, Paris, 1963), pp. 146 - 154
11) C.A. Colman-Porter, C.B. Monk, J. Chem. Soc., 1952, 4363
12) R. Näsänen, E. Uisitalo, Acta Chem. Scand., 8 (1954), 112 - 118
13) C.W. Davies, C.B. Monk, Trans. Faraday Soc., 50 (1954), 132

Electronically conducting phase	Intermediate species	Composition of the solution	Solvent	Temperature °C	Pressure Torr	Measuring method
a	b	c	d		e	d
Pb	$PbCO_3$, $BaCO_3$	$BaCl_2$	1	25		I
Pb	$PbCO_3$, $BaCO_3$	$BaCL_2$	7	25		I
Ba		$BaCl_2$: (0.9858 up to 1.0157) · 10^{-3} M; $K_3[Fe(CN)_6]$: (0.7945 up to 0.2041) · 10^{-3} M	1	25		III
Ba			1	25		V
Ba	BaO		1	25	760	IV
Ind.	BaO_2		1	25	760	IV
Ba		NaOH : (22 up to 47.8) · 10^{-3} M; $BaCl_2$: (60.6 up to 147.5) · 10^{-3} M	1	25		
Ba	$Ba(OH)_2$		1	25	760	IV
Ba	$Ba(OH)_2 \cdot 8H_2O$		1	25	760	IV
Ba			1	25		III
Ba			1	25		III
Ba	$Ba(S_2O_3)$	1) BaS_2O_3 : (0.21808 up to 0.64710) · 10^{-3} M; 2) BaS_2O_3 satd.	1	25		III, V
Ba	BaH_2		1	25	760	IV
Ind.	BaH_2		1	25	760	IV
Ind.	BaH_2, BaO		1	25	760	IV
Ind.	BaO, BaO_2		1	25	760	IV
Ba		Na acetate : (37.27 up to 72.13) · 10^{-3} M; $Ba(IO_3)_2$: satd.	1	25		V
Ba	Ba (adipate)		1	25		III, IV
Ba	Alanine	Alanine : (27.42 up to 73.19) · 10^{-3} M; $Ba(IO_3)_2$: satd.	1	25 ± 0.03		V
Ba			1	25		V
Ba		Na n-butyrate : (52.30 up to 77.39) · 10^{-3} M; $Ba(IO_3)_2$ satd.	1	25		V
Ba		Na formate : (51.10 up to 75.00) · 10^{-3} M; $Ba(IO_3)_2$ satd.	1	25		V
Ba	Ba (fumarate)		1	25		III
Glass	Ba (glutarate)		1	25		I
Ba	Glycine	Glycine : (30.36 up to 74.48) · 10^{-3} M; $Ba(IO_3)_2$: satd.	1	25 ± 0.03		V
1) Ba 2) Pt, H_2		1) Ba $(glycolate)^+$: (1.112 up to 1.209) · 10^{-3} M; 2) Na glycolate : (3.306 up to 33.25) · 10^{-3} M; $Ba(IO_3)_2$	1	25		V, I

Comparison electrode	Liquid junction	Electrode reaction	Standard value V	Uncertainty mV	Temperature Coefficient μV/°C	Notes	Reference	Electrode Reference Number
d			f		g	h		
3		$PbCO_3 + Ba^{2+} + 2e^- \rightleftharpoons Pb + BaCO_3$	-2.912		-395	1	1,6	1
3		$PbCO_3 + Ba^{2+} + 2e^- \rightleftharpoons Pb + BaCO_3$	-2.943			1	1	2
1		$Ba[Fe(CN)_6]^- + 2e^- \rightleftharpoons Ba + [Fe(CN)_6]^{3-}$	-2.997			2	2	3
1		$Ba[Fe(CN)_6]^{2-} + 2e^- \rightleftharpoons Ba + [Fe(CN)_6]^{4-}$	-3.027			3	3	4
		$BaO + 2H^+ + 2e^- \rightleftharpoons Ba + H_2O$	-1.509				4	5
		$BaO_2 + 4H^+ + 2e^- \rightleftharpoons Ba^{2+} + 2H_2O$	-2.419				4	6
		$Ba(OH)^+ + 2e^- \rightleftharpoons Ba + OH^-$	-2.937				5	7
		$Ba(OH)_2 + 2e^- \rightleftharpoons Ba + 2OH^-$	-2.81				6	8
		$Ba(OH)_2 \cdot 8H_2O + 2e^- \rightleftharpoons Ba + 2OH^- + 8H_2O$	-2.99		+380		6	9
1		$Ba(P_3O_9)^- + 2e^- \rightleftharpoons Ba + (P_3O_9)^{3-}$	-3.011			2	7	10
1		$Ba(P_4O_{12})^{2-} + 2e^- \rightleftharpoons Ba + (P_4O_{12})^{4-}$	-3.059			2	8	11
1		$Ba(S_2O_3) + 2e^- \rightleftharpoons Ba + (S_2O_3)^{2-}$	-2.979		-300	4	9	12
		$Ba + 2H^+ + 2e^- \rightleftharpoons BaH_2$	-0.685				4	13
		$Ba^{2+} + 2H^+ + 4e^- \rightleftharpoons BaH_2$	-1.110				4	14
		$BaO + 2H^+ + 2e^- \rightleftharpoons BaH_2 + H_2O$	-0.412				4	15
		$BaO_2 + 2H^+ + 2e^- \rightleftharpoons BaO + H_2O$	1.023				4	16
1		$Ba(acetate)^+ + 2e^- \rightleftharpoons Ba + (acetate)^-$	-2.924			5	10	17
1		$Ba(adipate) + 2e^- \rightleftharpoons Ba + (adipate)^{2+}$	-2.965			6	11,12	18
1		$Ba(alanine)^{2+} + 2e^- \rightleftharpoons Ba + (alanine)$	-2.935			5	13	19
1		$Ba(bromoacetate)^+ + 2e^- \rightleftharpoons Ba + (bromoacetate)^-$	-2.919			5	14	20
1		$Ba(n\text{-}butyrate)^+ + 2e^- \rightleftharpoons Ba + (n\text{-}butyrate)^-$	-2.912			5	10	21
1		$Ba(formate)^+ + 2e^- \rightleftharpoons Ba + (formate)^-$	-2.930			5	10	22
1		$Ba(fumarate) + 2e^- \rightleftharpoons Ba + (fumarate)^{2-}$	-2.959			2	11	23
		$Ba(glutarate) + 2e^- \rightleftharpoons Ba + (glutarate)^{2-}$	-2.972			7	12	24
1		$Ba(glycine)^{2+} + 2e^- \rightleftharpoons Ba + (glycine)$	-2.934			5	15	25
1) 1 2) 4		$Ba(glycolate)^+ + 2e^- \rightleftharpoons Ba + (glycolate)^-$	-2.942			8	10,11	26

Electronically conducting phase	Intermediate species	Composition of the solution	Solvent	Temperature °C	Pressure Torr	Measuring method
a	b	c	d		e	d
Glass	Ba (8-hydroxyquinoline-5-sulfonate)	8-hydroxyquinoline-5-sulfonate : $2.954 \cdot 10^{-3}$ M; NaOH : $1.4 \cdot 10^{-3}$ M; $BaCl_2$	1	25	743	I
1) Ba 2) Pt, H_2		1) Na lactate : (19.96 up to 59.88) $\cdot 10^{-3}$ M; $Ba(IO_3)_2$ satd. 2) Ba $(lactate)^+$: (3.144 up to 32.86) $\cdot 10^{-3}$ M	1	25		V, I
Ba	Ba (malate		1	25		III
Ba	Ba (maleate)		1	25		III
Ba	Ba (malonate)		1	25		III, V
Ba	Ba (iso-phthalate)		1	25		III
Ba		Na pivalate : (62.79 up to 70.40) $\cdot 10^{-3}$ M; $Ba(IO_3)_2$ satd.	1	25		V
Ba		Na propionate : (59.88 up to 88.30) $\cdot 10^{-3}$ M; $Ba(IO_3)_2$ satd.	1	25		V
Ba	Ba (salicylate)		1	25	760	IV
Ba	Ba (succinate)		1	25		III
Ba	Ba (tartrate)		1	25		III
Ba		Na valerate : (57.82 up to 70.40) $\cdot 10^{-3}$ M; $Ba(IO_3)_2$ satd.	1	25		V

Comparison electrode	Liquid junction	Electrode reaction	Standard value V	Uncertainty mV	Temperature Coefficient μV/°C	Notes	Reference	Electrode Reference Number
d			f		g	h		
4		Ba (8-hydroxyquinoline-5-sulfonate) + 2e$^-$ ⇌ Ba + (8-hydroxyquinoline-5-sulfonate)$^{2-}$	-2.980			7	16	27
1) 1 2) 4		Ba (lactate)$^+$ + 2e$^-$ ⇌ Ba + (lactate)$^-$	-2.933			8	10,17	28
1		Ba (malate) + 2e$^-$ ⇌ Ba + (malate)$^{2-}$	-2.971			2	11,18	29
1		Ba (maleate) + 2e$^-$ ⇌ Ba + (maleate)$^{2-}$	-2.979			2	11	30
1		Ba (malonate) + 2e$^-$ ⇌ Ba + (malonate)$^{2-}$	-2.968			9	12,19	31
1		Ba (iso-phthalate) + 2e$^-$ ⇌ Ba + (iso-phthalate)$^{2-}$	-2.958			2	11	32
1		Ba (pivalate)$^+$ + 2e$^-$ ⇌ Ba + (pivalate)$^-$	-2.914			5	10	33
1		Ba (propionate)$^+$ + 2e$^-$ ⇌ Ba + (propionate)$^-$	-2.916			5	10	34
		Ba (salicylate) + 2e$^-$ ⇌ Ba + (salicylate)$^{2-}$	-2.918				18	35
1		Ba (succinate) + 2e$^-$ ⇌ Ba + (succinate)$^{2-}$	-2.968			2	11,12	36
1		Ba (tartrate) + 2e$^-$ ⇌ Ba + (tartrate)$^{2-}$	-2.987			2	11	37
1		Ba (valerate)$^+$ + 2e$^-$ ⇌ Ba + (valerate)$^-$	-2.918			5	10	38

Standard value V	Solvent	Electrode Reference Number
-3.059	1	11
-3.027	1	4
-3.011	1	10
-2.997	1	3
-2.99	1	9
-2.987	1	37
-2.980	1	27
-2.979	1	30
-2.979	1	12
-2.972	1	24
-2.971	1	29
-2.968	1	31
-2.966	1	36
-2.965	1	18
-2.959	1	23
-2.958	1	32
-2.942	1	26
-2.937	1	7
-2.935	1	19
-2.934	1	25
-2.933	1	28
-2.930	1	22
-2.924	1	17
-2.919	1	20
-2.918	1	35
-2.918	1	38
-2.916	1	34
-2.914	1	33
-2.912	1	1
-2.912	1	21
-2.81	1	8
-2.419	1	6
-1.509	1	5
-1.110	1	14
-0.685	1	13
-0.412	1	15
1.023	1	16
-2.943	7	2

NOTES: Barium

1) Reaction equivalent to: $Ba^{2+} + 2e^- \rightleftharpoons Ba$
2) Conductometric data
3) Spectrophotometric data
4) Solubility and conductometric data
5) Solubility data
6) Conductometric and potentiometric data
7) Potentiometric pH measurements
8) Solubility and potentiometry data
9) Conductometry and colorimetry data

BIBLIOGRAPHY: Barium

1) B. Jakuszewski, S. Taniewska-Osinka, Rocz. Chem., 36 (1962), 329 - 334
2) C.W. Gibby, C.B. Monk, Trans. Faraday Soc., 48 (1952), 632 - 634
3) S.R. Cohen, Thesis, Cornell Univ. Ithaca, N.Y., 1956, Univ. Microfilms, 20,406
4) Data reported by M. Pourbaix, "Atlas d'Equilibres Electrochimiques à 25 °C" (Gauthier-Villars Editeur, Paris, 1963), pp. 146 - 154
5) R.P. Bell, M.H. Panekhurst, J. Chem. Soc., 1956, 2836 - 2840
6) Data reported by A.J. De Bethune, N.A. Swendeman-Loud, "Standard Aqueous Electrode Potentials and Temperature Coefficients at 25 °C" (Clifford A. Hampel, Ill.)
7) H.W. Jones, C.B. Monk, C.W. Davies, J. Chem. Soc., 1949, 2693
8) H.W. Jones, C.B. Monk, J. Chem. Soc., 1950, 3475
9) T.O. Denney, C.B. Monk, Trans. Faraday Soc., 47 (1951), 992 - 998
10) C.A. Colman-Porter, C.B. Monk, J. Chem. Soc., 1952, 4363 - 4368
11) P.B. Davies, C.B. Monk, Trans. Faraday Soc., 50 (1954), 128 - 132
12) J.M. Peacock, J.C. James, J. Chem. Soc., 1951, 2233
13) N.E. Topp, C.W. Davies, J. Chem. Soc., 1940, 87
14) C.W. Davies, P.A.H. Wyatt, Trans. Faraday Soc., 45 (1949), 770
15) C.B. Monk, Trans. Faraday Soc., 47 (1951), 1233 - 1240
16) R. Näsänen, E. Uisitalo, Acta Chem. Scand., 8 (112 - 118)
17) C.W. Davies, C.B. Monk, Trans. Faraday Soc., 50 (1954), 132 - 136
18) R.P. Bell, G.M. Waind, J. Chem. Soc., 1951, 2357
19) D.I. Stock, C.W. Davies, J. Chem. Soc., 1949, 1371

Electronically conducting phase	Intermediate species	Composition of the solution	Solvent	Temperature °C	Pressure Torr	Measuring method
a	b	c	d		e	d
Ra			1	25	760	IV
Ra			1	25	760	IV
Ra	RaO		1	25	760	IV

Comparison electrode	Liquid junction	Electrode reaction	Standard value	Uncertainty	Temperature Coefficient	Notes	Reference	Electrode Reference Number
			V	mV	μV/°C			
d			f		g	h		
		$Ra^{+} + e^{-} \rightleftharpoons Ra$	-4.0				1	1
		$Ra^{2+} + 2e^{-} \rightleftharpoons Ra$	-2.916		-5.90		2	2
		$RaO + 2H^{+} + 2e^{-} \rightleftharpoons Ra + H_2O$	-1.319				3	3

Standard value V	Solvent	Electrode Reference Number	Standard value V	Solvent	Electrode Reference Number	Standard value V	Solvent	Electrode Reference Number
-4.0	1	1						
-2.916	1	2						
-1.319	1	3						

BIBLIOGRAPHY: Radium

1) S.N. Pobedinskii, G.A. Krestov, L.L. Kuz'min, Izv. Vyssh. Uchebn. Zavedenii, Khim. i Khim. Tekhnol., 6 (1963), 5, 768 - 773

2) Data reported by A.J. De Bethune, N.A. Swendeman-Loud "Standard Aqueous Electrode Potentials and Temperature Coefficients at 25 °C" (Clifford A. Hampel, Ill.)

3) Data reported by M. Pourbaix "Atlas d'Equilibres Electrochimiques à 25 °C" (Gauthier-Villars Editeur, Paris, 1963), pp. 146 - 154

Electronically conducting phase	Intermediate species	Composition of the solution	Solvent	Temperature °C	Pressure Torr	Measuring method
a	b	c	d		e	d
Zn			1	25	760	IV
Zn (Hg)		1) $ZnBr_2$: 0.002 up to 0.1 M; 2) $ZnCl_2$	1	25		I
Zn		$ZnCl_2$: 0.1 N; Pb $(NO_3)_2$: 0.1 N; NH_4SCN : 20 mole %	2	-34± 0.5		I
Zn			15	25	760	IV
Zn			11	25	760	IV
Zn			12	25	760	IV
Zn		$ZnCl_2$: 0.01 N	16	25± 0.01		I
Zn		$ZnCl_2$: 0.00554, 0.0151 and 0.0370 M; KCl: satd.	18	25± 0.1		I
Zn		$ZnCl_2$: 0.01 N	23	25		I
Zn(Hg)	$ZnCl_2$	HCl	27	25± 0.5		I
Zn	$ZnCO_3$		1	25	760	IV
Zn			1	25		I
Zn	NH_3		1	25	760	IV
Zn	ZnO (ε)		1	25	760	IV
Zn	ZnO (amorphous)		1	25	760	IV
Zn			1	25	760	IV
Zn			1	25	760	IV
Zn	Zn $(OH)_2$		1	25	760	IV
Zn		NaOH : (0.413 up to 1167) · 10^{-3} M; Zn $(OH)_2$: (0 up to 20) · 10^{-3} M	1	25 ± 0.05		V
Zn		KOH(a_{OH^-}: 0.80, 1.10, 1.14, 1.75)	1	25± 0.2		V,I
Zn	ZnS		1	25	760	IV
Zn	Zn (S_2O_3)	$ZnCl_2$: (7.53 up to 14.92) · 10^{-3} M; BaS_2O_3 : (13.83 up to 15.88) · 10^{-3} M	1	25 ± 0.03		V
Glass	Alanine	$ZnSO_4$: (4.35 up to 4.53) · 10^{-3} M; alanine : (2.15 up to 34.95) · 10^{-3} M; NaOH : (0.82 up to 4.95) · ·10^{-3} N	1	25 ± 0.02		I
Glass	Alanine	$ZnSO_4$: (4.35 up to 4.53) · 10^{-3} M; alanine : (2.15 up to 34.95) · 10^{-3} M; NaOH : (0.82 up to 4.95) · 10^{-3} M	1	25± 0.02		I
Glass	Glycine	$ZnSO_4$: (3.67 up to 35.49) · 10^{-3} M; glycine : (64.02 up to 73.59) · 10^{-3} M; NaOH : (4.62 up to 6.14) · 10^{-3} M	1	25± 0.02		I
Glass	Glycine	$ZnSO_4$: (3.67 up to 35.49) · 10^{-3} M; glycine : (64.02 up to 73.59) · 10^{-3} M; NaOH : (4.62 up to 6.14) · 10^{-3} M	1	25± 0.02		I

Comparison electrode	Liquid junction	Electrode reaction	Standard value V	Uncertainty mV	Temperature Coefficient μV/°C	Notes	Reference	Electrode Reference Number
d			f		g	h		
		$HZnO_2^- + 3H^+ + 2e^- \rightleftharpoons Zn + 2H_2O$	0.054				1	1
1)5 2)4		$Zn^{2+} + 2e^- \rightleftharpoons Zn$	-0.7618	±0.8	+99		2,3	2
29		$Zn^{2+} + 2e^- \rightleftharpoons Zn$	-0.505	±4			4	3
		$Zn^{2+} + 2e^- \rightleftharpoons Zn$	-0.792				5	4
		$Zn^{2+} + 2e^- \rightleftharpoons Zn$	-0.853				5	5
		$Zn^{2+} + 2e^- \rightleftharpoons Zn$	-0.835				5	6
14		$Zn^{2+} + 2e^- \rightleftharpoons Zn$	-1.05		-600		6	7
25		$Zn^{2+} + 2e^- \rightleftharpoons Zn$	-0.757	±8	+140		7	8
7		$Zn^{2+} + 2e^- \rightleftharpoons Zn$	-0.74				8	9
1		$ZnCl_2 + 2e^- \rightleftharpoons Zn + 2Cl^-$	-0.788	±1			9	10
		$ZnCO_3 + 2e^- \rightleftharpoons Zn + CO_3^{2-}$	-1.06		-1164		10	11
		$ZnF^+ + 2e^- \rightleftharpoons Zn + F^-$	-0.798				11	12
		$Zn(NH_3)_4^{2+} + 2e^- \rightleftharpoons Zn + 4(NH_3)$	-1.04				10	13
		$ZnO + 2H^+ + 2e^- \rightleftharpoons Zn + H_2O$	-0.439				1	14
		$ZnO + 2H^+ + 2e^- \rightleftharpoons Zn + H_2O$	-0.400				1	15
		$ZnO_2^{2-} + 4H^+ + 2e^- \rightleftharpoons Zn + 2H_2O$	0.441				1	16
		$ZnO_2^{2-} + 2H_2O + 2e^- \rightleftharpoons Zn + 4(OH)^-$	-1.215				10	17
		$Zn(OH)_2 + 2e^- \rightleftharpoons Zn + 2OH^-$	-1.245		-1002		10	18
1		$Zn(OH)_3^- + 2e^- \rightleftharpoons Zn + 3(OH)^-$	-1.183			1	12	19
1		$Zn(OH)_4^{2-} + 2e^- \rightleftharpoons Zn + 4(OH)^-$	-1.214			2	13	20
		$ZnS + 2e^- \rightleftharpoons Zn + S^{2-}$	-1.405		-850		10	21
1		$Zn(S_2O_3) + 2e^- \rightleftharpoons Zn + (S_2O_3)^{2-}$	-0.831		-300	3	14	22
3		$Zn(alanine)^{2+} + 2e^- \rightleftharpoons Zn + alanine$	-0.915			4	15	23
3		$Zn(alanine)_2^{2+} + 2e^- \rightleftharpoons Zn + 2\ alanine$	-1.043			4	15	24
3		$Zn(glycine)^{2+} + 2e^- \rightleftharpoons Zn + glycine$	-0.925			4	15	25
3		$Zn(glycine)_2^{2+} + 2e^- \rightleftharpoons Zn + 2\ glycine$	-1.056			4	15	26

Electronically conducting phase	Intermediate species	Composition of the solution	Solvent	Temperature °C	Pressure Torr	Measuring method
a	b	c	d		e	d
Glass	Glycylglycine	$ZnSO_4$: (4.27 up to 39.07) · 10^{-3} M; glycylglycine : (13.05 up to 17.12) · 10^{-3} M; NaOH : (0.64 up to 8.34) · 10^{-3} M	1	25± 0.02		I
Glass	Glycylglycine	$ZnSO_4$: (4.27 up to 39.07) · 10^{-3} M; glycylglycine : (13.05 up to 17.12) · 10^{-3} M; NaOH : (0.64 up to 8.34) · 10^{-3} M	1	25± 0.02		I
Pt, H_2	Glycylglycylglycine	$ZnCl_2$: (11.52 up to 12.96) · 10^{-3} M; glycylglycylglycine : (21.50 up to 26.63) · 10^{-3} M; NaOH : (2.071 up to 4.203) · 10^{-3} M	1	25± 0.03		I
Pt, H_2	Glycylglycylglycine	$ZnCl_2$: (11.52 up to 12.96) · 10^{-3} M; glycylglycylglycine : (21.50 up to 26.63) · 10^{-3} M; NaOH : (2.071 up to 4.203) · 10^{-3} M	1	25± 0.03		I
Glass	Zn (8-hydroxyquinoline 5-sulfonate)	8-hydroxyquinoline 5-sulfonic acid : 2.954 · 10^{-3}M; Zn $(NO_3)_2$	1	25	743	I
Glass		8-hydroxyquinoline 5-sulfonic acid : 2.954 · 10^{-3}M; Zn $(NO_3)_2$	1	25	743	I
Zn	Zn (lactate)	Zn (lactate) : (3.175 up to 28.57) · 10^{-3} M	1	25		I
Zn	Zn (malate)	Zn (malate) : (1.7 up to 6.0) · 10^{-3} M	1	25		V
Zn		Zn (mandelate)$^-$: 0.0159 up to 0.0323 M	1	25± 0.01		V

Comparison electrode	Liquid junction	Electrode reaction	Standard value	Uncertainty	Temperature Coefficient	Notes	Reference	Electrode Reference Number
			V	mV	μV/°C			
d			f		g	h		
3		Zn (glycylglycine)$^{2+}$ + 2e$^-$ ⇌ Zn + glycylglycine	-0.874			4	15	27
3		Zn (glycylglycine)$_2^{2+}$ + 2e$^-$ ⇌ Zn + 2 glycylglycine	-0.955			4	15	28
4		Zn (glycylglycylglycine)$^{2+}$ + 2e$^-$ ⇌ Zn + glycylglycylglycine	-0.859			4	16	29
4		Zn (glycylglycylglycine)$_2^{2+}$ + 2e$^-$ ⇌ Zn + 2 glycylglycylglycine	-0.948			4	16	30
3	KCl satd.	Zn (8-hydroxyquinoline 5-sulfonate) + 2e$^-$ ⇌ Zn + 8-hydroxyquinoline 5-sulfonate $^{2-}$	-1.021			4	17	31
3	KCl satd.	Zn (8-hydroxyquinoline 5-sulfonate)$_2^{2-}$ + 2e$^-$ ⇌ Zn + 2 8-hydroxyquinoline 5-sulfonate $^{2-}$	-1.241			4	17	32
		Zn (lactate) + 2e$^-$ ⇌ Zn + lactate $^{2-}$	-0.826			4	18	33
1		Zn (malate) + 2e$^-$ ⇌ Zn + malate $^{2-}$	-0.859			5	12	34
1		Zn (mandelate)$^+$ + 2e$^-$ ⇌ Zn + mandelate$^-$	-0.828			5	19	35

Standard value V	Solvent	Electrode Reference Number	Standard value V	Solvent	Electrode Reference Number	Standard value V	Solvent	Electrode Reference Number
-1.405	1	21						
-1.245	1	18						
-1.241	1	32						
-1.215	1	17						
-1.214	1	20						
-1.183	1	19						
-1.06	1	11						
-1.056	1	26						
-1.043	1	24						
-1.04	1	13						
-1.021	1	31						
-0.955	1	28						
-0.948	1	30						
-0.925	1	25						
-0.915	1	23						
-0.874	1	27						
-0.859	1	29						
-0.859	1	34						
-0.831	1	22						
-0.828	1	35						
-0.826	1	33						
-0.798	1	12						
-0.7618	1	2						
-0.439	1	14						
-0.400	1	15						
0.054	1	1						
0.441	1	16						
-0.505	2	3						
-0.792	15	4						
-0.853	11	5						
-0.835	12	6						
-1.05	16	7						
-0.757	18	8						
-0.74	23	9						
-0.788	27	10						

NOTES: Zinc

1) Solubility data

2) Solubility and potentiometry data

3) Solubility and spectrophotometry data

4) Potentiometric pH measurements

5) Kinetics data

BIBLIOGRAPHY: Zinc

1) Data reported by M. Pourbaix, "Atlas d'Equilibres Electrochimiques à 25 ^{o}C (Gauthier-Villars Editeur, Paris, 1963), pp. 406 - 413

2) R.M. Stokes, I.M. Stokes, Trans. Faraday Soc., 41 (1945), 688 - 695

3) G. Carraro, H.L. Stephens, J. Electrochem. Soc., 104 (1957), 512 - 515

4) F.E. Rosztoczy, C.W. Tobias, Electrochim. Acta, 11 (1966), 857 - 861

5) N.E. Khomutov, Zh. Fiz. Khim., 42 (1968), 2223 - 2229

6) V.A. Pleskov, Acta Physicochim. U.S.S.R., 21 (1946), 41 - 54

7) T. Pavlopoulos, M. Strehlow, Z. Physik. Chem. (N.F.), 2 (1954), 89 - 103

8) V.A. Pleskov, Zh. Fiz. Khim., 22 (1948), 351 - 361

9) L.M. Mukerjee, J.J. Kelly, W. Baranetzky, J. Sica, J. Phys. Chem., 72 (1968), 3410 - 3415

10) Data reported by A.J. De Bethune, N.A. Swendeman-Loud, "Standard Aqueous Electrode Potentials and Temperature Coefficients at 25 ^{o}C" (Clifford A. Hampel, Ill.)

11) A.D. Paul, Thesis, Univ. California, Berkeley, Calif., 1955 , UCRL - 2926

12) J.W. Fulton, D.F. Swinehart, J. Am. Chem. Soc., 76 (1954), 864

13) T.P. Dirkse, J. Electrochem. Soc., 101 (1954), 328 - 331

14) T.O. Denney, C.B. Monk, Trans. Faraday Soc., 47 (1951), 992 - 998

15) B. Monk, Trans. Faraday Soc., 47 (1951), 297 - 302

16) J.I. Evans, C.B. Monk, Trans. Faraday Soc., 51 (1955), 1244 - 1250

17) R. Näsänen, E. Uisitalo, Acta Chem. Scand., 8 (1954), 112 - 118

18) C.W. Davies, C.B. Monk, Trans. Faraday Soc., 50 (1954), 132 - 133

19) R.P. Bell, G.M. Waind, J. Chem. Soc., 1951, 2357 - 2362

Electronically conducting phase	Intermediate species	Composition of the solution	Solvent	Temperature °C	Pressure Torr	Measuring method
a	b	c	d		e	d
Cd		1) $CdCl_2$: (1.089 up to 21.645) · 10^{-3} m; 2) Cd $(ClO_4)_2$: 10^{-3} M; $HClO_4$: pH 3 - 4	1	25		I
Cd		$CdCl_2$: 0.1 N; Pb $(NO_3)_2$: 0.1 N; NH_4 SCN: 20 mol %	2	-34 ± 0.5		I
Cd-Cd(Hg)		$CdCl_2$: 0.0005 up to 0.1 M	7	25		I
Cd(Hg)		Cd $(NO_3)_2$: 0.098 up to 0.908 M; NH_4NO_3 : 0.01 M	8	25 ± 0.3		II
Cd		$CdCl_2$: 0.01 N	16	25.00 ± 0.01		I
Cd(Hg)		1) $CdCl_2$: satd ; KCl: 0.05 up to 0.5 N; 2) $CdCl_2$, KCl	18	25		I
Cd		CdI_2 : 0.01 N	23	25		I
Cd(Hg)			1	25		II
Cd(Hg)	Cd $(Br)_2$		1	25		II
Cd(Hg)			1	25		II
Cd(Hg)			1	25		II
Cd(Hg)		$Cd(NO_3)_2$: 0.01 up to 0.1 M; NH_4 Br : 0.03 up to 0.2 M	8	25 ± 0.3		II
Cd(Hg)		Cd $(ClO_4)_2$: 0.01 M; NaCl : 0 up to 0.5 N; $HClO_4$: 0.01 N	1	25± 0.05		I
Cd(Hg)	Cd $(Cl)_2$	Cd $(ClO_4)_2$: 0.01 M; NaCl : 0 up to 0.5 N; $HClO_4$: 0.01 N	1	25 ± 0.05		I
Cd(Hg)		Cd $(ClO_4)_2$: 0.01 M; NaCl : 0 up to 0.5 N; $HClO_4$: 0.01 N	1	25± 0.05		I
Cd			1	25	760	IV
Cd	$CdCO_3$		1	25	760	IV
Cd(Hg)	$Cd_2[Fe(CN)_6]$	$Li_4[Fe(CN)_6]$: 0.001, 0.01 and 0.05 M	1	25		I
Cd	NH_3		1	25	760	IV
Cd	CdO		1	25	760	IV
Cd	Cd $(OH)_2$		1	25	760	IV
Cd	CdS		1	25	760	IV
Cd(Hg)	Cd $(SCN)_2$	Cd $(NO_3)_2$: 0.01 up to 0.1 M; NH_4 Br: 0.3 up to 0.2 M	8	25 ± 0.3		II
Cd	$CdSO_4$	H_2SO_4 : 0.2 M	3	25 ± 0.1		I
Glass	Cd (S_2O_3)	$CdCl_2$: (4.68 up to 11.71) · 10^{-3} N; BaS_2O_3 : (0.2181 up to 0.6471) · 10^{-3} N	1	25± 0.03		I III
Glass		$CdCl_2$: (4.69 up to 11.71) · 10^{-3} N; $Na_2S_2O_3$: (0.2181 up to 0 6471) · 10^{-3} N	1	25± 0.03		I III

Comparison electrode	Liquid junction	Electrode reaction	Standard value V	Uncertainty mV	Temperature Coefficient μV/°C	Notes	Reference	Electrode Reference Number
d			f		g	h		
4 3		$Cd^{2+} + 2e^- \rightleftharpoons Cd$	-0.4030	±0.5	-93		1,2,13	1
29		$Cd^{2+} + 2e^- \rightleftharpoons Cd$	-0.189	±4			3	2
4		$Cd^{2+} + 2e^- \rightleftharpoons Cd$	-0.410	±2			4	3
3		$Cd^{2+} + 2e^- \rightleftharpoons Cd$	-0.340				5	4
14		$Cd^{2+} + 2e^- \rightleftharpoons Cd$	-0.75		-700		6	5
4		$Cd^{2+} + 2e^- \rightleftharpoons Cd$	-0.414	±6	-592		7,8	6
7		$Cd^{2+} + 2e^- \rightleftharpoons Cd$	-0.47				9	7
3		$Cd(Br)^+ + 2e^- \rightleftharpoons Cd + Br^-$	-0.469				10	8
3		$Cd(Br)_2 + 2e^- \rightleftharpoons Cd + 2\,Br^-$	-0.492				10	9
3		$Cd(Br)_3^- + 2e^- \rightleftharpoons Cd + 3\,Br^-$	-0.482				10	10
3		$Cd(Br)_4^{2-} + 2e^- \rightleftharpoons Cd + 4\,Br^-$	-0.489				10	11
3		$Cd(Br)_3^- + 2e^- \rightleftharpoons Cd + 3\,Br^-$	-0.779				11	12
26		$Cd(Cl)^+ + 2e^- \rightleftharpoons Cd + Cl^-$	-0.463	±0.5			12	13
26		$Cd(Cl)_2 + 2e^- \rightleftharpoons Cd + 2\,Cl^-$	-0.483	±0.5			12	14
26		$Cd(Cl)_3^- + 2e^- \rightleftharpoons Cd + 3\,Cl^-$	-0.468	±0.5			12	15
		$Cd(CN)_4^{2-} + 2e^- \rightleftharpoons Cd + 4\,CN^-$	-1.028				13	16
		$CdCO_3 + 2e^- \rightleftharpoons Cd + CO_3^{2-}$	-0.74		-1232		13	17
10		$Cd_2[Fe(CN)_6] + 4e^- \rightleftharpoons 2\,Cd + [Fe(CN)_6]^{4-}$	-0.6205				14	18
		$Cd(NH_3)_4^{2+} + 2e^- \rightleftharpoons Cd + 4\,NH_3$ (aq)	-0.613				13	19
		$CdO + 2H^+ + 2e^- \rightleftharpoons Cd + H_2O$	0.063				15	20
		$Cd(OH)_2 + 2e^- \rightleftharpoons Cd + 2OH^-$	-0.809		-1014		13	21
		$CdS + 2e^- \rightleftharpoons Cd + S^{2-}$	-1.175		-870		15	22
		$Cd(SCN)_2 + 2e^- \rightleftharpoons Cd + 2\,SCN^-$	-0.740				11	23
1		$CdSO_4 + 2e^- \rightleftharpoons Cd + SO_4^{4-}$	-0.246				16	24
3		$Cd(S_2O_3) + 2e^- \rightleftharpoons Cd + S_2O_3^{2-}$	-0.520			1	17	25
3		$Cd(S_2O_3)_2^{2-} + 2e^- \rightleftharpoons Cd + 2\,S_2O_3^{2-}$	-0.594			1	17	26

Electronically conducting phase	Intermediate species	Composition of the solution	Solvent	Temperature °C	Pressure Torr	Measuring method
a	b	c	d		e	d
Cd			1	25	760	IV
Cd	CdH		1	25	760	IV
Pt-H_2	Glycine	Cd $(ClO_4)_2$: (3.546 up to 7.770) · 10^{-3} M; glycine : 0.1099 up to 0.1525 M	1	25 ± 0.03		I
Pt-H_2	Glycine	Cd $(ClO_4)_2$: (3.546 up to 7.770) · 10^{-3} M; glycine : 0.1099 up to 0.1525 M	1	25± 0.03		I
Pt-H_2	Glycylglycine	Cd $(ClO_4)_2$: (5.222 up to 7.244) · 10^{-3} M; glycylglycine : (28.32 up to 62.1) · 10^{-3} M	1	25 ± 0.03		I
Pt-H_2	Glycylglycine	Cd $(ClO_4)_2$: (5.222 up to 7.244) · 10^{-3} M; glycylglycine: (28.32 up to 62.1) · 10^{-3} M	1	25± 0.3		I
Pt-H_2	Glycylglycylglycine	$CdCl_2$: (6.593 up to 8.161) · 10^{-3} M; glycylglycylglycine : (22.46 up to 27.80) · 10^{-3} M	1	25± 0.03		I
Pt-H_2	Glycylglycylglycine	$CdCl_2$: (6.593 up to 8.161) · 10^{-3} M; glycylglycylglycine : (22.46 up to 27.80) · 10^{-3} M	1	25± 0.03		I
Glass	Cd (8-hydroquinoline-5-sulfonate)	8-hydroquinoline-5-sulfonic acid : 2.954 · 10^{-3} M; Cd $(ClO_4)_2$	1	25	743	I
Glass		8-hydroxyquinoline-5-sulfonic acid : 2.954 · 10^{-3} M; Cd $(ClO_4)_2$	1	25	743	I
Cd	Cd (oxalate)	Cd^{2+} : (0.16 up to 0.3) · 10^{-3} M; K_2 oxalate : (up to 0.01) · 10^{-3} m	1	25		V
Cd		Cd^{2+} : (0.16 up to 0.3) · 10^{-3} m; K_2 oxalate : (up to 0.01) · 10^{-3} m	1	25		V

Comparison electrode	Liquid junction	Electrode reaction	Standard value	Uncertainty	Temperature Coefficient	Notes	Reference	Electrode Reference Number
			V	mV	μV/°C			
d			f		g	h		
		$HCdO_2^- + 3H^+ + 2e^- \rightleftharpoons Cd + 2H_2O$	0.583				15	27
		$Cd + H^+ + e^- \rightleftharpoons CdH$	-2.417				15	28
4		$Cd(glycine)^{2+} + 2e^- \rightleftharpoons Cd + glycine$	-0.547			2	18	29
4		$Cd(glycine)_2^{2+} + 2e^- \rightleftharpoons Cd + 2\ glycine$	-0.667			2	18	30
4		$Cd(glycylglycine)^{2+} + 2e^- \rightleftharpoons Cd + glycylglycine$	-0.502			2	18	31
4		$Cd(glycylglycine)_2^{2+} + 2e^- \rightleftharpoons Cd + 2\ glycylglycine$	-0.578			2	18	32
4		$Cd(glycyglycylglycine)^{2+} + 2e^- \rightleftharpoons$ Cd + glycyglycylglycine	-0.501			2	18	33
4		$Cd(glycylglycylglycine)_2^{2+} + 2e^- \rightleftharpoons$ $Cd + 2\ glycylglycylglycine^-$	-0.577			2	18	34
4		$Cd(8\text{-hydroquinoline-5-sulfonate}) + 2e^- \rightleftharpoons$ $Cd + 8\text{-hydroquinoline-5-sulfonate}^{2-}$	-0.633			2	19	35
4		$Cd(8\text{-hydroxyquinoline-5-sulfonate})_2^{2-} + 2e^- \rightleftharpoons$ $Cd + 2\ 8\text{-hydroxyquinoline-5-sulfonate}^{2-}$	-0.828			2	19	36
		$Cd(oxalate) + 2e^- \rightleftharpoons Cd + oxalate^{2-}$	-0.522			3	20	37
		$Cd(oxalate)_2^{2-} + 2e^- \rightleftharpoons Cd + 2\ oxalate^{2-}$	-0.572			3	21	38

Standard value V	Solvent	Electrode Reference Number	Standard value V	Solvent	Electrode Reference Number	Standard value V	Solvent	Electrode Reference Number
-2.417	1	28	-0.75	16	5			
-1.175	1	22	-0.414	18	6			
-1.028	1	16	-0.47	23	7			
-0.828	1	36						
-0.809	1	21						
-0.74	1	17						
-0.667	1	30						
-0.633	1	35						
-0.6205	1	18						
-0.613	1	19						
-0.594	1	26						
-0.578	1	32						
-0.577	1	34						
-0.572	1	38						
-0.547	1	29						
-0.522	1	37						
-0.520	1	25						
-0.502	1	31						
-0.501	1	33						
-0.492	1	9						
-0.489	1	11						
-0.483	1	14						
-0.482	1	10						
-0.469	1	8						
-0.468	1	15						
-0.463	1	13						
-0.4030	1	1						
0.063	1	20						
0.583	1	27						
-0.189	2	2						
-0.246	3	24						
-0.410	7	3						
-0.779	8	12						
-0.740	8	23						
-0.340	8	4						

NOTES: Cadmium

1) Potentiometric pH measurements and conductometric data
2) Potentiometric pH measurements
3) Solubility data

BIBLIOGRAPHY: Cadmium

1) W.B. Treumann, L.M. Ferris, J. Am. Chem. Soc., 80 (1952), 5048 - 5050
2) J.L. Burnett, M.H. Zirin, J. Inorg. Nucl. Chem., 28 (1966), 902 - 904
3) F.E. Rosztoczy, C.W. Tobias, Electrochim. Acta, 11 (1966), 857 - 861
4) J.D. Hefley, E.S. Amis, J. Electrochem. Soc., 112 (1965), 336 - 340
5) Ya. I. Turyan, Zh. Fiz. Khim., 28 (1954), 2152 - 2155
6) V.A. Pleskov, Acta Physicochim. U.S.S.R., 21 (1946), 41 - 54
7) R.W.C. Broadbank, B.L. Muju, K.W. Morcom, Trans. Faraday Soc., 64 (1968), 3318 - 3320
8) M. De Rossi, G. Pecci, B. Scrosati, Ric. Sci., 37 (1957), 342 - 346
9) V.A. Pleskov, Zh. Fiz. Khim., U.S.S.R., 22 (1948), 351 - 361
10) P. Kivalo, P. Ekari, Suomen Kem., 30 B (1957), 116
11) Ya. I. Turyan, Zh. Anal. Khim., 11 (1956), 71 - 76
12) C.E. Vanderzee, H.J. Dawson, Jr. J. Am. Chem. Soc., 75 (1953), 5659 - 5663
13) Data reported by A.J. De Bethune, N.A Swendeman-Loud, "Standard Aqueous Electrode Potentials and Temperature Coefficients at 25 °C" (Clifford A. Hampel, Ill.)
14) A. Basinski, S. Poczpko, Roczn. Chem., 31 (1957), 449 - 455
15) Data reported by M. Pourbaix, "Atlas d'Equilibres Electrochimiques à 25 °C" (Gauthier-Villars Editeur, Paris, 1963), pp. 414 - 420
16) C. Furlani, Ann. Chim., 45 (1955), 264 - 273
17) T.O. Denney, C.B. Monk, Trans. Faraday Soc., 51 (1947), 992
18) J.I. Evans, C.B. Monk, Trans. Faraday Soc., 51 (1955), 1244 - 1250
19) R. Näsänen, E. Uisitalo, Acta Chem. Scand., 8 (1954), 112 - 118
20) J.E. Barney, W.J. Argensinger, Jr., C.A. Reynolds, J. Am. Chem. Soc., 73 (1951), 3785 - 3788
21) W.C. Vosburgh, J.F. Beckman, J. Am. Chem. Soc., 62 (1940), 1028

Electronically conducting phase	Intermediate species	Composition of the solution	Solvent	Temperature °C	Pressure Torr	Measuring method
a	b	c	d		e	d
Hg		1) $Hg_2(NO_3)_2$: 0.0455 up to 0.728 M; 2) $Hg_2(ClO_4)_2$: 0.01, 0.05 and 0.1012 M; 3) Hg_2Cl_2 : satd.	1	25		I
Hg			7	25	760	IV
Hg			8	25	760	IV
Hg			11	25	760	IV
Hg			12	25	760	IV
Hg			15	25	760	IV
Hg		HgHCOO : 0.01 N	16	25 ± 0.01		I
Hg		$HgBr_2$: 0.01 N	23	25		I
Ind.			1	25	760	IV
Hg	Hg_2Br_2	HBr : 0.005 up to 0.2 N	1	25		I
Hg			1	25	760	IV
Hg	Hg_2Cl_2	HCl : 0.004 up to 1 m	1	25		I
Hg	Hg_2Cl_2	KCl: satd.	17	25		I
Hg	Hg_2Cl_2	KCl: satd.	18	25		I
Hg			1	25	760	IV
Hg	$(Hg_2)_3[Co(CN)_6]_2$	$K_3[Co(CN)_6]$: 0.128 and 0.05 m	1	25± 0.005		I
Hg			1	25		I
Hg	Hg_2HPO_4	H_3PO_4 : 0.01 up to 0.1 m	1	25		I
Hg	Hg_2I_2		1	25	760	IV
Hg			1	25	760	IV
Hg	$Hg_2(IO_3)_2$	KIO_3	1	25± 0.05		I
Hg	$Hg_2(N_3)_2$		1	25		I
Hg	HgO (red)		1	25	760	IV
Hg	HgO (yellow)		1	25	760	IV
Hg	HgO (red)	NaOH : 5, 10, 20 and 30 %	1	25		I
Ind.	HgO (red)		1	25	760	IV
Ind	HgO (yellow)		1	25	760	IV
Hg	$Hg(OH)_2$		1	25	760	IV
Ind.	$Hg(OH)_2$		1	25	760	IV
Hg	HgS (black)		1	25	760	IV

Comparison electrode	Liquid junction	Electrode reaction	Standard value	Uncertainty	Temperature Coefficient	Notes	Reference	Electrode Reference Number
			V	mV	μV/°C			
d			f		g	h		
1) 4 2,3)1		$Hg_2^{2+} + 2e^- \rightleftharpoons 2Hg$	0.7973				1,2,3	1
		$Hg_2^{2+} + 2e^- \rightleftharpoons 2Hg$	0.746				4	2
		$Hg_2^{2+} + 2e^- \rightleftharpoons 2Hg$	0.828				4	3
		$Hg_2^{2+} + 2e^- \rightleftharpoons 2Hg$	0.677				4	4
		$Hg_2^{2+} + 2e^- \rightleftharpoons 2Hg$	0.716				4	5
		$Hg_2^{2+} + 2e^- \rightleftharpoons 2Hg$	0.716				4	6
14		$Hg_2^{2+} + 2e^- \rightleftharpoons 2Hg$	0.180	± 5	-460		5	7
7		$Hg_2^{2+} + 2e^- \rightleftharpoons 2Hg$	0.25				6	8
		$2Hg^{2+} + 2e^- \rightleftharpoons Hg_2^{2+}$	0.920				7	9
1		$Hg_2Br_2 + 2e^- \rightleftharpoons 2Hg + 2Br^-$	0.13923	± 0.03	-154		8	10
		$Hg(Br)_4^{2-} + 2e^- \rightleftharpoons Hg + 4\,Br^-$	0.223		-420		7	11
1)4; 2,3, 4,5)1 6) 33		$Hg_2Cl_2 + 2e^- \rightleftharpoons 2Hg + 2Cl^-$	0.26808	± 0.13	-317		9,10 11,12 13,14	12
1		$Hg_2Cl_2 + 2e^- \rightleftharpoons 2Hg + 2Cl^-$	0.27				15	13
4		$Hg_2Cl_2 + 2e^- \rightleftharpoons 2Hg + 2Cl^-$	0.2451	± 0.7			16	14
		$Hg(CN)_4^{2-} + 2e^- \rightleftharpoons Hg + 4\,CN^-$	-0.37		+780		7	15
3		$(Hg_2)_3[Co(CN)_6]_2 + 6e^- \rightleftharpoons 6Hg + 2[Co(CN)_6]^{3-}$	-0.427	± 1		1	17	16
		$HgF^+ + 2e^- \rightleftharpoons Hg + F^-$	0.876				18	17
1		$Hg_2HPO_4 + 2e^- \rightleftharpoons 2Hg + HPO_4^{2-}$	0.6359				19	18
		$Hg_2I_2 + 2e^- \rightleftharpoons 2Hg + 2I^-$	-0.0405		+19		7	19
		$Hg(I)_4^{2-} + 2e^- \rightleftharpoons Hg + 4\,I^-$	-0.038		+40		7	20
		$Hg_2(IO_3)_2 + 2e^- \rightleftharpoons 2Hg + 2IO_3^-$	0.3942				20	21
		$Hg_2(N_3)_2 + 2e^- \rightleftharpoons 2Hg + 2N_3^-$	-0.26				21	22
		$HgO + 2H^+ + 2e^- \rightleftharpoons Hg + H_2O$	0.926				22	23
		$HgO + 2H^+ + 2e^- \rightleftharpoons Hg + H_2O$	0.927				22	24
9		$HgO + H_2O + 2e^- \rightleftharpoons Hg + 2OH^-$	0.0977				23	25
		$2HgO + 4H^+ + 2e^- \rightleftharpoons Hg_2^{2+} + 2H_2O$	1.064				22	26
		$2HgO + 4H^+ + 2e^- \rightleftharpoons Hg_2^{2+} + 2H_2O$	1.065				22	27
		$Hg(OH)_2 + 2H^+ + 2e^- \rightleftharpoons Hg + 2H_2O$	1.034				22	28
		$2Hg(OH)_2 + 4H^+ + 2e^- \rightleftharpoons Hg_2^{2+} + 4H_2O$	1.279				22	29
		$HgS + 2e^- \rightleftharpoons Hg + S^{2-}$	-0.69		-790		7	30

Electronically conducting phase	Intermediate species	Composition of the solution	Solvent	Temperature °C	Pressure Torr	Measuring method
a	b	c	d		e	d
Hg			1	25		I
Hg			1	25		I
Hg	Hg_2SO_4	H_2SO_4 : 0.1 up to 8.0 m	1	25		I
Hg	Hg_2SO_4	H_2SO_4 : 0.00025 up to 0.4800 m	7	25		I
Hg			1	25		I
Hg			1	25		I
Hg			1	25		I
Hg			1	25	760	IV
Ind.			1	25	760	IV
Hg	HgH		1	25	760	IV
Hg	Hg_2(acetate)$_2$	1) CH_3COOH: 0.4 and 1 M 2) CH_3COOH + (Na or K) CH_3COO; 3) CH_3COOH : 0.2 up to 2 m; KNO_3 : 0.1 m	1	25		I
Hg	Hg (acetate)$_2$	$NaCH_3COO$	17	25		I
Hg	Hg_2(benzoate)$_2$	Benzoic acid :0.01 up to 0 .06 m	1	25 ± 0.1		I
Hg	Hg_2 (formate)$_2$	NaHCOO : 0.005 up to 4 m	1	25± 0.005		I
Hg	Hg_2 (oxalate)	Oxalic acid : satd.	1	25		I
Hg	Hg_2 (propionate)$_2$	NaC_2H_5COO : 0.0005 up to 3 M	1	25 ± 0.05		I

Comparison electrode	Liquid junction	Electrode reaction	Standard value V	Uncertainty mV	Temperature Coefficient μV/°C	Notes	Reference	Electrode Reference Number
d			f		g	h		
		$Hg(SCN)_4^{2-} + 2e^- \rightleftharpoons Hg + 4\ SCN^-$	0.267				24	31
		$Hg(SeCN)_4^{2-} + 2e^- \rightleftharpoons Hg + 4\ SeCN^-$	0.025				24	32
1		$Hg_2SO_4 + 2e^- \rightleftharpoons 2Hg + SO_4^{2-}$	0.6125	±0.15			25	33
1		$Hg_2SO_4 + 2e^- \rightleftharpoons 2Hg + SO_4^{2-}$	0.5392	±0.5	-802		26	34
		$Hg(S_2O_3)_2^{2-} + 2e^- \rightleftharpoons Hg + 2\ S_2O_3^{2-}$	0.028				3	35
		$Hg(S_2O_3)_3^{4-} + 2e^- \rightleftharpoons Hg + 3(S_2O_3)^{2-}$	-0.057				24	36
		$Hg(S_2O_3)_4^{6-} + 2e^- \rightleftharpoons Hg + 4(S_2O_3)^{2-}$	-0.100				24	37
		$HHgO_2^- + 3H^+ + 2e^- \rightleftharpoons Hg + 2H_2O$	1.474				22	38
		$2\ HHgO_2^- + 6H^+ + 2e^- \rightleftharpoons Hg_2^{2+} + 4H_2O$	2.159				22	39
		$Hg + H^+ + e^- \rightleftharpoons HgH$	-2.281				22	40
1,2)1 3) 33	3)satd. KNO_3	$Hg_2(acetate)_2 + 2e^- \rightleftharpoons 2Hg + 2\ acetate^-$	0.51163	±0.3	-580		27,28 29	41
33		$Hg(acetate)_2 + 2e^- \rightleftharpoons Hg + 2\ acetate^-$	-0.1599	±0.8			30	42
1		$Hg_2(benzoate)_2 + 2e^- \rightleftharpoons 2Hg + 2\ benzoate^-$	0.4263	±0.1			31	43
3	Agar gel	$Hg_2(formate)_2 + 2e^- \rightleftharpoons 2Hg + 2\ formate^-$	0.567	±2			32	44
13		$Hg_2(oxalate) + 2e^- \rightleftharpoons 2Hg + oxalate^{2-}$	0.4158				33	45
3		$Hg_2(propionate)_2 + 2e^- \rightleftharpoons 2Hg + 2\ propionate^-$	0.499	±2			32	46

Mercury 1

Standard value V	Solvent	Electrode Reference Number
-2.281	1	40
-0.69	1	30
-0.427	1	16
-0.37	1	15
-0.26	1	22
-0.100	1	37
-0.057	1	36
-0.0405	1	19
-0.038	1	20
0.025	1	32
0.028	1	35
0.0977	1	25
0.13923	1	10
0.223	1	11
0.267	1	31
0.26808	1	12
0.3942	1	21
0.4158	1	45
0.4263	1	43
0.499	1	46
0.51163	1	41
0.567	1	44
0.6125	1	33
0.6359	1	18
0.7973	1	1
0.876	1	17
0.920	1	9
0.926	1	23
0.927	1	24
1.034	1	28
1.064	1	26
1.065	1	27
1.279	1	29
1.474	1	38
2.159	1	39
0.5392	7	34
0.746	7	2
0.828	8	3
0.677	11	4
0.716	12	5
0.716	15	6
0.180	16	7
0.2451	18	14
-0.1599	18	42
0.27	18	13
0.25	23	8

NOTES: Mercury

1) Liquid junction electric tension eliminated by K amalgam

BIBLIOGRAPHY: Mercury

1) O.D. Bonner, F.A. Unietis, J. Am. Chem. Soc., 75 (1953), 5111 - 5113

2) S.E.S. Elwakkad, T.M. Sabem, J. Phys. Colloid Chem., 54 (1950), 1371 - 1383

3) J.G. Murgulescu, E. Constantinescu, Rev. Chim. Acad. Rep. Populaire Roumaine, 8 (1963), 5 - 12

4) N.E. Khomutov, Zh. Fiz. Khim., 42(1968), 2223 - 2229

5) V.A. Pleskov, Acta Physicochim. U.S.S.R., 21 (1946), 41 - 54

6) V.A. Pleskov, Zh. Fiz. Khim., 22 (1948), 351 - 361

7) Data reported by A.J. De Bethune, N.A. Swendeman-Loud, "Standard Aqueous Electrode Potentials and Temperature Coefficients at 25 °C" (Clifford A. Hampel, Ill.)

8) S.R. Gupta, G.J. Hills, D.J.G. Ives, Trans. Faraday Soc., 59 (1963), 1886 - 1891

9) J. Pouradier, H. Chateau, Compt. Rend., 237 (1953), 711 - 713

10) A.K. Covington, J.V. Dobson, L.W. Jones, Electrochim. Acta, 12 (1967), 513 - 523

11) S.R. Gupta, G.J. Hillis, D.J.G. Ives, Trans. Faraday Soc., 59 (1963), 1874 - 1885

12) A.K. Gozybowski, J. Phys. Chem., 62 (1958), 550 - 552

13) K. Schwabe, E. Ferse, Ber. Bunsen-Ges.Phys. Chem., 69(1965), 383 - 391

14) L. Sharma, G. Sahu, B. Prasad, J. Indian Chem. Soc., 45 (1968), 580 - 582

15) S. Tutundžic, P. Putanov, Glasnik Hem. Društva, Beograd, 21 (1956), 257 - 267

16) M. De Rossi, G. Pecci, B. Scrosati, Ric. Sci., 37 (1967), 342 - 346

BIBLIOGRAPHY: Mercury

17) P.A. Rock, Inorg. Chem., 4 (1965), 1667 - 1668

18) A.D. Paul, Thesis,Universisty of California, Berkeley, Calif., 1955, UCRL - 2926

19) W.D. Larson, J. Phys. Colloid Chem., 54 (1950), 310 - 315

20) I. Takács, Magyar Chém. Folyoirat, 49 (1943), 33 - 51, 100 - 108

21) S. Suzuki, J. Chem. Soc. Japan, Pure Chem. Sect., 73 (1952), 278 - 280

22) Data reported by M. Pourbaix, "Atlas d'Equilibres Electrochimiques à 25 ^{o}C" (Gauthier-Villars Editeur, Paris, 1963), pp. 421 - 427

23) W.J. Hamer, D.N. Craig, J. Electrochem. Soc., 104 (1957), 206 - 211

24) V.F. Tacopova, Zh. Neorg. Khim., 1 (1956), 243

25) A.K. Covington, J.V. Dobson, L.W. Jones, Trans. Faraday Soc., 61 (1965), 2050 - 2056

26) W. Kanning, M.G. Bowman, J. Am. Chem. Soc., 68 (1946), 2042 - 2046

27) W.D. Larson, J. Phys. Chem., 67 (1963), 937 - 938

28) A.K. Covington, P.K. Talukdev, H.R. Thirsk, Trans. Faraday Soc., 60 (1964), 412 - 416

29) Li Shu-Chen, K. Pan, J. Chinese Chem. Soc., (Taipei), 15 (1968), 106 - 112

30) B. Jakuszewski, J.B. Jedraejwska, Rocz. Chem., 39 (1965), 907 - 911

31) J. Bertram, S.J. Bone, Trans. Faraday Soc., 63 (1967), 415

32) J. Chauchard, J. Gauthier, Bull. Soc. Chim. France, 1966 , 2635 - 2639

33) D.T. Ferrell, Jr., I. Blackburn, W.C. Vosburgh, J. Am. Chem. Soc., 70 (1948), 3812 - 3815

Group III

Electronically conducting phase	Intermediate species	Composition of the solution	Solvent	Temperature °C	Pressure Torr	Measuring method
a	b	c	d		e	d
B	BO (gas)		1	25	760	IV
B	B_2O_3		1	25	760	IV
B			1	25	760	IV
B			1	25	760	IV
B			1	25	760	IV
B			1	25	760	IV
B			1	25	760	IV
B	H_3BO_3 (s)		1	25	760	IV
B	H_3BO_3 (aq)		1	25	760	IV
B			1	25	760	IV
B	BH		1	25	760	IV
B	BH, H_3BO_3 (s)		1	25	760	IV
B	B_2H_6		1	25	760	IV
Ind.	B_2H_6		1	25	760	IV
Ind.	B_2H_6		1	25	760	IV
Ind.	B_2H_6		1	25	760	IV
Ind.	B_2H_6		1	25	760	IV
Ind.	B_2H_6 ; H_3BO_3 (s)		1	25	760	IV
Ind.	B_2H_6 ; H_3BO_3 (aq)		1	25	760	IV
Ind.	B_2H_6		1	25	760	IV
B	B_5H_9		1	25	760	IV
Ind.	B_5H_9, H_3BO_3(s)		1	25	760	IV
B	$B_{10}H_{14}$ (s)		1	25	760	IV
B	$B_{10}H_{14}$ (gas)		1	25	760	IV
Hg(DME)	$B_{10}H_{14}$(s)	$B_{10}H_{14}$: 0.05 M; R_4NClO_4: 0.1 up to 0.4 M ($R = C_2H_5$ or C_4H_9)	23	24 ± 2		II
Ind.	$B_{10}H_{14}$ (s)		1	25	760	IV
Ind.	$B_{10}H_{14}$ (s)		1	25	760	IV
Ind.	$B_{10}H_{14}$ (s)		1	25	760	IV
Ind.	$B_{10}H_{14}$ (s)		1	25	760	IV
Ind.	$B_{10}H_{14}$(s), H_3BO_3(s)		1	25	760	IV
Ind.	$B_{10}H_{14}$(s), H_3BO_3(aq)		1	25	760	IV
Ind.	$B_{10}H_{14}$ (gas), H_3BO_3 (s)		1	25	760	IV

Comparison electrode	Liquid junction	Electrode reaction	Standard value V	Uncertainty mV	Temperature Coefficient µV/°C	Notes	Reference	Electrode Reference Number
d			f		g	h		
		$BO + 2H^+ + 2e^- \rightleftharpoons B + H_2O$	0.806				1	1
		$B_2O_3 + 6H^+ + 6e^- \rightleftharpoons 2B + 3H_2O$	-0.841				1	2
		$BO_3^{3-} + 6H^+ + 3e^- \rightleftharpoons B + 3H_2O$	-0.165				1	3
		$B_4O_7^{2-} + 14H^+ + 12e^- \rightleftharpoons 4B + 7H_2O$	-0.792				1	4
		$HBO_3^{2-} + 5H^+ + 3e^- \rightleftharpoons B + 3H_2O$	-0.437				1	5
		$H_2BO_3^- + 4H^+ + 3e^- \rightleftharpoons B + 3H_2O$	-0.687				1	6
		$H_2BO_3^- + H_2O + 3e^- \rightleftharpoons B + 4OH^-$	-1.79		-1147		2	7
		$H_3BO_3 + 3H^+ + 3e^- \rightleftharpoons B + 3H_2O$	-0.869				1	8
		$H_3BO_3 + 3H^+ + 3e^- \rightleftharpoons B + 3H_2O$	-0.8698		-481		2	9
		$HB_4O_7^- + 13H^+ + 12e^- \rightleftharpoons 4B + 7H_2O$	-0.836				1	10
		$B + H^+ + e^- \rightleftharpoons BH$	-4.883				1	11
		$H_3BO_3 + 4H^+ + 4e^- \rightleftharpoons BH + 3H_2O$	-1.873				1	12
		$2B + 6H^+ + 6e^- \rightleftharpoons B_2H_6$	-0.143				1	13
		$2BO_3^{3-} + 18H^+ + 12e^- \rightleftharpoons B_2H_6 + 6H_2O$	-0.154				1	14
		$B_4O_7^{2-} + 26H^+ + 24e^- \rightleftharpoons 2B_2H_6 + 7H_2O$	-0.467				1	15
		$2HBO_3^{2-} + 16H^+ + 12e^- \rightleftharpoons B_2H_6 + 6H_2O$	-0.289				3	16
		$2H_2BO_3^- + 14H^+ + 12e^- \rightleftharpoons B_2H_6 + 6H_2O$	-0.415				1	17
		$2H_3BO_3 + 12H^+ + 12e^- \rightleftharpoons B_2H_6 + 6H_2O$	-0.506				1	18
		$2H_3BO_3 + 12H^+ + 12e^- \rightleftharpoons B_2H_6 + 6H_2O$	-0.506				1	19
		$HB_4O_7^- + 25H^+ + 24e^- \rightleftharpoons 2B_2H_6 + 7H_2O$	-0.490				1	20
		$5B + 9H^+ + 9e^- \rightleftharpoons B_5H_9$	-0.189				1	21
		$5H_3BO_3 + 24H^+ + 24e^- \rightleftharpoons B_5H_9 + 15H_2O$	-0.614				1	22
		$10B + 14H^+ + 14e^- \rightleftharpoons B_{10}H_{14}$	-0.201				1	23
		$10B + 14H^+ + 14e^- \rightleftharpoons B_{10}H_{14}$	-0.220				1	24
3			-0.78	± 20		1	3	25
		$10B_3^{3-} + 74H^+ + 44e^- \rightleftharpoons B_{10}H_{14} + 30H_2O$	-0.176				1	26
		$5B_4O_7^{2-} + 98H^+ + 88e^- \rightleftharpoons 2B_{10}H_{14} + 35H_2O$	-0.610				1	27
		$10HBO_3^{2-} + 64H^+ + 44e^- \rightleftharpoons B_{10}H_{14}$	-0.362				1	28
		$10H_2BO_3^- + 54H^+ + 44e^- \rightleftharpoons B_{10}H_{14} + 30H_2O$	-0.536				1	29
		$10H_3BO_3 + 44H^+ + 44e^- \rightleftharpoons B_{10}H_{14} + 30H_2O$	-0.657				1	30
		$10H_3BO_3 + 44H^+ + 44e^- \rightleftharpoons B_{10}H_{14} + 30H_2O$	-0.656				1	31
		$10H_3BO_3 + 44H^+ + 44e^- \rightleftharpoons B_{10}H_{14} + 30H_2O$	-0.662				1	32

Electronically conducting phase	Intermediate species	Composition of the solution	Solvent	Temperature °C	Pressure Torr	Measuring method
a	b	c	d		e	d
Ind.	$B_{10}H_{14}$ (s)		1	25	760	IV
Ind.	BO (gas), H_3BO_3 (s)		1	25	760	IV

Comparison electrode	Liquid junction	Electrode reaction	Standard value V	Uncertainty mV	Temperature Coefficient μV/°C	Notes	Reference	Electrode Reference Number
d			f		g	h		
		$5HB_4O_7^- + 93H^+ + 88e^- \rightleftharpoons 2B_{10}H_{14} + 35H_2O$	-0.642				1	33
		$H_3BO_3 + H^+ + e^- \rightleftharpoons BO + 2H_2O$	-4.219				1	34

Boron 1

Standard value V	Solvent	Electrode Reference Number
-4.883	1	11
-4.219	1	34
-1.873	1	12
-1.79	1	7
-0.8698	1	9
-0.869	1	8
-0.841	1	2
-0.836	1	10
-0.792	1	4
-0.687	1	6
-0.662	1	32
-0.657	1	30
-0.656	1	31
-0.642	1	33
-0.614	1	22
-0.610	1	27
-0.536	1	29
-0.506	1	19
-0.506	1	18
-0.490	1	20
-0.467	1	15
-0.437	1	5
-0.415	1	17
-0.362	1	28
-0.289	1	16
-0.220	1	24
-0.201	1	23
-0.189	1	21
-0.176	1	26
-0.165	1	3
-0.154	1	14
-0.143	1	13
0.806	1	1
-0.78	23	25

Standard value V	Solvent	Electrode Reference Number

Standard value V	Solvent	Electrode Reference Number

NOTES: Boron

1) Liquid junction containing R_4NClO_4

BIBLIOGRAPHY: Boron

1) Data reported by M. Pourbaix, "Atlas d'Equilibres Electrochimiques à 25 °C" (Gauthier-Villars Editeur, Paris, 1963), 158 - 167

2) Data reported by A.J. De Bethune, N.A. Swendeman-Loud: "Standard Aqueous Electrode Potentials and Temperature Coefficients at 25 °C" (Clifford A. Hampel, Ill.)

3) J.A. Chambus, A.D. Norma, J. Am. Chem. Soc., 90 (1968), 6056 - 6062

Electronically conducting phase	Intermediate species	Composition of the solution	Solvent	Temperature °C	Pressure Torr	Measuring method
a	b	c	d		e	d
Al			1	25	760	IV
			7	25		I
Quinhy-drone			1	25		I
Al			1	25	760	IV
Al			1	25	760	IV
Al	$Al_2O_3 \cdot 3H_2O$ (bayerite)		1	25	760	IV
Al	$Al_2O_3 \cdot H_2O\gamma$ (boehmite)		1	25	760	IV
Al	$Al_2O_3\ \alpha$ (Corundum)		1	25	760	IV
Al	$Al_2O_3 \cdot 3H_2O$ (hydrargillitc)		1	25	760	IV
Al	$Al(OH)_3$		1	25	760	IV
Al	$Al(OH)_3$		1	25	760	IV
Al			1	25	760	IV

Comparison electrode	Liquid junction	Electrode reaction	Standard value V	Uncertainty mV	Temperature Coefficient μV/°C	Notes	Reference	Electrode Reference Number
d			f		g	h		
		$Al^{3+} + 3e^- \rightleftharpoons Al$	-1.662				1	1
1		$Al^{3+} + 3e^- \rightleftharpoons Al$	-0.632				2	2
		$Al(Cl)^{2+} + 3e^- \rightleftharpoons Al + Cl^-$	-1.802			1 2	3 4	3
		$Al(F)_6^{3-} + 3e^- \rightleftharpoons Al + 6F^-$	-2.069		-200		1	4
		$AlO_2^- + 4H^+ + 3e^- \rightleftharpoons Al + 2H_2O$	-1.262		+504		5	5
		$Al_2O_3 \cdot 3H_2O + 6H^+ + 6e^- \rightleftharpoons 2Al + 6H_2O$	-1.535				5	6
		$Al_2O_3 \cdot H_2O + 6H^+ + 6e^- \rightleftharpoons 2Al + 4H_2O$	-1.505				5	7
		$Al_2O_3 + 6H^+ + 6e^- \rightleftharpoons 2Al + 3H_2O$	-1.494				5	8
		$Al_2O_3 \cdot 3H_2O + 6H^+ + 6e^- \rightleftharpoons 2Al + 6H_2O$	-1.550				5	9
		$Al(OH)_3 + 3H^+ + 3e^- \rightleftharpoons Al + 3H_2O$	-1.471				5	10
		$Al(OH)_3 + 3e^- \rightleftharpoons Al + 3OH^-$	-2.30		-930		1	11
		$H_2AlO_3^- + H_2O + 3e^- \rightleftharpoons Al + 4OH^-$	-2.33				1	12

Aluminum 1

Standard value V	Solvent	Electrode Reference Number
-2.33	1	12
-2.30	1	11
-2.069	1	4
-1.802	1	3
-1.662	1	1
-1.550	1	9
-1.535	1	6
-1.505	1	7
-1.494	1	8
-1.471	1	10
-1.262	1	5
-0.632	7	2

Standard value V	Solvent	Electrode Reference Number

Standard value V	Solvent	Electrode Reference Number

NOTES: Aluminum

1) Potentiometric pH measurements

2) The value is an average of those given by two authors; they all are in the same range of uncertainty

BIBLIOGRAPHY: Aluminum

1) Data reported by A.J. De Bethune, N.A. Swendeman-Loud "Standard Aqueous Electrode Potentials and Temperature Coefficients at 25 °C" (Clifford A. Hampel, Ill.)

2) B. Jakuszewski, Z. Kozlowskii, Rocz. Chem., 38 (1964), 93

3) C. Brosset, Diss., Stockholm, 1942

4) C. Brosset, J. Orring, Swensk Kem. Tidskr., 55 (1943), 101

5) Data reported by M. Pourbaix "Atlas d'Equilibres Electrochimiques à 25 °C" (Gauthier-Villars Editeur, Paris, 1963), 168 - 176

Electronically conducting phase	Intermediate species	Composition of the solution	Solvent	Temperature °C	Pressure Torr	Measuring method
a	b	c	d		e	d
Ga		a) $GaCl_3$; HCl : 0.005 up to 0.03 N; b) Ga $(ClO_4)_3$; $HClO_4$: 0.005 up to 0.03 N	1	25		I
Ind.			1	25	760	IV
Ga		$GaCl_3$: 0.15 m; HCl : up to 7 m	1	25 ± 3		V
Ga		$GaCl_3$: 0.15 m; HCl : up to 7 m	1	25 ± 3		V
Ga	Ga $(Cl)_3$	$GaCl_3$: 0.15 m; HCl : up to 7 m	1	25 ± 3		V
Ga		$GaCl_3$: 0.15 m; HCl : up to 7 m	1	25 ± 3		V
			1	25		I
Ga	Ga_2O		1	25	760	IV
Ga			1	25	760	IV
Ind.			1	25	760	IV
Ga			1	25	760	IV
Ind.			1	25	760	IV
Ga			1	25	760	IV
Ind.			1	25	760	IV
Ga	Ga_2O_3		1	25	760	IV
Ga			1	25	760	IV
Ind.			1	25	760	IV
Ga	Ga $(OH)_3$		1	25	760	IV
Ga		KOH, KGa $(OH)_4$	1	25		I
Ga			1	25	760	IV
Ga			1	25	760	IV
Ind.			1	25	760	IV
Ind.	Ga_2O, Ga_2O_3		1	25	760	IV
Ind.	Ga_2O, Ga $(OH)_2$		1	25	760	IV

Comparison electrode	Liquid junction	Electrode reaction	Standard value V	Uncertainty mV	Temperature Coefficient μV/°C	Notes	Reference	Electrode Reference Number
d			f		g	h		
1		$Ga^{3+} + 3e^- \rightleftharpoons Ga$	-0.560	± 5			1	1
		$Ga^{3+} + e^- \rightleftharpoons Ga^{2+}$	-0.677				2	2
1		$Ga(Cl)^{2+} + 3e^- \rightleftharpoons Ga + Cl^-$	-0.548			1	3	3
1		$Ga(Cl)_2^+ + 3e^- \rightleftharpoons Ga + 2\,Cl^-$	-0.514			1	3	4
1		$Ga(Cl)_3 + 3e^- \rightleftharpoons Ga + 3\,Cl^-$	-0.471			1	3	5
1		$Ga(Cl)_4^- + 3e^- \rightleftharpoons Ga + 4\,Cl^-$	-0.447			1	3	6
		$Ga(F)^{2+} + 3e^- \rightleftharpoons Ga + F^-$	-0.676				4	7
		$Ga_2O + 2H^+ + 2e^- \rightleftharpoons 2\,Ga + H_2O$	-0.401				2	8
		$GaO^+ + 2H^+ + 3e^- \rightleftharpoons Ga + H_2O$	-0.415				2	9
		$GaO^+ + 2H^+ + e^- \rightleftharpoons Ga^{2+} + H_2O$	-0.334				2	10
		$GaO_2^- + 4H^+ + 3e^- \rightleftharpoons Ga + 2H_2O$	-0.114				2	11
		$GaO_2^- + 4H^+ + e^- \rightleftharpoons Ga^{2+} + 2H_2O$	0.567				2	12
		$GaO_3^{3-} + 6H^+ + 3e^- \rightleftharpoons Ga + 3H_2O$	0.319				2	13
		$GaO_3^{3-} + 6H^+ \rightleftharpoons Ga^{3+} + 3H_2O$	1.868				2	14
		$Ga_2O_3 + 6H^+ + 6e^- \rightleftharpoons 2\,Ga + 3H_2O$	-0.485				2	15
		$Ga(OH)^{2+} + H^+ + 3e^- \rightleftharpoons Ga + H_2O$	-0.479				2	16
		$Ga(OH)^{2+} + H^+ + e^- \rightleftharpoons Ga^{2+} + H_2O$	-0.525				2	17
		$Ga(OH)_3 + 3H^+ + 3e^- \rightleftharpoons Ga + 3H_2O$	-0.419				2	18
3		$Ga(OH)_4^- + 3e^- \rightleftharpoons Ga + 4\,OH^-$	-1.326	± 1			5	19
		$HGaO_3^{2-} + 5H^+ + 3e^- \rightleftharpoons Ga + 3H_2O$	0.088				2	20
		$H_2GaO_3^- + H_2O + 3e^- \rightleftharpoons Ga + 4OH^-$	-1.219				6	21
		$HGaO_3^{2-} + 5H^+ + e^- \rightleftharpoons Ga^{2+} + 3H_2O$	1.174				2	22
		$Ga_2O_3 + 4H^+ + 4e^- \rightleftharpoons Ga_2O + 2H_2O$	-0.527				2	23
		$2Ga(OH)_3 + 4H^+ + 4e^- \rightleftharpoons Ga_2O + 5H_2O$	-0.428				2	24

Gallium 1

Standard value V	Solvent	Electrode Reference Number	Standard value V	Solvent	Electrode Reference Number	Standard value V	Solvent	Electrode Reference Number
-1.326	1	19						
-1.219	1	21						
-0.677	1	2						
-0.676	1	7						
-0.560	1	1						
-0.548	1	3						
-0.527	1	23						
-0.525	1	17						
-0.514	1	4						
-0.485	1	15						
-0.479	1	16						
-0.471	1	5						
-0.447	1	6						
-0.428	1	24						
-0.419	1	18						
-0.415	1	9						
-0.401	1	8						
-0.334	1	10						
-0.114	1	11						
0.088	1	20						
0.319	1	13						
0.567	1	12						
1.174	1	22						
1.868	1	14						

NOTES : Gallium

1) Anion exchange data

BIBLIOGRAPHY: Gallium

1) W.M.Saltman, N.H. Nachtrieb, J. Electrochem. Soc., 100 (1953), 126 - 130

2) Data reported by M. Pourbaix "Atlas d'Equilibres Electrochimiques à 25 °C" (Gauthier-Villars Editeur, Paris, 1963), 428 - 435

3) K.A. Kraus, F. Nelson, G.W. Smith, J. Phys. Chem., 58 (1954), 11 - 17

4) L.M. Yates; H.W. Dodgen, Am. Chem. Soc., 122nd Meeting, Sept. 1952, Abstract 18P

5) A.T. Van Vagramian, T.I. Leshawa, Z. Physik. Chem., 234 (1967), 57 - 60

6) Data reported by A.J. De Bethune, N.A. Swendeman-Loud, "Standard Aqueous Electrode Potentials and Temperature Coefficients at 25 °C" (Clifford A. Hampel, Ill.)

Electronically conducting phase	Intermediate species	Composition of the solution	Solvent	Temperature °C	Pressure Torr	Measuring method
a	b	c	d		e	d
In		$InClO_4$; $HClO_4$: 0.1 m; Ce $(ClO_4)_3$: I = 0.83 m	1	25 ± 0.1		V
In		$InCl_3$: 2 · 10^{-3} up to 2 · 10^{-2} M; HCl : 2 · 10^{-2} M	1	25 ± 0.05		I
Ind.		$InClO_4$; $HClO_4$: 0.1 m; Ce $(ClO_4)_3$: I = 0.83 m	1	25± 0.1		V
Ind.			1	25	760	IV
Ind.		$InClO_4$; $HClO_4$: 0.1 m; Ce $(ClO_4)_3$; I = 0.83 m	1	25 ± 0.1		V
Pt		In(ClO_4); Fe $(ClO_4)_3$: 0.5 · 10^{-3} m; Fe $(ClO_4)_2$: 0.5 · 10^{-3} m; $HClO_4$; NaF: (0.374 up to 16.993) · 10^{-3} m	1	25± 0.01		I
In			1	25	760	IV
In	In_2O_3		1	25	760	IV
Ind.			1	25	760	IV
Ind.	In_2O_3		1	25	760	IV
In			1	25	760	IV
Ind.			1	25	760	IV
In	In $(OH)_3$		1	25	760	IV
In	In $(OH)_3$		1	25	760	IV
Ind.	In $(OH)_3$		1	25	760	IV
In	InH (gas)		1	25	760	IV

Comparison electrode	Liquid junction	Electrode reaction	Standard value V	Uncertainty mV	Temperature Coefficient μV/°C	Notes	Reference	Electrode Reference Number
d			f		g	h		
1		$In^{+} + e^{-} \rightleftharpoons In$	-0.14			1	1	1
4		$In^{3+} + 3e^{-} \rightleftharpoons In$	-0.3382	±3	-40		2	2
1		$In^{2+} + e^{-} \rightleftharpoons In^{+}$	-0.40			1	1	3
		$In^{3+} + 2e^{-} \rightleftharpoons In^{+}$	-0.443				3	4
1		$In^{3+} + e^{-} \rightleftharpoons In^{2+}$	-0.49			1	1	5
1		$In(F)^{2+} + 3e^{-} \rightleftharpoons In + F^{-}$	-0.429			2	4	6
		$InO_2^{-} + 4H^{+} + 3e^{-} \rightleftharpoons In + 2H_2O$	0.146				3	7
		$In_2O_3 + 6H^{+} + 6e^{-} \rightleftharpoons 2\,In + 3H_2O$	-0.190				3	8
		$InO_2^{-} + 4H^{+} + 2e^{-} \rightleftharpoons In^{+} + 2H_2O$	0.262				3	9
		$In_2O_3 + 6H^{+} + 4e^{-} \rightleftharpoons 2\,In^{+} + 3H_2O$	-0.216				3	10
		$In(OH)^{2+} + H^{+} + 3e^{-} \rightleftharpoons In + H_2O$	-0.266				3	11
		$In(OH)^{2+} + H^{+} + 2e^{-} \rightleftharpoons In^{+} + H_2O$	-0.330				3	12
		$In(OH)_3 + 3H^{+} + 3e^{-} \rightleftharpoons In + 3H_2O$	-0.172				3	13
		$In(OH)_3 + 3e^{-} \rightleftharpoons In + 3OH^{-}$	-1.00		-970		5	14
		$In(OH)_3 + 3H^{+} + 2e^{-} \rightleftharpoons In^{+} + 3H_2O$	-0.189				3	15
		$In + H^{+} + e^{-} \rightleftharpoons InH$	-1.951				3	16

Indium 1

Standard value V	Solvent	Electrode Reference Number
-1.951	1	16
-1.00	1	14
-0.49	1	5
-0.443	1	4
-0.429	1	6
-0.40	1	3
-0.3382	1	2
-0.330	1	12
-0.266	1	11
-0.216	1	10
-0.190	1	8
-0.189	1	15
-0.172	1	13
-0.14	1	1
0.146	1	7
0.262	1	9

NOTES: Indium

1) Equilibrium data

2) Cell junction constituted by agar-agar containing 1.6 m $NaClO_4$

BIBLIOGRAPHY: Indium

1) L.G. Heplet, Z.Z. Hugusti, J. Am. Chem. Soc., 75 (1953), 5652 - 5654

2) A.K. Cavington, M.A. Hakeem, W.F.K. Wynne-Jones, J. Chem. Soc., 1963, 4394 - 4401

3) Data reported by M. Pourbaix "Atlas d'Equilibres Electrochimiques à 25 °C" (Gauthier-Villars Editeur, Paris, 1963), 436 - 442

4) L.G. Heplet, Z.Z. Hugusti, Jr., J.W. Kury, J. Phys. Chem., 58 (1954), 26 - 28

5) Data reported by A.J. De Bethune, N.A. Swendeman-Loud "Standard Aqueous Electrode Potentials and Temperature Coefficients at 25 °C" (Clifford A. Hampel, Ill.)

Electronically conducting phase	Intermediate species	Composition of the solution	Solvent	Temperature °C	Pressure Torr	Measuring method
a	b	c	d		e	d
Tl			1	25	760	IV
Tl		$TlNO_3$: 0.00293, 0.00820 and 0.0194 M	18	25 ± 0.1		I
Tl(Hg)		$TlCl_4$: 10^{-4} up to 10^{-3} M, Et_4NClO_4 : 0.01 M	23	25		I
Tl			29	25		I
Ind.			1	25	760	IV
Tl	TlBr		1	25	760	IV
Tl	Tl (Br)		1	25		V
Tl			1	25		V
Tl	TlCl		1	25	760	IV
Tl	Tl (Cl)	1) Tl (Cl) : (0.697 up to 3.881) · 10^{-3} M 2) TlCl : (1.5 up to 7.5) · 10^{-3} M 3) TlCl : (1.52 up to 3.70) · 10^{-3} M; HCl : up to 0.3 M	1	25		III, V
Tl		TlCl : (1.52 up to 3.70) · 10^{-3} M; HCl: up to 0.2 M	1	25		V
	Tl_2CrO_4		1	25		I
Tl	Tl (F)	Tl (F) : (1.919 up to 2.185) · 10^{-3} M	1	25		V
Tl	TlI		1	25	760	IV
Tl	$TlIO_3$	Tl (IO_3) : (0.650 up to 4.108) · 10^{-3} M	1	25		V
	TlN_3		1	25		I
Tl	Tl (N_3)	Tl (N_3) : (0.043 up to 0.233) · 10^{-3} M	1	25		V
Tl	Tl (NO_3)	Tl (NO_3) : (0.040 up to 0.204) · 10^{-3} M	1	25		V
Tl	Tl_2O		1	25	760	IV
Ind	Tl_2O_3		1	25	760	IV
Tl	TlOH		1	25	760	IV
Tl	TlOH		1	25		I
Tl	Tl (OH)	1) Tl (OH) : (0.690 up to 4.313) · 10^{-3} M 2) Tl (OH) : (21.1 up to 50.7) · 10^{-3} M 3) Tl (OH) and $C_2H_5NO_2$: 0.0211 up to 0.0507 m	1	25		V
Ind.	$Tl(OH)_3$		1	25	760	IV
Tl	Tl_2S		1	25	760	IV
Hg (D.M.E.)	TlSCN	Tl (SCN) : 2 · 10^{-3} m; KSCN : 1 and 10 m		25		I, II

Comparison electrode	Liquid junction	Electrode reaction	Standard value	Uncertainty	Temperature Coefficient	Notes	Reference	Electrode Reference Number
			V	mV	μV/°C			
d			f		g	h		
		$Tl^+ + e^- \rightleftharpoons Tl$	-0.336				1	1
25	KCl satd.	$Tl^+ + e^- \rightleftharpoons Tl$	-0.34	± 14	-850		2	2
7	Et_4NClO_4 0.1 M	$Tl^+ + e^- \rightleftharpoons Tl$	-0.410				3	3
1		$Tl^+ + e^- \rightleftharpoons Tl$	-0.3817		+191		17	4
		$Tl^{3+} + 2e^- \rightleftharpoons Tl^+$	1.252				1	5
		$TlBr + e^- \rightleftharpoons Tl + Br^-$	-0.658		-413		4	6
1		$Tl(Br) + e^- \rightleftharpoons Tl + Br^-$	-0.392			1	5	7
1		$Tl(Br)_2^- + e^- \rightleftharpoons Tl + 2\,Br^-$	-0.395			1	5	8
		$TlCl + e^- \rightleftharpoons Tl + Cl^-$	-0.5568		-560		4	9
1		$Tl(Cl) + e^- \rightleftharpoons Tl + Cl^-$	-0.372		+200	2 4	6,7,8	10
1		$Tl(Cl)_2^- + e^- \rightleftharpoons Tl + 2\,Cl^-$	-0.346				8	11
		$Tl_2CrO_4 + 2e^- \rightleftharpoons 2\,Tl + CrO_4^{2-}$	-1.056				9	12
1		$Tl(F) + e^- \rightleftharpoons Tl + F^-$	-0.342			1	6	13
		$TlI + e^- \rightleftharpoons Tl + I^-$	-0.752		-150		4	14
1		$TlIO_3 + e^- \rightleftharpoons Tl + IO_3^-$	-0.666			1,4	6	15
		$TlN_3 + e^- \rightleftharpoons Tl + N_3^-$	-0.554				10	16
1		$Tl(N_3)^- + e^- \rightleftharpoons Tl + N_3^-$	-0.359		+130	1	11	17
1		$Tl(NO_3) + e^- \rightleftharpoons Tl + NO_3^-$	-0.356		+130	1	11	18
		$Tl_2O + 2H^+ + 2e^- \rightleftharpoons 2\,Tl + H_2O$	0.512				1	19
		$Tl_2O_3 + 6H^+ + 4e^- \rightleftharpoons 2\,Tl^+ + 3H_2O$	1.329				1	20
		$TlOH + H^+ + e^- \rightleftharpoons Tl + H_2O$	0.485				1	21
		$TlOH + e^- \rightleftharpoons Tl + OH^-$	-0.34				12	22
1		$Tl(OH) + e^- \rightleftharpoons Tl + OH^-$	-0.385		+130	3,4	6 15 16	23
		$Tl(OH)_3 + 3H^+ + 2e^- \rightleftharpoons Tl^+ + 3H_2O$	1.189				1	24
		$Tl_2S + 2e^- \rightleftharpoons 2\,Tl + S^{2-}$	-0.90		-940		4	25
3	KCl satd.	$TlSCN + e^- \rightleftharpoons Tl + SCN^-$	-0.56				13,14	26

Electronically conducting phase	Intermediate species	Composition of the solution	Solvent	Temperature °C	Pressure Torr	Measuring method
a	b	c	d		e	d
Tl	Tl (SCN)	Tl (SCN) : (0.693 up to 3.885) · 10^{-3} M	1	25		V
Tl	$Tl(SO_4)^-$	$Tl(SO_4)^-$: (0.043 up to 0.752) · 10^{-3} M	1	25		V
Tl	TlH		1	25	760	IV
Ind.	Tl_2O, Tl_2O_3		1	25	760	IV
Ind.	TlOH, $Tl(OH)_3$		1	25	760	IV
Tl	$Tl(OH)_3$, TlOH		1	25	760	IV

Comparison electrode	Liquid junction	Electrode reaction	Standard value V	Uncertainty mV	Temperature Coefficient µV/°C	Notes	Reference	Electrode Reference Number
d			f		g	h		
1		$Tl(SCN) + e^- \rightleftharpoons Tl + SCN^-$	-0.384		+735	1,4	6	27
1		$Tl(SO_4)^- + e^- \rightleftharpoons Tl + SO_4^{2-}$	-0.418		+66	1	6	28
		$Tl + H^+ + e^- \rightleftharpoons TlH$	-1.865				1	29
		$Tl_2O_3 + 4H^+ + 4e^- \rightleftharpoons Tl_2O + 2H_2O$	0.905				1	30
		$Tl(OH)_3 + 2H^+ + 2e^- \rightleftharpoons TlOH + 2H_2O$	0.778				1	31
		$Tl(OH)_3 + 2e^- \rightleftharpoons TlOH + 2OH^-$	-0.05		-940		4	32

Thallium 1

Standard value V	Solvent	Electrode Reference Number	Standard value V	Solvent	Electrode Reference Number	Standard value V	Solvent	Electrode Reference Number
-1.865	1	29						
-1.056	1	12						
-0.90	1	25						
-0.752	1	14						
-0.666	1	15						
-0.658	1	6						
-0.56	1	26						
-0.5568	1	9						
-0.554	1	16						
-0.418	1	28						
-0.395	1	8						
-0.392	1	7						
-0.385	1	23						
-0.384	1	27						
-0.372	1	10						
-0.359	1	17						
-0.356	1	18						
-0.346	1	11						
-0.342	1	13						
-0.34	1	22						
-0.336	1	1						
-0.05	1	32						
0.485	1	21						
0.512	1	19						
0.778	1	31						
0.905	1	30						
1.189	1	24						
1.252	1	5						
1.329	1	20						
-0.34	18	2						
-0.410	23	3						
-0.3817	29	4						

NOTES: Thallium

1) Solubility data

2) Solubility and conductometric data

3) Solubility and kinetic data

4) The value is an average of those given by several authors; they all are in the same range of uncertainty

BIBLIOGRAPHY: Thallium

1) Data reported by M. Pourbaix, "Atlas d'Equilibres Electrochimiques à 25 °C" (Gauthier-Villars Editeur, Paris, 1963), 443 - 448

2) T. Pavlopoulos, H. Strehlow, Z. Physik. Chem. (Leipzig), 2 (1954), 89 - 103

3) T.F. Coetzee, J.J. Campion, J. Am. Chem. Soc., 89 (1967), 2513 - 2517

4) Data reported by A.J. De Bethune, N.A. Swendeman-Loud, "Standard Aqueous Electrode Potentials and Temperature Coefficients at 25 °C" (Clifford A. Hampel, Ill.)

5) L.G. Carpenter, Thesis, Columbia University, 1956, Univ. Microfilms, Order n. 17044

6) R.P. Bell, J.H.B. George, Trans. Faraday Soc., 49 (1953), 619 - 627

7) A.G. Garrett, S.J. Vellenga, J. Am. Chem. Soc., 67 (1945), 225 - 228

8) K.H. Hu, A.B. Scott, J. Am. Chem. Soc., 77 (1955), 1380 - 1382

9) S. Suzuki, J. Chem. Soc. Japan, 74 (1953), 219

10) S. Suzuki, J. Chem. Soc. Japan, 73 (1952), 150

11) V.S.K. Nair, G.H. Nancollas, J. Chem. Soc., 1957, 318 - 323

12) S. Suzuki, J. Chem. Soc. Japan, 72 (1951), 265 - 267

13) S. Suzuki, J. Chem. Soc. Japan, 73 (1952), 153 - 155

14) J.N. Grenier, L. Meites, Anal. Chim. Acta, 14 (1956), 482 - 494

15) R.P. Bell, M.H. Panckhurst, J. Chem. Soc., 1956, 2836

16) R.P. Bell, M.N. Panckhurst, Rec. Trav. Chim. Pays Bas, 75 (1956), 725 - 729

17) D.P. Boden, Diss. Abstr., 31 B (1970), 3223

Electronically conducting phase	Intermediate species	Composition of the solution	Solvent	Temperature °C	Pressure Torr	Measuring method
a	b	c	d		e	d
Sc			1	25	760	IV
			1	25		I
			1	25		I
	$Sc(F)_3$		1	25		I
Sc	Sc_2O_3		1	25	760	IV
Sc			1	25	760	IV
Sc	$Sc(OH)_3$		1	25	760	IV
Sc	$Sc(OH)_3$		1	25	760	IV

Comparison electrode	Liquid junction	Electrode reaction	Standard value V	Uncertainty mV	Temperature Coefficient μV/°C	Notes	Reference	Electrode Reference Number
d			f		g	h		
		$Sc^{3+} + 3e^- \rightleftharpoons Sc$	-2.077				1	1
		$Sc(F)^{2+} + 3e^- \rightleftharpoons Sc + F^-$	-2.239				2	2
		$Sc(F)_2^+ + 3e^- \rightleftharpoons Sc + 2F^-$	-2.333				2	3
		$Sc(F)_3 + 3e^- \rightleftharpoons Sc + 3F^-$	-2.428				2	4
		$Sc_2O_3 + 6H^+ + 6e^- \rightleftharpoons 2Sc + 3H_2O$	-1.591				1	5
		$Sc(OH)^{2+} + 3e^- \rightleftharpoons Sc + H_2O$	-1.980				1	6
		$Sc(OH)_3 + 3H^+ + 3e^- \rightleftharpoons Sc + 3H_2O$	-1.784				1	7
		$Sc(OH)_3 + 3e^- \rightleftharpoons Sc + 3OH^-$	-2.61				3	8

Standard value V	Solvent	Electrode Reference Number
-2.61	1	8
-2.428	1	4
-2.333	1	3
-2.239	1	2
-2.077	1	1
-1.980	1	6
-1.784	1	7
-1.591	1	5

Standard value V	Solvent	Electrode Reference Number

Standard value V	Solvent	Electrode Reference Number

BIBLIOGRAPHY: Scandium

1) Data reported by M. Pourbaix, "Atlas d'Equilibres Electrochimiques à 25 °C" (Gauthier-Villars Editeur, Paris, 1963), 177 - 182

2) A.D. Paul, Thesis, University of California, Berkekey, 1955, UCRL-2926

3) Data reported by A.J. De Bethune, N.A. Swendeman-Loud, "Standard Aqueous Electrode Potentials and Temperature Coefficients at 25 °C" (Clifford A. Hampel, Ill.)

Electronically conducting phase	Intermediate species	Composition of the solution	Solvent	Temperature °C	Pressure Torr	Measuring method
a	b	c	d		e	d
Y			1	25	760	IV
Y	Y_2O_3		1	25	760	IV
Y	$Y(OH)_3$ (s)		1	25	760	IV
Y	$Y(OH)_3$ (s)		1	25	760	IV
		$Y_2(SO_4)_3$: up to 1.2357 N	1	25		V

Comparison electrode	Liquid junction	Electrode reaction	Standard value V	Uncertainty mV	Temperature Coefficient μV/°C	Notes	Reference	Electrode Reference Number
d			f		g	h		
		$Y^{3+} + 3e^- \rightleftharpoons Y$	-2.372				1	1
		$Y_2O_3 + 6H^+ + 6e^- \rightleftharpoons 2Y + 3H_2O$	-1.676				1	2
		$Y(OH)_3 + 3H^+ + 3e^- \rightleftharpoons Y + 3H_2O$	-1.981				1	3
		$Y(OH)_3 + 3e^- \rightleftharpoons Y + 3OH^-$	-2.81		-950		2	4
1		$Y(SO_4)^+ + 3e^- \rightleftharpoons Y + SO_4^{2-}$	-2.872			1	3	5

Yttrium 1

Standard value V	Solvent	Electrode Reference Number
-2.872	1	5
-2.81	1	4
-2.372	1	1
-1.981	1	3
-1.676	1	2

Standard value V	Solvent	Electrode Reference Number

Standard value V	Solvent	Electrode Reference Number

NOTES: Yttrium

1) Solubility data

BIBLIOGRAPHY: Yttrium

1) Data reported by M. Pourbaix, "Atlas d'Equilibres Electrochimiques à 25 °C" (Gauthier-Villars Editeur, Paris, 1963), 177 - 182

2) Data reported by A.J. De Bethune, N.A. Swendeman-Loud, "Standard Aqueous Electrode Potentials and Temperature Coefficients at 25 °C" (Clifford A. Hampel, Ill.)

3) F.H. Spedding, S. Jaffe, J. Am. Chem. Soc., 76 (1954), 882 - 884

Group IV

Electronically conducting phase	Intermediate species	Composition of the solution	Solvent	Temperature °C	Pressure Torr	Measuring method
a	b	c	d		e	d
C	CCl_4		1	25	760	IV
C	CO		1	25	760	IV
C	CO_2		1	25	760	IV
C			1	25	760	IV
C			1	25	760	IV
C	H_2CO_3		1	25	760	IV
C			1	25	760	IV
C	HCOOH		1	25	760	IV
C	CH_4		1	25	760	IV
Ind.	CH_4, CH_3OH		1	25	760	IV
Ind.	CH_4, CO		1	25	760	IV
Ind.	CH_4, CO_2		1	25	760	IV
Ind.	C_2H_4, C_2H_2		1	25	760	IV
Ind.	C_2H_6, C_2H_4		1	25	760	IV
Ind.	CH_3OH		1	25	760	IV
Ind.	CH_3OH		1	25	760	IV
Ind.	CH_3OH, H_2CO_3		1	25	760	IV
Ind.	CH_3OH, HCOH		1	25	760	IV
Ind.	CH_3OH		1	25	760	IV
Ind.	CH_3OH, HCOOH		1	25	760	IV
Ind.			1	25	760	IV
Ind.	$(CN)_2$, HCNO		1	25	760	IV
Ind.	$(CNS)_2$		1	25	760	IV
Ind.	CO, CO_2		1	25	760	IV
Ind.			1	25	760	IV
Ind.			1	25	760	IV
Ind.	H_2CO_3		1	25	760	IV
Ind.	HCN, $(CN)_2$		1	25	760	IV
Ind.	HCOH		1	25	760	IV
Ind.	HCOH		1	25	760	IV
Ind.	HCOH, H_2CO_3		1	25	760	IV
Ind.	HCOH		1	25	760	IV
Ind.	HCOH, HCOOH		1	25	760	IV

Comparison electrode	Liquid junction	Electrode reaction	Standard value V	Uncertainty mV	Temperature Coefficient µV/°C	Notes	Reference	Electrode Reference Number
d			f		g	h		
		$CCl_4 + 4e^- \rightleftharpoons C + 4Cl^-$	1.18		-645		1	1
		$CO + 2H^+ + 2e^- \rightleftharpoons C + H_2O$	0.518				2	2
		$CO_2 + 4H^+ + 4e^- \rightleftharpoons C + 2H_2O$	0.207				2	3
		$CO_3^{2-} + 6H^+ + 4e^- \rightleftharpoons C + 3H_2O$	0.475				2	4
		$HCO_3^- + 5H^+ + 4e^- \rightleftharpoons C + 3H_2O$	0.323				2	5
		$H_2CO_3 + 4H^+ + 4e^- \rightleftharpoons C + 3H_2O$	0.228				2	6
		$HCOO^- + 3H^+ + 2e^- \rightleftharpoons C + 2H_2O$	0.724				2	7
		$HCOOH + 2H^+ + 2e^- \rightleftharpoons C + 2H_2O$	0.627				2	8
		$C + 4H^+ + 4e^- \rightleftharpoons CH_4$	-0.1316		-209		1	9
		$CH_3OH + 2H^+ + 2e^- \rightleftharpoons CH_4 + H_2O$	0.588		-35		1	10
		$CO + 6H^+ + 6e^- \rightleftharpoons CH_4 + H_2O$	0.497				2	11
		$CO_2 + 8H^+ + 8e^- \rightleftharpoons CH_4 + 2H_2O$	0.169				2	12
		$C_2H_2 + 2H^+ + 2e^- \rightleftharpoons C_2H_4$	0.731		-580		1	13
		$C_2H_4 + 2H^+ + 2e^- \rightleftharpoons C_2H_6$	0.52		-625		1	14
		$CO_3^{2-} + 8H^+ + 6e^- \rightleftharpoons CH_3OH + 2H_2O$	0.209				2	15
		$HCO_3^- + 7H^+ + 6e^- \rightleftharpoons CH_3OH + 2H_2O$	0.107				2	16
		$H_2CO_3 + 6H^+ + 6e^- \rightleftharpoons CH_3OH + 2H_2O$	0.044				2	17
		$HCOH + 2H^+ + 2e^- \rightleftharpoons CH_3OH$	0.232				2	18
		$HCOO^- + 5H^+ + 4e^- \rightleftharpoons CH_3OH + H_2O$	0.199				2	19
		$HCOOH + 4H^+ + 4e^- \rightleftharpoons CH_3OH + H_2O$	0.145				2	20
		$CNO^- + H_2O + 2e^- \rightleftharpoons CN^- + 2OH^-$	-0.970		-1211		1	21
		$2HCNO + 2H^+ + 2e^- \rightleftharpoons (CN)_2 + 2H_2O$	-0.330		-588		1	22
		$(CNS)_2 + 2e^- \rightleftharpoons 2CNS^-$	0.77				1	23
		$CO_2 + 2H^+ + 2e^- \rightleftharpoons CO + H_2O$	-0.103				2	24
		$2CO_3^{2-} + 4H^+ + 2e^- \rightleftharpoons C_2O_4^{2-} + 2H_2O$	0.441				2	25
		$2HCO_3^- + 2H^+ + 2e^- \rightleftharpoons C_2O_4^{2-} + 2H_2O$	-0.170				2	26
		$2H_2CO_3 + 2e^- \rightleftharpoons C_2O_4^{2-} + 2H_2O$	-0.547				2	27
		$(CN)_2 + 2H^+ + 2e^- \rightleftharpoons 2HCN$	0.373		-596		1	28
		$CO_3^{2-} + 6H^+ + 4e^- \rightleftharpoons HCOH + 2H_2O$	0.197				2	29
		$HCO_3^- + 5H^+ + 4e^- \rightleftharpoons HCOH + 2H_2O$	0.044				2	30
		$H_2CO_3 + 4H^+ + 4e^- \rightleftharpoons HCOH + 2H_2O$	-0.050				2	31
		$HCOO^- + 3H^+ + 2e^- \rightleftharpoons HCOH + H_2O$	0.167				2	32
		$HCOOH + 2H^+ + 2e^- \rightleftharpoons HCOH + H_2O$	0.056				2	33

Electronically conducting phase	Intermediate species	Composition of the solution	Solvent	Temperature °C	Pressure Torr	Measuring method
a	b	c	d		e	d
Ind.			1	25	760	IV
Ind.			1	25	760	IV
Ind.	H_2CO_3		1	25	760	IV
Ind.			1	25	760	IV
Ind.			1	25	760	IV
Ind.	HCOOH, CO_2		1	25	760	IV
Ind.	HCOOH, H_2CO_3		1	25	760	IV
Ind.	HCOOH		1	25	760	IV
Ind.	HCOOH, $H_2C_2O_4$		1	25	760	IV
Ind.	H_2CO_3		1	25	760	IV
Ind.	$H_2C_2O_4$, H_2CO_3		1	25	760	IV

Comparison electrode	Liquid junction	Electrode reaction	Standard value V	Uncertainty mV	Temperature Coefficient μV/°C	Notes	Reference	Electrode Reference Number
d			f		g	h		
		$CO_3^{2-} + 3H^+ + 2e^- \rightleftharpoons HCOO^- + H_2O$	0.227				2	34
		$HCO_3^- + 2H^+ + 2e^- \rightleftharpoons HCOO^- + H_2O$	-0.078				2	35
		$H_2CO_3 + H^+ + 2e^- \rightleftharpoons HCOO^- + H_2O$	-0.267				2	36
		$C_2O_4^{2-} + 2H^+ + 2e^- \rightleftharpoons 2HCOO^-$	0.013				2	37
		$HC_2O_4^- + H^+ + 2e^- \rightleftharpoons 2HCOO^-$	-0.111				2	38
		$CO_2 + 2H^+ + 2e^- \rightleftharpoons HCOOH$	-0.199		-936		1	39
		$H_2CO_3 + 2H^+ + 2e^- \rightleftharpoons HCOOH + H_2O$	-0.156				2	40
		$HC_2O_4^- + 3H^+ + 2e^- \rightleftharpoons 2HCOOH$	0.110				2	41
		$H_2C_2O_4 + 2H^+ + 2e^- \rightleftharpoons 2HCOOH$	0.074				2	42
		$2H_2CO_3 + H^+ + 2e^- \rightleftharpoons HC_2O_4^- + 2H_2O$	-0.423				2	43
		$2H_2CO_3 + 2H^+ + 2e^- \rightleftharpoons H_2C_2O_4 + 2H_2O$	-0.386				2	44

Carbon 1

Standard value V	Solvent	Electrode Reference Number	Standard value V	Solvent	Electrode Reference Number	Standard value V	Solvent	Electrode Reference Number
-0.1316	1	9	0.497	1	11			
-0.970	1	21	0.518	1	2			
-0.547	1	27	0.52	1	14			
-0.423	1	43	0.588	1	10			
-0.386	1	44	0.627	1	8			
-0.330	1	22	0.724	1	7			
-0.267	1	36	0.731	1	13			
-0.199	1	39	0.77	1	23			
-0.170	1	26	1.18	1	1			
-0.156	1	40						
-0.111	1	38						
-0.103	1	24						
-0.078	1	35						
-0.050	1	31						
0.013	1	37						
0.044	1	17						
0.044	1	30						
0.056	1	33						
0.074	1	42						
0.107	1	16						
0.110	1	41						
0.145	1	20						
0.167	1	32						
0.169	1	12						
0.197	1	29						
0.199	1	19						
0.207	1	3						
0.209	1	15						
0.227	1	34						
0.228	1	6						
0.232	1	18						
0.323	1	5						
0.373	1	28						
0.441	1	25						
0.475	1	4						

BIBLIOGRAPHY: Carbon

1) Data reported by A.J. De Bethune, N.A. Swendeman-Loud, "Standard Aqueous Electrode Potentials and Temperature Coefficients at 25 °C" (Clifford A. Hampel, Ill.)

2) Data reported by M. Pourbaix, "Atlas d'Equilibres Electrochimiques à 25 °C" (Gauthier-Villars Editeur, Paris, 1963), 449 - 456

Electronically conducting phase	Intermediate species	Composition of the solution	Solvent	Temperature °C	Pressure Torr	Measuring method
a	b	c	d		e	d
Si			1	25	760	IV
Si			1	25	760	IV
Si			1	25	760	IV
Si	SiO_2 (cristobalite)		1	25	760	IV
Si	$SiO_2 \cdot XH_2O$		1	25	760	IV
Si	SiO_2 (lechatelierite)		1	25	760	IV
Si	SiO_2 (quartz)		1	25	760	IV
Si	SiO_2 (trydimite)		1	25	760	IV
Si			1	25	760	IV
Si			1	25	760	IV
Ind.	SiH_4 (gas)		1	25	760	IV
Ind.	SiH_4 (gas)		1	25	760	IV
Si	SiH_4 (gas)		1	25	760	IV
Ind.	SiH_4 (gas), SiO_2 (cristobalite)		1	25	760	IV
Ind.	SiH_4 (gas), $SiO_2 \cdot XH_2O$		1	25	760	IV
Ind.	SiH_4 (gas), SiO_2 (lechatelierite)		1	25	760	IV
Ind.	SiH_4 (gas), SiO_2 (quartz)		1	25	760	IV
Ind.	SiH_4 (gas), SiO_2 (trydimite)		1	25	760	IV
Ind.	SiH_4 (gas)		1	25	760	IV

Comparison electrode	Liquid junction	Electrode reaction	Standard value	Uncertainty	Temperature Coefficient	Notes	Reference	Electrode Reference Number
			V	mV	μV/°C			
d			f		g	h		
		$HSiO_3^- + 5H^+ + 4e^- \rightleftharpoons Si + 3H_2O$	-0.632				1	1
		$H_2SiO_3 + 4H^+ + 4e^- \rightleftharpoons Si + 3H_2O$	-0.780				1	2
		$SiF_6^{2-} + 4e^- \rightleftharpoons Si + 6F^-$	-1.24		-650		2	3
		SiO_2 (cristobalite) $+ 4H^+ + 4e^- \rightleftharpoons Si + 2H_2O$	-0.853				1	4
		$SiO_2 \cdot XH_2O + 4H^+ + 4e^- \rightleftharpoons Si + (X+2)\ H_2O$	-0.807				1	5
		SiO_2 (lechatelierite) $+ 4H^+ + 4e^- \rightleftharpoons Si + 2H_2O$	-0.840				1	6
		SiO_2 (quartz) $+ 4H^+ + 4e^- \rightleftharpoons Si + 2H_2O$	-0.857		-374		2	7
		SiO_2 (trymidite) $+ 4H^+ + 4e^- \rightleftharpoons Si + 2H_2O$	-0.851				1	8
		$SiO_3^{2-} + 3H_2O + 4e^- \rightleftharpoons Si + 6OH^-$	-1.697				2	9
		$SiO_3^{2-} + 6H^+ + 4e^- \rightleftharpoons Si + 3H_2O$	-0.455				1	10
		$HSiO_3^- + 9H^+ + 8e^- \rightleftharpoons SiH_4 + 3H_2O$	-0.265				1	11
		$H_2SiO_3 + 8H^+ + 8e^- \rightleftharpoons SiH_4 + 3H_2O$	-0.339				1	12
		$Si + 4H^+ + 4e^- \rightleftharpoons SiH_4$	-0.102		-197		2	13
		SiO_2 (cristobalite) $+ 8H^+ + 8e^- \rightleftharpoons SiH_4 + 2H_2O$	-0.376				1	14
		$SiO_2 \cdot XH_2O + 8H^+ + 8e^- \rightleftharpoons SiH_4 + (X+2)H_2O$	-0.353				1	15
		SiO_2 (lechatelierite) $+ 8H^+ + 8e^- \rightleftharpoons SiH_4 + 2H_2O$	-0.369				1	16
		SiO_2 (quartz) $+ 8H^+ + 8e^- \rightleftharpoons SiH_4 + 2H_2O$	-0.377				1	17
		SiO_2 (trydimite) $+ 8H^+ + 8e^- \rightleftharpoons SiH_4 + 2H_2O$	-0.375				1	18
		$SiO_3^{2-} + 10H^+ + 8e^- \rightleftharpoons SiH_4 + 3H_2O$	-0.176				1	19

Silicon 1

Standard value V	Solvent	Electrode Reference Number
-1.697	1	9
-1.24	1	3
-0.857	1	7
-0.853	1	4
-0.851	1	8
-0.840	1	6
-0.807	1	5
-0.780	1	2
-0.632	1	1
-0.455	1	10
-0.377	1	17
-0.376	1	14
-0.375	1	18
-0.369	1	16
-0.353	1	15
-0.339	1	12
-0.265	1	11
-0.176	1	19
-0.102	1	13

Standard value V	Solvent	Electrode Reference Number

Standard value V	Solvent	Electrode Reference Number

BIBLIOGRAPHY: Silicon

1) Data reported by M. Pourbaix, "Atlas d'Equilibres Electrochimiques à 25 °C" (Gauthier-Villars Editeur, Paris, 1963), 458 - 463

2) Data reported by A.J. De Bethune, N.A. Swendeman-Loud, "Standard Aqueous Electrode Potentials and Temperature Coefficient at 25 °C" (Clifford A. Hampel, Ill.)

Electronically conducting phase	Intermediate species	Composition of the solution	Solvent	Temperature °C	Pressure Torr	Measuring method
a	b	c	d		e	d
Ge		H_2SO_4 : 1 M	1	25		III
Ge		H_2SO_4 : 1 M	1	25		III
Ge		H_2SO_4 : 1 M	1	25		III
Ge	GeO (brown)		1	25	760	IV
Ge	GeO (yellow)		1	25	760	IV
Ge	GeO_2 (white ,hexag.)		1	25	760	IV
Ge	GeO_2 (white, tetrag.)		1	25	760	IV
Ge			1	25	760	IV
Ge			1	25	760	IV
Ge			1	25	760	IV
Ind.			1	25	760	IV
Ge	H_2GeO_3		1	25	760	IV
Ind.	H_2GeO_3		1	25	760	IV
Ge	GeH_4 (gas)		1	25	760	IV
Ind.	GeO (brown), GeO_2 (white, hexag.)	HCl : 0.026, 0.1042 and 1.06 m	1	25 ± 0.01		I
Ind.	GeO (brown), GeO_2 (white, tetrag.)		1	25	760	IV
Ind.	GeO (yellow) , GeO_2 (white, hexag.)		1	25	760	IV
Ind.	GeO (yellow), GeO_2 (white, tetrag.)		1	25	760	IV
Ind.	GeO (brown)		1	25	760	IV
Ind.	GeO (yellow)		1	25	760	IV
Ind.	GeO (brown)		1	25	760	IV
Ind.	GeO (yellow)		1	25	760	IV
Ind.	GeO (brown), H_2GeO_3		1	25	760	IV
Ind.	GeO (yellow), H_2GeO_3		1	25	760	IV

Comparison electrode	Liquid junction	Electrode reaction	Standard value V	Uncertainty mV	Temperature Coefficient μV/°C	Notes	Reference	Electrode Reference Number
d			f		g	h		
19		$Ge^{2+} + 2e^- \rightleftharpoons Ge$	0.24			1	1	1
19		$Ge^{4+} + 4e^- \rightleftharpoons Ge$	0.124			1	1	2
19		$Ge^{4+} + 2e^- \rightleftharpoons Ge^{2+}$	0.00			1	1	3
		GeO (brown) $+ 2H^+ + 2e^- \rightleftharpoons Ge + H_2O$	-0.286				2	4
		GeO (yellow) $+ 2H^+ + 2e^- \rightleftharpoons Ge + H_2O$	-0.130				2	5
		GeO_2 (white, hexag.) $+ 4H^+ + 4e^- \rightleftharpoons Ge + 2H_2O$	-0.202				2	6
		GeO_2 (white, tetrag.) $+ 4H^+ + e^- \rightleftharpoons Ge + 2H_2O$	-0.246				2	7
		$GeO_3^{2-} + 6H^+ + 4e^- \rightleftharpoons Ge + 3H_2O$	0.132				2	8
		$HGeO_3^- + 2H_2O + 4e^- \rightleftharpoons Ge + 5OH^-$	-1.03		-1290		3	9
		$HGeO_3^- + 5H^+ + 4e^- \rightleftharpoons Ge + 3H_2O$	-0.056				2	10
		$HGeO_3^- + 5H^+ + 2e^- \rightleftharpoons Ge^{2+} + 3H_2O$	-0.111				2	11
		$H_2GeO_3 + 4H^+ + 4e^- \rightleftharpoons Ge + 3H_2O$	-0.182				2	12
		$H_2GeO_3 + 4H^+ + 2e^- \rightleftharpoons Ge^{2+} + 3H_2O$	-0.363				2	13
		$Ge + 4H^+ + 4e^- \rightleftharpoons GeH_4$	-0.867				2	14
3		GeO_2 (white, hexag.) $+ 2H^+ + 2e^- \rightleftharpoons GeO$ (brown) $+ H_2O$	-0.118	± 10			4	15
		GeO_2 (white, tetrag.) $+ 2H^+ + 2e^- \rightleftharpoons GeO$ (brown) $+ H_2O$	-0.206				2	16
		GeO_2 (white, hexag.) $+ 2H^+ + 2e^- \rightleftharpoons GeO$ (yellow) $+ H_2O$	-0.273				2	17
		GeO_2 (white, tetrag.) $+ 2H^+ + 2e^- \rightleftharpoons GeO$ (yellow) $+ H_2O$	-0.362				2	18
		$GeO_3^{2-} + 4H^+ + 2e^- \rightleftharpoons GeO$ (brown) $+ 2H_2O$	0.550				2	19
		$GeO_3^{2-} + 4H^+ + 2e^- \rightleftharpoons GeO$ (yellow) $+ 2H_2O$	0.394				2	20
		$HGeO_3^- + 3H^+ + 2e^- \rightleftharpoons GeO$ (brown) $+ 2H_2O$	0.177				2	21
		$HGeO_3^- + 3H^+ + 2e^- \rightleftharpoons GeO$ (yellow) $+ 2H_2O$	0.019				2	22
		$H_2GeO_3 + 2H^+ + 2e^- \rightleftharpoons GeO$ (brown) $+ 2H_2O$	-0.074				2	23
		$H_2GeO_3 + 2H^+ + 2e^- \rightleftharpoons GeO$ (yellow) $+ 2H_2O$	-0.232				2	24

Standard value V	Solvent	Electrode Reference Number
-1.03	1	9
-0.867	1	14
-0.363	1	13
-0.362	1	18
-0.286	1	4
-0.273	1	17
-0.246	1	7
-0.232	1	24
-0.206	1	16
-0.202	1	6
-0.182	1	12
-0.130	1	5
-0.118	1	15
-0.111	1	11
-0.074	1	23
-0.056	1	10
0.00	1	3
0.019	1	22
0.124	1	2
0.132	1	8
0.177	1	21
0.24	1	1
0.394	1	20
0.550	1	19

NOTES: Germanium

1) Anodic dissolution data

BIBLIOGRAPHY: Germanium

1) W.E. Reid, Jr., J. Phys. Chem., 69 (1965), 3168 - 3171

2) Data reported by M. Pourbaix, "Atlas d'Equilibres Electrochimiques à 25 ^{o}C" (Gauthier-Villars Editeur, Paris, 1963), 464 - 474

3) Data reported by A.J. De Bethune, N.A. Swendeman-Loud, "Standard Aqueous Electrode Potentials and Temperature Coefficients at 25 ^{o}C" (Clifford A. Hampel, Ill.)

4) W.L. Jolly, W.M. Latimer, J. Am. Chem. Soc., 74 (1952), 5751 - 5752

Electronically conducting phase	Intermediate species	Composition of the solution	Solvent	Temperature °C	Pressure Torr	Measuring method
a	b	c	d		e	d
Ind.	SnO_2 (white)		1	25	760	IV
Ind.			1	25	760	IV
Ind.	Sn $(OH)_4$ (white)		1	25	760	IV
Ind.			1	25	760	IV
Sn			1	25	760	IV
Sn			1	25	760	IV
Sn		Sn $(ClO_4)_2$: (0.849 up to 16.999) $\cdot$ 10^{-3} M; $HClO_4$: 0.01 up to 0.05 M	1	25		I
Sn		LiCl : 0.01 m	7	25		I
Ind.			1	25	760	IV
Sn(Hg)			1	25		II
Sn			1	25	760	IV
Sn	SnO (black)		1	25	760	IV
Sn	SnO_2 (white)		1	25	760	IV
Ind.	SnO_2 (white)		1	25	760	IV
Ind.			1	25	760	IV
Sn	$Sn(OH)_2$ (white)		1	25	760	IV
Sn	Sn $(OH)_4$ (white)		1	25	760	IV
Ind.	Sn $(OH)_4$ (white)		1	25	760	IV
Sn	SnS		1	25	760	IV
Sn	SnH_4	H_2SO_4 (pH 0.25)	1	25		III
Ind.	SnO (black), SnO_2 (white)		1	25	760	IV
Ind.	SnO		1	25	760	IV
Ind.	SnO (black), Sn $(OH)_4$ (white)		1	25	760	IV
Ind.	Sn $(OH)_2$ (white), SnO_2 (white)		1	25	760	IV
Ind.	Sn $(OH)_2$ (white)		1	25	760	IV
Ind.	Sn $(OH)_2$ (white), Sn $(OH)_4$ (white)		1	25	760	IV

Comparison electrode	Liquid junction	Electrode reaction	Standard value	Uncertainty	Temperature Coefficient	Notes	Reference	Electrode Reference Number
			V	mV	μV/°C			
d			f		g	h		
		$SnO_2 + H^+ + 2e^- \rightleftharpoons HSnO_2^-$	-0.546				1	1
		$SnO_3^{2-} + 3H^+ + 2e^- \rightleftharpoons HSnO_2^- + H_2O$	0.374				1	2
		$Sn(OH)_4 + H^+ + 2e^- \rightleftharpoons HSnO_2^- + 2H_2O$	-0.349				1	3
		$Sn(OH)_6^{2-} + 2e^- \rightleftharpoons HSnO_2^- + H_2O + 3OH^-$	-0.93				2	4
		$HSnO_2^- + 3H^+ + 2e^- \rightleftharpoons Sn + 2H_2O$	0.333				1	5
		$HSnO_2^- + H_2O + 2e^- \rightleftharpoons Sn + 3OH^-$	-0.909				2	6
1		$Sn^{2+} + 2e^- \rightleftharpoons Sn$	-1.375	± 0.5			3	7
1		$Sn^{2+} + 2e^- \rightleftharpoons Sn$	-0.532				4	8
		$Sn^{4+} + 2e^- \rightleftharpoons Sn^{2+}$	0.151				1	9
		$Sn(F)^+ + 2e^- \rightleftharpoons Sn + F^-$	-1.510				5	10
		$SnF_6^{2-} + 4e^- \rightleftharpoons Sn + 6F^-$	-0.25				2	11
		$SnO + 2H^+ + 2e^- \rightleftharpoons Sn + H_2O$	-0.104				1	12
		$SnO_2 + 4H^+ + 4e^- \rightleftharpoons Sn + 2H_2O$	-0.106				1	13
		$SnO_2 + 4H^+ + 2e^- \rightleftharpoons Sn^{2+} + 2H_2O$	-0.77				1	14
		$SnO_3^{2-} + 6H^+ + 2e^- \rightleftharpoons Sn^{2+} + 3H_2O$	0.844				1	15
		$Sn(OH)_2 + 2H^+ + 2e^- \rightleftharpoons Sn + 2H_2O$	-0.091				1	16
		$Sn(OH)_4 + 4H^+ + 4e^- \rightleftharpoons Sn + 4H_2O$	-0.008				1	17
		$Sn(OH)_4 + 4H^+ + 2e^- \rightleftharpoons Sn^{2+} + 4H_2O$	0.120				1	18
		$SnS + 2e^- \rightleftharpoons Sn + S^{2-}$	-0.87				2	19
28		$Sn + 4H^+ + e^- \rightleftharpoons SnH_4$	-1.07	± 50		1	6	20
		$SnO_2 + 2H^+ + 2e^- \rightleftharpoons SnO + H_2O$	-0.108				1	21
		$SnO_3 + 4H^+ + 2e^- \rightleftharpoons SnO + 2H_2O$	0.812				1	22
		$Sn(OH)_4 + 2H^+ + 2e^- \rightleftharpoons SnO + 3H_2O$	0.088				1	23
		$SnO_2 + 2H^+ + 2e^- \rightleftharpoons Sn(OH)_2$	-0.121				1	24
		$SnO_3^{2-} + 4H^+ + 2e^- \rightleftharpoons Sn(OH)_2 + H_2O$	0.791				1	25
		$Sn(OH)_4 + 2H^+ + 2e^- \rightleftharpoons Sn(OH)_2 + 2H_2O$	0.075				1	26

Tin 1

Standard value V	Solvent	Electrode Reference Number	Standard value V	Solvent	Electrode Reference Number	Standard value V	Solvent	Electrode Reference Number
-1.510	1	10						
-1.375	1	7						
-1.07	1	20						
-0.93	1	4						
-0.909	1	6						
-0.87	1	19						
-0.77	1	14						
-0.546	1	1						
-0.349	1	3						
-0.25	1	11						
-0.121	1	24						
-0.108	1	21						
-0.106	1	13						
-0.104	1	12						
-0.091	1	16						
-0.008	1	17						
0.075	1	26						
0.088	1	23						
0.120	1	18						
0.151	1	9						
0.333	1	5						
0.374	1	2						
0.791	1	25						
0.812	1	22						
0.844	1	15						
-0.532	7	8						

NOTES: Tin

1) Electrolytic data

BIBLIOGRAPHY: Tin

1)Data reported by M. Pourbaix, "Atlas d'Equilibres Electrochimiques à 25 °C" (Gauthier-Villars Editeur, Paris, 1963), 475 - 484

2) Data reported by A.J. De Bethune, N.A. Swendeman-Loud, "Standard Aqueous Electrode Potentials and Temperature Coefficients at 25 °C" (Clifford A. Hampel, Ill.)

3) S.E.S. El Wakkad, T.M. Salem, J.A. Al Sayed, J. Chem. Soc., 1957, 3770 - 3776

4) B. Jakuszewski, Z. Kozlowskii, Rocz. Chem., 38 (1964), 93

5) A.D. Paul, Thesis, University of California, Berkeley 1955, UCRL-2926

6) N. De Zoubov, E. Deltombe, Proc. 7th. Intern. Comm. Electrochem. Thermodynam. and Kinet., 1957, 240 - 243

Electronically conducting phase	Intermediate species	Composition of the solution	Solvent	Temperature °C	Pressure Torr	Measuring method
a	b	c	d		e	d
Ind.	PbO_2		1	25	760	IV
Ind.			1	25	760	IV
Ind.	Pb_3O_4		1	25	760	IV
Pb			1	25	760	IV
Hg (D.M.E.)		NaOH : 02 up to 1.94 M	1	25		II
Pb		1) $PbCl_2$: 0.0002 up to 0.0330 m (pH 5 - 6.1); 2) $Pb(ClO_4)_2$: 0.001026 up to 0.1026 m; $NaClO_4$: 0.005 up to 0.1 m	1	25		I
Pb			7	25	760	IV
Hg (D.M.E.)		NH_4NO_3 : 0.01 M	8	25		II
Pb			11	25	760	IV
Pb			12	25	760	IV
Pb			15	25	760	IV
Pb		$Pb(HCOO)_2$: 0.01 N	16	25 ± 0.01		I
Pb		$Pb(NO_3)_2$: 0.000171 up to 0.0130 M	34	25		I
Pb		$Pb(ClO_4)_2$: 0.01 N	23	25		I
Pt		Pb (acetate)$_2$; Pb (acetate)$_4$; $HClO_4$: 1.1 up to 5.8 N	1	25		I
Pb		1) $PbBr_2$: (0.15755 up to 0.36200) · 10^{-3} M; 2) $Pb(ClO_4)_2$: (4 up to 12) · 10^{-3} M; KBr : 2 · · 10^{-3} M	1	25		III V
Pb	$PbBr_2$		1	25	760	IV
Pb (Hg)	$PbCl_2$	KCl : 10^{-5} up to 0.4 M	1	25		I
Pb	$PbCl_2$	KCl	7	25		I
Pb(Hg)	$PbCl_2$	$PbCl_2$ satd.	13	25± 0.01		I
Glass	Pb (Cl) (OH)	$PbCl_2$: 0.01 and 0.1 M; NaOH	1	25		I, V
Pb	$PbCO_3$		1	25	760	IV
Pb (Hg) (two phases)	PbF_2	NaF: 0.05 up to 0.90 m	1	25		I
Pb	$PbHPO_4$	NaH_2PO_4 : 0.001 up to 0.1 M; CH_3COOH, CH_3COONa, pH = 4.6	1	25		I
Pb		$Pb(ClO_4)_2$: (4 up to 12) · 10^{-3} M; KI : 2 ·10^{-3} M	1	25		V
Pb	PbI_2		1	25	760	IV

Comparison electrode	Liquid junction	Electrode reaction	Standard value V	Uncertainty mV	Temperature Coefficient μV/°C	Notes	Reference	Electrode Reference Number
d			f		g	h		
		$PbO_2 + H^+ + 2e^- \rightleftharpoons HPbO_2^-$	0.621				1	1
		$PbO_3^{2-} + 3H^+ + 2e^- \rightleftharpoons HPbO_2^- + H_2O$	1.547				1	2
		$PbO_3O_4 + 2H_2O + 2e^- \rightleftharpoons 3HPbO_2^-$	-0.390				1	3
		$HPbO_2^- + 3H^+ + 2e^- \rightleftharpoons Pb + 2H_2O$	0.702				1	4
2		$HPbO_2^- + H_2O + 2e^- \rightleftharpoons Pb + 3OH^-$	-0.537	±2			2	5
1) 4 2) 3		$Pb^{2+} + 2e^- \rightleftharpoons Pb$	-0.1262	±1.2		1	3, 4	6
		$Pb^{2+} + 2e^- \rightleftharpoons Pb$	-0.113				5	7
3		$Pb^{2+} + 2e^- \rightleftharpoons Pb$	-0.145				6	8
		$Pb^{2+} + 2e^- \rightleftharpoons Pb$	-0.172				5	9
		$Pb^{2+} + 2e^- \rightleftharpoons Pb$	-0.068				5	10
		$Pb^{2+} + 2e^- \rightleftharpoons Pb$	-0.155				5	11
14		$Pb^{2+} + 2e^- \rightleftharpoons Pb$	-0.72		-540		7	12
25	KCl satd.	$Pb^{2+} + 2e^- \rightleftharpoons Pb$	-0.193	±12	0	1	8,9	13
7		$Pb^{2+} + 2e^- \rightleftharpoons Pb$	-0.12				10	14
3		$Pb^{4+} + 2e^- \rightleftharpoons Pb^{2+}$	1.670				11	15
		$Pb(Br)^+ + 2e^- \rightleftharpoons Pb + Br^-$	-0.174			1 2	12 13	16
		$PbBr_2 + 2e^- \rightleftharpoons Pb + 2Br^-$	-0.284		-341		14	17
3		$PbCl_2 + 2e^- \rightleftharpoons Pb + 2Cl^-$	-0.2675	±0.9			15	18
3		$PbCl_2 + 2e^- \rightleftharpoons Pb + 2Cl^-$	-0.2873				16	19
4		$PbCl_2 + 2e^- \rightleftharpoons Pb + 2Cl^-$	0.4428	±0.1			17	20
4		$Pb(Cl)(OH) + 2e^- \rightleftharpoons Pb + Cl^- + OH^-$	-0.532			3	18	21
		$PbCO_3^{2-} + 2e^- \rightleftharpoons Pb + CO_3^{2-}$	-0.509		-1294		14	22
32		$PbF_2 + 2e^- \rightleftharpoons Pb + 2F^-$	-0.3444		1025		19	23
3		$PbHPO_4 + 2e^- \rightleftharpoons Pb + HPO_4^{2-}$	-0.465		0	4	20	24
		$Pb(I)^+ + 2e^- \rightleftharpoons Pb + I^-$	-0.183			5	13	25
		$PbI_2 + 2e^- \rightleftharpoons Pb + 2I^-$	-0.365		-124		14	26

Electronically conducting phase	Intermediate species	Composition of the solution	Solvent	Temperature °C	Pressure Torr	Measuring method
a	b	c	d		e	d
Pb(Hg)	$Pb(N_3)_2$		1	25		I
Pb		$Pb(NO_3)_2$: (0.16163 up to 0.46815) · 10^{-3} M	1	25		III
Pb	PbO (yellow)		1	25	760	IV
Pb	PbO (red)		1	25	760	IV
Pb	PbO (red)		1	25	760	IV
Ind.	PbO_2		1	25	760	IV
Ind.			1	25	760	IV
Ind.	Pb_3O_4		1	25	760	IV
Pb	$Pb(OH)_2$		1	25	760	IV
Pb	$Pb_3(PO_4)_2$	Na_3PO_4 : 0.001 up to 0.005 M; NaOH : 0 up to 0.05 N	1	25		I
Pb (Hg)		$K_4P_2O_7$; $NaNO_3$; Pb^{2+}: 3 · 10^{-5} g. ion dm^{-3}	1	25		II
Pb	PbS		1	25	760	IV
Pb	$PbSeO_4$		1	25		V
Pb	$PbSO_4$		1	25	760	IV
Pb	PbH_2 (gas)		1	25	760	IV
Ind.	PbO (red); PbO_2		1	25	760	IV
Ind.	PbO (red)		1	25	760	IV
Ind.	PbO (yellow)		1	25	760	IV
Ind.	PbO (red), Pb_3O_4		1	25	760	IV
Ind.	PbO (yellow), Pb_3O_4		1	25	760	IV
Ind.	Pb_2O_3 , PbO_2		1	25	760	IV
Ind.	Pb_3O_4 , PbO_2		1	25	760	IV
Ind.	Pb_3O_4		1	25	760	IV
Ind.	Pb_3O_4 , Pb_2O_3		1	25	760	IV
Ind.	$Pb(OH)_2$		1	25	760	IV
Ind.	$Pb(OH)_2$, Pb_3O_4		1	25	760	IV
Pt	PbO_2 , $PbSO_4$	H_2SO_4 : 0.1 up to 8.272 m	1	25		I
Glass	Alanine	Alanine : (31.99 up to 35.61) · 10^{-3} M; NaOH: (0.92 up to 8.43) · 10^{-3} M; $Pb(NO_3)_2$:(4.82 up to 5.28) · 10^{-3} M	1	25 ± 0.02		I
Glass	Alanine	Alanine : (31.99 up to 35.61) · 10^{-3} M; NaOH : (0.92 up to 8.43) · 10^{-3} M; $Pb(NO_3)_2$: (4.82 up to 5.28) · 10^{-3} M	1	25± 0.02		I
	Pb (citrate)		1	25		I

Comparison electrode	Liquid junction	Electrode reaction	Standard value	Uncertainty	Temperature Coefficient	Notes	Reference	Electrode Reference Number
			V	mV	μV/°C			
d			f		g	h		
		$Pb(N_3)_2 + 2e^- \rightleftharpoons Pb + 2N_3^-$	-0.380				21	27
		$Pb(NO_3)^+ + 2e^- \rightleftharpoons Pb + NO_3^-$	-0.161			6	12	28
		$PbO + 2H^+ + 2e^- \rightleftharpoons Pb + H_2O$	0.252				1	29
		$PbO + 2H^+ + 2e^- \rightleftharpoons Pb + H_2O$	0.248				1	30
		$PbO + H_2O + 2e^- \rightleftharpoons Pb + 2OH^-$	-0.580		-1163		14	31
		$PbO_2 + 4H^+ + 2e^- \rightleftharpoons Pb^{2+} + 2H_2O$	1.455		-238		14	32
		$PbO_3^{2-} + 6H^+ + 2e^- \rightleftharpoons Pb^{2+} + 3H_2O$	2.375				1	33
		$Pb_3O_4 + 8H^+ + 2e^- \rightleftharpoons 3Pb^{2+} + 4H_2O$	2.094				1	34
		$Pb(OH)_2 + 2H^+ + 2e^- \rightleftharpoons Pb + 2H_2O$	0.277				1	35
3		$Pb_3(PO_4)_2 + 6e^- \rightleftharpoons 3Pb + 2PO_4^{3-}$	-0.54				20	36
3		$Pb(P_2O_7)^{2-} + 2e^- \rightleftharpoons Pb + P_2O_7^{4-}$	-0.633				22	37
		$PbS + 2e^- \rightleftharpoons Pb + S^{2-}$	-0.93		-900		14	38
		$PbSeO_4 + 2e^- \rightleftharpoons Pb + SeO_4^{2-}$	-0.329			7	23	39
		$PbSO_4 + 2e^- \rightleftharpoons Pb + SO_4^{2-}$	-0.3588		-1015		14	40
		$Pb + 2H^+ + 2e^- \rightleftharpoons PbH_2$	-1.507				1	41
		$PbO_2 + H_2O + 2e^- \rightleftharpoons PbO + 2OH^-$	0.247		-1194		14	42
		$PbO_3^{2-} + 4H^+ + 2e^- \rightleftharpoons PbO + 2H_2O$	2.001				1	43
		$PbO_3^{2-} + 4H^+ + 2e^- \rightleftharpoons PbO + 2H_2O$	1.997				1	44
		$Pb_3O_4 + 2H^+ + 2e^- \rightleftharpoons 3PbO + H_2O$	0.972				1	45
		$Pb_3O_4 + 2H^+ + 2e^- \rightleftharpoons 3PbO + H_2O$	0.959				1	46
		$2PbO_2 + 2H^+ + 2e^- \rightleftharpoons Pb_2O_3 + H_2O$	1.093				1	47
		$3PbO_2 + 4H^+ + 4e^- \rightleftharpoons Pb_3O_4 + 2H_2O$	1.127				1	48
		$3PbO_3^{2-} + 10H^+ + 4e^- \rightleftharpoons Pb_3O_4 + 5H_2O$	2.515				1	49
		$3Pb_2O_3 + 2H^+ + 2e^- \rightleftharpoons 2Pb_3O_4 + H_2O$	1.228				1	50
		$PbO_3^{2-} + 4H^+ + 2e^- \rightleftharpoons Pb(OH)_2 + H_2O$	1.972				1	51
		$Pb_3O_4 + 2H_2O + 2H^+ + 2e^- \rightleftharpoons 3Pb(OH)_2$	0.885				1	52
1		$PbO_2 + SO_4^{2-} + 4H^+ + 2e^- \rightleftharpoons PbSO_4 + 2H_2O$	1.6913	± 5		1	24, 25, 26	53
		$Pb(alanine)^{2+} + 2e^- \rightleftharpoons Pb + alanine$	-0.274			8	27	54
		$Pb(alanine)_2^{2+} + 2e^- \rightleftharpoons Pb + 2\ alanine$	-0.370			8	27	55
		$Pb(citrate) + 2e^- \rightleftharpoons Pb + citrate^{2+}$	-0.302				28	56

Electronically conducting phase	Intermediate species	Composition of the solution	Solvent	Temperature °C	Pressure Torr	Measuring method
a	b	c	d		e	d
Glass	Glycine	Glycine : (22.19 up to 28.77) · 10^{-3} M; NaOH : (0.87 up to 9.98) · 10^{-3} M; Pb $(NO_3)_2$: (5.41 up to 7.02) · 10^{-3} M	1	25 ± 0.02		I
Glass	Glycine	Glycine : (22.19 up to 28.77) · 10^{-3} M; NaOH: (0.87 up to 9.98) · 10^{-3} M; Pb $(NO_3)_2$: (5.41 up to 7.02) · 10^{-3} M	1	25± 0.02		I
Glass	Glycylglycine	Glycylglycine : (23.25 up to 27.29) · 10^{-3} M; NaOH : (0.60 up to 80.2) · 10^{-3} M; Pb $(NO_3)_2$: (5.18 up to 6.08) · 10^{-3} M	1	25± 0.02		I
Glass	Glycylglycine	Glycylglycine : (23.25 up to 27.29) · 10^{-3} M; NaOH : (0.60 up to 80.2) · 10^{-3} M; Pb $(NO_3)_2$: (5.18 up to 6.08) · 10^{-3} M	1	25± 0.02		I
Glass	Glycylglycylglycine	Glycylglycylglycine : (33.87 up to 37.15) · 10^{-3} M; Na(OH) : (1.072 up to 4.307) · 10^{-3} M; Pb $(NO_3)_2$: (5.311 up to 6.363) · 10^{-3} M	1	25± 0.03		I
Glass	Glycylglycylglycine	Glycylglycylglycine : (33.87 up to 37.15) · 10^{-3} M; Na (OH) : (1.072 up to 4.307) · 10^{-3} M; Pb$(NO_3)_2$ (5.311 up to 6.363) · 10^{-3} M	1	25± 0.03		I
Glass			1	25		I
Glass	Pb (8-hydroxyquinoli-ne-5-sulfonate)	8-hydroxyquinoline-5-sulfonic acid : 2.954 · 10^{-3} M; Pb $(NO_3)_2$	1	25		I
Glass		8-hydroxyquinoline-5-sulfonic acid : 2.954 · 10^{-3} M; Pb $(NO_3)_2$	1	25		I

Comparison electrode	Liquid junction	Electrode reaction	Standard value V	Uncertainty mV	Temperature Coefficient μV/°C	Notes	Reference	Electrode Reference Number
d			f		g	h		
		Pb (glycine)$^{2+}$ + 2e$^-$ ⇌ Pb + glycine	-0.288			8	27	57
		Pb (glycine)$_2^{2+}$ + 2e$^-$ ⇌ Pb + 2 glycine	-0.388			8	27	58
		Pb (glycylglycine)$^{2+}$ + 2e$^-$ ⇌ Pb + glycylglycine	-0.222			8	27	59
		Pb (glycylglycine)$_2^{2+}$ + 2e$^-$ ⇌ Pb + 2 glycylglycine	-0.302			8	27	60
		Pb (glycylglycylglycine)$^{2+}$ + 2e$^-$ ⇌ Pb + glycylglycylglycine	-0.215			8	29	61
		Pb (glycylglycylglycine)$_2^{2+}$ + 2e$^-$ ⇌ Pb + 2 glycylglycylglycine	-0.296			8	29	62
		Pb (8-hydroxyquinoline)$^{2+}$ + 2e$^-$ ⇌ Pb + 8-hydroxyquinoline	-0.393			8	30	63
4		Pb (8-hydroxyquinoline-5-sulfonate) + 2e$^-$ ⇌ Pb + 8-hydroxyquinoline-5-sulfonate^{2-}	-0.379			8	31	64
4		Pb (8-hydroxyquinoline-5-sulfonate)$_2^{2-}$ + 2e$^-$ ⇌ Pb + 2 8-hydroxyquinoline-5-sulfonate^{2-}	-0.601			8	31	65

Lead 1

Standard value V	Solvent	Electrode Reference Number	Standard value V	Solvent	Electrode Reference Number	Standard value V	Solvent	Electrode Reference Number
-1.507	1	41	0.252	1	29			
-0.93	1	38	0.277	1	35			
-0.633	1	37	0.621	1	1			
-0.601	1	65	0.702	1	4			
-0.580	1	31	0.885	1	52			
-0.54	1	36	0.959	1	46			
-0.537	1	5	0.972	1	45			
-0.532	1	21	1.093	1	47			
-0.509	1	22	1.127	1	48			
-0.465	1	24	1.228	1	50			
-0.393	1	63	1.455	1	32			
-0.390	1	3	1.547	1	2			
-0.388	1	58	1.670	1	15			
-0.380	1	27	1.6813	1	53			
-0.379	1	64	1.972	1	51			
-0.370	1	55	1.997	1	44			
-0.365	1	26	2.001	1	43			
-0.3588	1	40	2.094	1	34			
-0.3444	1	23	2.375	1	33			
-0.329	1	39	2.515	1	49			
-0.302	1	56	-0.2873	7	19			
-0.302	1	60	-0.113	7	7			
-0.296	1	62	-0.145	8	8			
-0.288	1	57	-0.172	11	9			
-0.284	1	17	-0.068	12	10			
-0.274	1	54	0.4428	13	20			
-0.2675	1	18	-0.155	15	11			
-0.222	1	59	-0.72	16	12			
-0.215	1	61	-0.193	34	13			
-0.183	1	25	-0.12	23	14			
-0.174	1	16						
-0.161	1	28						
-0.1262	1	6						
0.247	1	42						
0.248	1	30						

NOTES: Lead

1) The value is an average of those given by several authors; they all are in the same range of uncertainty

2) Conductometric and spectrophotometric data

3) Potentiometric pH measurements and solubility data

4) We report only the most recent value. Another author gives for the same electrode the value: -0.2507.V (20°), which is estimated to be equally reliable.

5) Spectrophotometric data

6) Conductometric data

7) Solubility data

8) Potentiometric pH measurements

BIBLIOGRAPHY: Lead

1) Data reported by M. Pourbaix, "Atlas d'Equilibres Electrochimiques à 25 ^{o}C" (Gauthier-Villars Editeur, Paris, 1963), 485 - 492

2) A.A. Vleck, Chem. Listy, 48 (1954), 1474 - 1484

3) A.R. Tourchy, A.A. Abdul-Azim, L.A. Shalaby, J. Chem. U.A.R., 10 (1967), 253 - 266

4) K. Farayo, Acta Lit. Sci. Regiae Univ. Hung. Francisco-Josephinae Sect. Chem. Phys., 5 (1937), 193 - 210

5) N.E. Khomutov, Zh. Fiz. Khim., 42 (1968), 2223 - 2229

6) Ya. I. Turyan, Zh. Fiz. Khim., 28 (1954), 2129 - 2136

7) V. Pleskov, Acta Physicochim. U.S.S.R., 21 (1946), 41 - 54

8) T. Pavlopoulos, H. Strehlow, Z. Physik. Chem. (Leipzig), 2 (1954), 89 - 103

9) L. Fischer, G. Winkler, G. Jander, Z. Elektrochem., 62 (1958), 1 - 28

10) V.A. Pleskov, J. Phys. Chem., 22 (1948), 351 - 361

11) A. Berka, V. Dvorak, I. Nemec, J. Zyka, J. Electroanal. Chem., 4 (1952), 150 - 159

12) G.H. Nancollas, J. Chem. Soc., 1955 , 1458 - 1462

13) A.I. Biggs, H.N. Parton, R.A. Robinson, J. Am. Chem. Soc., 77 (1955), 5844 - 5848

14) Data reported by A.J. De Bethune, N.A. Swendeman-Loud, "Standard Aqueous Electrode Potentials and Temperature coefficients at 25 ^{o}C" (Clifford A. Hampel, Ill.)

15) V.I. Altynov, B.V. Ptitsyn, Zh. Neorgan. Khim., 9 (1964), 2407 - 2410

16) C.L. Carlson, E.W. Malmberg, D.I. Brovon, J. Phys. Chem., 54 (1950), 322 - 325

17) A.G. Barret, R. Bryant, G.F. Kiefer, J. Am. Chem. Soc., 65 (1943), 1095 - 1097

18) P. Deschamps, B. Charreton, Compt. Rend., 232 C (1951), 162 - 163

19) R.W. Iwett, T. De Vries, J. Am. Chem. Soc., 63 (1941), 2021 - 2025

20) S.A. Awad, Z.A. Elhady, J. Electroanal. Chem., 20 (1969), 79 - 87

20')C.M. Mason, W.M. Blum, J. Am. Chem. Soc., 69 (1947), 1246 - 1250

21) S. Suzuki, J. Chem. Soc. Japan, Pure Chem. Sect., 73 (1952), 378 - 380

22) I. N. Popova, A.G. Stromberg, Elektrokhim. 4 (1968), 1147 - 1152

23) N.M. Selivanova, R.Ya. Boguslavskii, Zh. Fiz. Khim., 29 (1955), 128

24) S.J. Bone, K.P. Singh, W.F.K. Wynne-Jones, Electrochim. Acta, 4 (1961), 288 - 293

25) A.K. Covington, J.V. Dobson, W.F.K. Wynne-Jones, Trans. Faraday Soc., 61 (1965), 2050 - 2056

26) W.H. Beck, K.S. Singh, W.F.K. Wynne-Jones, Trans. Faraday Soc., 55 (1959), 331 - 338

27) C.B. Monk, Trans. Faraday Soc., 47 (1951), 297 - 302

28) S. Suzuki, J. Chem. Soc. Japan, Pure Chem. Sect., 73 (1952), 92

BIBLIOGRAPHY: Lead

29) J.I. Evans, C.B. Monk, Trans. Faraday Soc., 51 (1955), 1244 - 1250

30) R. Näsänen, Suomen Kem., 26 B (1953), 11

31) R. Näsänen, E. Uisitalo, Acta Chem. Scand., 8 (1954), 112 - 118

Electronically conducting phase	Intermediate species	Composition of the solution	Solvent	Temperature °C	Pressure Torr	Measuring method
a	b	c	d		e	d
Ind.			1	25	760	IV
Ind.			1	25	760	IV
Ti			1	25	760	IV
Ind.			1	25	760	IV
Pt		HCl : 0.05 M	20	25	760	I
Pt		HCl : 0.20 M	27	25	760	I
Ti			1	25	760	IV
Ti	TiO		1	25	760	IV
Ti			1	25	760	IV
Ind.			1	25	760	IV
Ind.			1	25	760	IV
Ind.	TiO_2		1	25	760	IV
Ind.	TiO_2		1	25	760	IV
Ind.	Ti_2O_3		1	25	760	IV
Ind.	$Ti(OH)_3$		1	25	760	IV
Ind.	$Ti(OH)_4$		1	25	760	IV
Hg (D.M.E.)		HCl or $HClO_4$: 0.5 M	1	25		II
Ind.	$Ti(OH)_4$		1	25	760	IV
Hg (D.M.E.)		HBr + KBr : 1 M; $Ti^{4+} + Ti^{3+}$: $5 \cdot 10^{-3}$ M	1	25		II
Ind.	TiO, Ti_2O_3		1	25	760	IV
Ind.			1	25	760	IV
Ind.	TiO, $Ti(OH)_3$		1	25	760	IV
Ind.	TiO_2		1	25	760	IV
Ind.	Ti_2O_3		1	25	760	IV
Ind.	Ti_2O_3, TiO_2		1	25	760	IV
Ind.	Ti_2O_3, Ti_3O_5		1	25	760	IV
Ind.	Ti_2O_3, $Ti(OH)_4$		1	25	760	IV
Ind.	Ti_3O_5, TiO_2		1	25	760	IV
Ind.	Ti_3O_5, $Ti(OH)_4$		1	25	760	IV
Ind.	$Ti(OH)_3$		1	25	760	IV
Ind.	$Ti(OH)_3$, TiO_2		1	25	760	IV
Ind.	$Ti(OH)_3$, Ti_3O_5		1	25	760	IV

Comparison electrode	Liquid junction	Electrode reaction	Standard value	Uncertainty	Temperature Coefficient	Notes	Reference	Electrode Reference Number
			V	mV	μV/°C			
d			f		g	h		
		$TiO_2^{2+} + H_2O + 2e^- \rightleftharpoons HTiO_3^- + H^+$	1.303				1	1
		$HTiO_3^- + 5H^+ + 2e^- \rightleftharpoons Ti^{2+} + 3H_2O$	0.362				1	2
		$Ti^{2+} + 2e^- \rightleftharpoons Ti$	-1.630				1	3
		$Ti^{3+} + e^- \rightleftharpoons Ti^{2+}$	-0.368				1	4
3		$Ti^{4+} + e^- \rightleftharpoons Ti^{3+}$	0.000			1	2	5
3		$Ti^{4+} + e^- \rightleftharpoons Ti^{3+}$	-0 190				2	6
		$TiF_6^{2-} + 4e^- \rightleftharpoons Ti + 6F^-$	-1.191		-965		3	7
		$TiO + 2H^+ + 2e^- \rightleftharpoons Ti + H_2O$	-1.306				1	8
		$TiO^{2+} + 2H^+ + 4e^- \rightleftharpoons Ti + H_2O$	-0.882				3	9
		$TiO^{2+} + 2H^+ + 2e^- \rightleftharpoons Ti^{2+} + H_2O$	-0.135				1	10
		$TiO^{2+} + 2H^+ + e^- \rightleftharpoons Ti^{3+} + H_2O$	0.100				1	11
		$TiO_2 + 4H^+ + 2e^- \rightleftharpoons Ti^{2+} + 2H_2O$	-0.502				1	12
		$TiO_2 + 4H^+ + e^- \rightleftharpoons Ti^{3+} + 2H_2O$	-0.666				1	13
		$Ti_2O_3 + 6H^+ + 6e^- \rightleftharpoons 2Ti^{2+} + 3H_2O$	-0.478				1	14
		$Ti(OH)_3 + 3H^+ + e^- \rightleftharpoons Ti^{2+} + 3H_2O$	-0.248				1	15
		$Ti(OH)_4 + 4H^+ + 2e^- \rightleftharpoons Ti^{2+} + 4H_2O$	-0.169				1	16
3		$Ti(OH)^{3+} + H^+ + e^- \rightleftharpoons Ti^{3+} + H_2O$	-0.055	± 10			4	17
		$Ti(OH)_4 + 4H^+ + e^- \rightleftharpoons Ti^{3+} + 4H_2O$	0.029				1	18
3		$Ti(OH)(Br_2) + H^+ + e^- \rightleftharpoons Ti^{3+} + 2Br^- + H_2O$	0.235				5	19
		$Ti_2O_3 + 2H^+ + 2e^- \rightleftharpoons 2TiO + H_2O$	-1.123				1	20
		$TiO_2^{2+} + 2H^+ + 2e^- \rightleftharpoons TiO^{2+} + H_2O$	1.800				1	21
		$Ti(OH)_3 + H^+ + e^- \rightleftharpoons TiO + 2H_2O$	-0.894				1	22
		$TiO_2^{2+} + 2e^- \rightleftharpoons TiO_2$	2.182				1	23
		$2HTiO_3^- + 4H^+ + 2e^- \rightleftharpoons Ti_2O_3 + 3H_2O$	-0.601				1	24
		$2TiO_2 + 2H^+ + 2e^- \rightleftharpoons Ti_2O_3 + H_2O$	-0.556				1	25
		$2Ti_3O_5 + 2H^+ + 2e^- \rightleftharpoons 3Ti_2O_3 + H_2O$	-0.490				1	26
		$2Ti(OH)_4 + 2H^+ + 2e^- \rightleftharpoons Ti_2O_3 + 5H_2O$	-0.139				1	27
		$3TiO_2 + 2H^+ + 2e^- \rightleftharpoons Ti_3O_5 + H_2O$	-0.589				1	28
		$3Ti(OH)_4 + 2H^+ + 2e^- \rightleftharpoons Ti_3O_5 + 7H_2O$	-0.453				1	29
		$HTiO_3^- + 2H^+ + e^- \rightleftharpoons Ti(OH)_3$	0.973				1	30
		$TiO_2 + H_2O + H^+ + e^- \rightleftharpoons Ti(OH)_3$	-0.786				1	31
		$Ti_3O_5 + 4H_2O + H^+ + e^- \rightleftharpoons 3Ti(OH)_3$	-1.178				1	32

Electronically conducting phase	Intermediate species	Composition of the solution	Solvent	Temperature °C	Pressure Torr	Measuring method
a	b	c	d		e	d
Ind.	Ti $(OH)_3$, Ti $(OH)_4$		1	25	760	IV
Ind.	Ti $(OH)_4$		1	25	760	IV
Hg (D.M.E.)		$HClO_4$: 0.5 M	1	25		II

Comparison electrode	Liquid junction	Electrode reaction	Standard value V	Uncertainty mV	Temperature Coefficient μV/°C	Notes	Reference	Electrode Reference Number
d			f		g	h		
		$Ti(OH)_4 + H^+ + e^- \rightleftharpoons Ti(OH)_3 + H_2O$	-0.091				1	33
		$TiO_2^{2+} + 2H_2O + 2e^- \rightleftharpoons Ti(OH)_4$	1.835				1	34
3		$Ti(SCN)(OH)^{2+} + H^+ + e^- \rightleftharpoons Ti(SCN)^{2+} + H_2O$	-0.03	± 20			4	35

Titanium 1

Standard value V	Solvent	Electrode Reference Number	Standard value V	Solvent	Electrode Reference Number	Standard value V	Solvent	Electrode Reference Number
-1.630	1	3						
-1.306	1	8						
-1.191	1	7						
-1.178	1	32						
-1.123	1	20						
-0.894	1	22						
-0.882	1	9						
-0.786	1	31						
-0.666	1	13						
-0.601	1	24						
-0.589	1	28						
-0.556	1	25						
-0.502	1	12						
-0.490	1	26						
-0.478	1	14						
-0.453	1	29						
-0.368	1	4						
-0.248	1	15						
-0.169	1	16						
-0.139	1	27						
-0.135	1	10						
-0.091	1	33						
-0.055	1	17						
-0.03	1	35						
0.029	1	18						
0.100	1	11						
0.235	1	19						
0.362	1	2						
0.973	1	30						
1.303	1	1						
1.800	1	21						
1.835	1	34						
2.182	1	23						
0.000	20	5						
-0.190	27	6						

NOTES: Titanium

1) Liquid junction: KCl satd. in methanol

BIBLIOGRAPHY: Titanium

1) Data reported by M. Pourbaix, "Atlas d'Equilibres Electrochimiques à 25 °C" (Gauthier-Villars Editeur, Paris, 1963), 213 - 222

2) Z. Hladky, J. Vrestal, Collect. Czech. Chem. Commun., 34(1969), 984 - 991

3) Data reported by A.J. De Bethune, N.A. Swendeman-Loud, "Standard Aqueous Electrode Potentials and Temperature Coefficients at 25 °C" (Clifford A. Hampel, Ill.)

4) S. Tribalat, D. Delafosse, Anal. Chim. Acta, 19 (1958), 74 - 89

5) A.I. Kartushinskaya, A.G. Stromberg, Zh. Neorgan. Khim., 7 (1962), 291 - 297

Electronically conducting phase	Intermediate species	Composition of the solution	Solvent	Temperature °C	Pressure Torr	Measuring method
a	b	c	d		e	d
Zr			1	25	760	IV
Zr	H_2ZrO_3		1	25	760	IV
Zr			1	25	760	IV
Zr			1.	25 ± 0.05		V
Zr			1	25	760	IV
Zr	ZrO_2		1	25	760	IV
Zr	$Zr(OH)_4$		1	25	760	IV
Zr	$ZrO(OH)_2$		1	25	760	IV

Comparison electrode	Liquid junction	Electrode reaction	Standard value	Uncertainty	Temperature Coefficient	Notes	Reference	Electrode Reference Number
			V	mV	μV/°C			
d			f		g	h		
		$HZrO_3^- + 5H^+ + 4e^- \rightleftharpoons Zr + 3H_2O$	-1.276				1	1
		$H_2ZrO_3 + H_2O + 4e^- \rightleftharpoons Zr + 4OH^-$	-2.36		-1110		2	2
		$Zr^{4+} + 4e^- \rightleftharpoons Zr$	-1.539				1	3
		$Zr(F)^{3+} + 4e^- \rightleftharpoons Zr + F^-$	-1.685			1	3	4
		$ZrO^{2+} + 2H^+ + 4e^- \rightleftharpoons Zr + H_2O$	-1.570				1	5
		$ZrO_2 + 4H^+ + 4e^- \rightleftharpoons Zr + 2H_2O$	-1.456				1	6
		$Zr(OH)_4 + 4H^+ + 4e^- \rightleftharpoons Zr + 4H_2O$	-1.553				1	7
		$ZrO(OH)_2 + 4H^+ + 4e^- \rightleftharpoons Zr + 3H_2O$	-1.533				1	8

Zirconium 1

Standard value V	Solvent	Electrode Reference Number	Standard value V	Solvent	Electrode Reference Number	Standard value V	Solvent	Electrode Reference Number
-2.36	1	2						
-1.685	1	4						
-1.570	1	5						
-1.539	1	3						
-1.553	1	7						
-1.533	1	8						
-1.456	1	6						
-1.276	1	1						

NOTES: Zirconium

1) Distribution between two phases

BIBLIOGRAPHY: Zirconium

1) Data reported by M. Pourbaix, "Atlas d'Equilibres Electrochimiques à 25 °C" (Gauthier-Villars Editeur, Paris, 1963), 223 - 229

2) Data reported by A.J. De Bethune, N.A. Swendeman-Loud, "Standard Aqueous Electrode Potentials and Temperature Coefficients at 25 °C" (Clifford A. Hampel, Ill.)

3) R.E. Connick, W.H. McVey, J. Am. Chem. Soc., 71 (1949), 3182 - 3191

Electronically conducting phase	Intermediate species	Composition of the solution	Solvent	Temperature °C	Pressure Torr	Measuring method
a	b	c	d		e	d
Hf			1	25	760	IV
Hf			1	25	760	IV
Hf	HfO_2		1	25	760	IV
Hf	$HfO(OH)_2$		1	25	760	IV
Hf	$HfO(OH)_2$		1	25	760	IV

Comparison electrode	Liquid junction	Electrode reaction	Standard value V	Uncertainty mV	Temperature Coefficient μV/°C	Notes	Reference	Electrode Reference Number
d			f		g	h		
		$Hf^{4+} + 4e^- \rightleftharpoons Hf$	-1.700				1	1
		$HfO^{2+} + 2H^+ + 2e^- \rightleftharpoons Hf + H_2O$	-1.724				1	2
		$HfO_2 + 4H^+ + 4e^- \rightleftharpoons Hf + 2H_2O$	-1.505				1	3
		$HfO(OH)_2 + 4H^+ + 4e^- \rightleftharpoons Hf + 3H_2O$	-1.685				1	4
		$HfO(OH)_2 + H_2O + 4e^- \rightleftharpoons Hf + 4OH^-$	-2.50				2	5

Hafnium 1

Standard value V	Solvent	Electrode Reference Number	Standard value V	Solvent	Electrode Reference Number	Standard value V	Solvent	Electrode Reference Number
-2.50	1	5						
-1.724	1	2						
-1.700	1	1						
-1.685	1	4						
-1.505	1	3						

BIBLIOGRAPHY: Hafnium

1) Data reported by M. Pourbaix, "Atlas d'Equilibres Electrochimiques at 25 ^{o}C" (Gauthier-Villars Editeur, Paris, 1963), 230 - 233

2) Data reported by A.J. De Bethune, N.A. Swendeman-Loud, "Standard Aqueous Electrode Potentials and Temperature Coefficients at 25 ^{o}C" (Clifford A. Hampel, Ill.)

Group V

Electronically conducting phase	Intermediate species	Composition of the solution	Solvent	Temperature °C	Pressure Torr	Measuring method
a	b	c	d		e	d
Ind.			1	25	760	IV
Ind.	$(CN)_2$, HCNO		1	25	760	IV
Ind.	$(CNS)_2$		1	25	760	IV
Ind.	HCN (aq), $(CN)_2$		1	25	760	IV
Ind.	HN_3 (gas), N_2		1	25	760	IV
Ind.	HN_3 (aq), N_2		1	25	760	IV
Ind.	HNO_2, HNO_3		1	25	760	IV
Ind.	HNO_2, NO_2		1	25	760	IV
Ind.	HNO_2		1	25	760	IV
Ind.	N_2O_4, HNO_2		1	25	760	IV
Ind.	$H_2N_2O_2$, HNO_2		1	25	760	IV
Ind.	$H_2N_2O_2$, NO		1	25	760	IV
Ind.	N_2, HNO_2		1	25	760	IV
Ind.	N_2, $H_2N_2O_2$		1	25	760	IV
Ind.	N_2, HNO_3		1	25	760	IV
Ind.	N_2, NO		1	25	760	IV
Ind.	N_2, N_2O		1	25	760	IV
Ind.	N_2, NO_2		1	25	760	IV
Ind.	N_2		1	25	760	IV
Ind.	N_2		1	25	760	IV
Ind.	N_2, N_2O_4		1	25	760	IV
Ind.	HNO_2, NH_3 (gas)		1	25	760	IV
Ind.	N_2, NH_3 (gas)		1	25	760	IV
Ind.	NH_3 (gas)		1	25	760	IV
Ind.	N_2, HN_3 (aq.)		1	25	760	IV
Ind.	HN_3 (aq)		1	25	760	IV
Ind.	HNO_2		1	25	760	IV
Ind.	N_2		1	25	760	IV
Ind.			1	25	760	IV
Ind.			1	25	760	IV
Ind.	NO		1	25	760	IV
Ind.	N_2O		1	25	760	IV
Ind.			1	25	760	IV

Comparison electrode	Liquid junction	Electrode reaction	Standard value V	Uncertainty mV	Temperature Coefficient µV/°C	Notes	Reference	Electrode Reference Number
d			f		g	h		
		$CNO^{-} + H_2O + 2e^{-} \rightleftharpoons CN^{-} + 2OH^{-}$	-0.970		-1211		1	1
		$2HCNO + 2H^{+} + 2e^{-} \rightleftharpoons (CN)_2 + 2H_2O$	-0.330		-588		1	2
		$(CNS)_2 + 2e^{-} \rightleftharpoons 2CNS^{-}$	0.77				1	3
		$(CN)_2 + 2H^{+} + 2e^{-} \rightleftharpoons 2HCN$	0.373		-596		1	4
		$3N_2 + 2H^{+} + 2e^{-} \rightleftharpoons 2HN_3$	-3.40		-1193		1	5
		$3N_2 + 2H^{+} + 2e^{-} \rightleftharpoons 2HN_3$	-3.09		-1570		1	6
		$HNO_3 + 2H^{+} + 2e^{-} \rightleftharpoons HNO_2 + H_2O$	0.934				2	7
		$NO_2 + H^{+} + e^{-} \rightleftharpoons HNO_2$	1.093				2	8
		$NO_3^{-} + 3H^{+} + 2e^{-} \rightleftharpoons HNO_2 + H_2O$	0.934		-800		2,1	9
		$N_2O_4 + 2H^{+} + 2e^{-} \rightleftharpoons 2HNO_2$	1.065		-726		2,1	10
		$2HNO_2 + 4H^{+} + 4e^{-} \rightleftharpoons H_2N_2O_2 + 2H_2O$	0.86		-510		1	11
		$2NO + 2H^{+} + 2e^{-} \rightleftharpoons H_2N_2O_2$	0.712		-1732		1	12
		$2HNO_2 + 6H^{+} + 6e^{-} \rightleftharpoons N_2 + 4H_2O$	1.454				2	13
		$H_2N_2O_2 + 2H^{+} + 2e^{-} \rightleftharpoons N_2 + 2H_2O$	2.65		-90		1	14
		$2HNO_3 + 10H^{+} + 10e^{-} \rightleftharpoons N_2 + 6H_2O$	1.246				2	15
		$2NO + 4H^{+} + 4e^{-} \rightleftharpoons N_2 + 2H_2O$	1.678				2	16
		$N_2O + 2H^{+} + 2e^{-} \rightleftharpoons N_2 + H_2O$	1.766				2	17
		$2NO_2 + 8H^{+} + 8e^{-} \rightleftharpoons N_2 + 4H_2O$	1.363				2	18
		$2NO_2^{-} + 8H^{+} + 6e^{-} \rightleftharpoons N_2 + 4H_2O$	1.520				2	19
		$2NO_3^{-} + 12H^{+} + 10e^{-} \rightleftharpoons N_2 + 6H_2O$	1.246				2	20
		$N_2O_4 + 8H^{+} + 8e^{-} \rightleftharpoons N_2 + 4H_2O$	1.357				2	21
		$HNO_2 + 6H^{+} + 6e^{-} \rightleftharpoons NH_3 + 2H_2O$	0.755				2	22
		$N_2 + 6H^{+} + 6e^{-} \rightleftharpoons 2NH_3$	0.057				2	23
		$NO_2^{-} + 7H^{+} + 6e^{-} \rightleftharpoons NH_3 + 2H_2O$	0.789				2	24
		$HN_3 + 3H^{+} + 2e^{-} \rightleftharpoons NH_4^{+} + N_2$	1.96		-140		1	25
		$HN_3 + 11H^{+} + 8e^{-} \rightleftharpoons 3NH_4^{+}$	0.695		-498		1	26
		$HNO_2 + 7H^{+} + 6e^{-} \rightleftharpoons NH_4^{+} + 2H_2O$	0.864				2	27
		$N_2 + 8H^{+} + 6e^{-} \rightleftharpoons 2NH_4^{+}$	0.275				2	28
		$N_2H_5^{+} + 3H^{+} + 2e^{-} \rightleftharpoons 2NH_4^{+}$	1.275		-1440		1	29
		$NH_3OH^{+} + 2H^{+} + 2e^{-} \rightleftharpoons NH_4^{+} + H_2O$	1.35		-530		1	30
		$NO + 6H^{+} + 5e^{-} \rightleftharpoons NH_4^{+} + H_2O$	0.836				2	31
		$N_2O + 10H^{+} + 8e^{-} \rightleftharpoons 2NH_4^{+} + H_2O$	0.647				2	32
		$NO_2^{-} + 8H^{+} + 6e^{-} \rightleftharpoons NH_4^{+} + 2H_2O$	0.897				2	33

Electronically conducting phase	Intermediate species	Composition of the solution	Solvent	Temperature °C	Pressure Torr	Measuring method
a	b	c	d		e	d
Ind.	N_2H_4, NH_2OH		1	25	760	IV
Ind.	N_2		1	25	760	IV
Ind.			1	25	760	IV
Ind.	$H_2N_2O_2$		1	25	760	IV
Ind.	NH_4OH, N_2		1	25	760	IV
Ind.	NH_4OH, NO		1	25	760	IV
Ind.	NH_4OH, N_2O		1	25	760	IV
Ind.	NH_4OH		1	25	760	IV
Pt	NO, HNO_2	H_2SO_4 : 0.1 up to 4 M; $NaNO_2$: 10^{-5} up to 10^{-2} M	1	25		I
Ind.	NO, HNO_3		1	25	760	IV
Pt	NO	1) H_2SO_4 : 50 and 80 %; NO 2) H_2SO_4 : 0.1 up to 4 M; NO; $NaNO_2$: 10^{-5} up to 10^{-2} M	1	25		III I
Ind.	NO, NO_2		1	25	760	IV
Ind.	NO		1	25	760	IV
Ind.	NO		1	25	760	IV
Ind.	NO, N_2H_4		1	25	760	IV
Pt	N_2O_4	H_2SO_4 : 50 and 80 %; N_2O_4, N_2O_3	1	25		III
Ind.	N_2O, HNO_2		1	25	760	IV
Ind.	N_2O, HNO_3		1	25	760	IV
Ind.	N_2O, NO		1	25	760	IV
Ind.	N_2O, NO_2		1	25	760	IV
Ind.	N_2O		1	25	760	IV
Ind.	N_2O		1	25	760	IV
Ind.	NO_2, HNO_3		1	25	760	IV
Ind.	NO_2		1	25	760	IV
Ind.	NO_2		1	25	760	IV
Ind.			1	25	760	IV
Ind.	N_2O_4, HNO_3		1	25	760	IV
Ind.	N_2O_4		1	25	760	IV
Ind.	N_2O_4		1	25	760	IV

Comparison electrode	Liquid junction	Electrode reaction	Standard value V	Uncertainty mV	Temperature Coefficient μV/°C	Notes	Reference	Electrode Reference Number
d			f		g	h		
		$2NH_2OH + 2e^- \rightleftharpoons N_2H_4 + 2OH^-$	0.73		-1810		1	34
		$N_2 + 5H^+ + 4e^- \rightleftharpoons N_2H_5^+$	-0.23		-840		1	35
		$2NH_3OH^+ + H^+ + 2e^- \rightleftharpoons N_2H_5^+ + 2H_2O$	1.42		-880		1	36
		$H_2N_2O_2 + 6H^+ + 4e^- \rightleftharpoons 2NH_3OH^+$	0.387		-44		1	37
		$N_2 + 2H_2O + 6H^+ + 6e^- \rightleftharpoons 2NH_4OH$	0.092				2	38
		$NO + 5H^+ + 5e^- \rightleftharpoons NH_4OH$	0.727				2	39
		$N_2O + H_2O + 8H^+ + 8e^- \rightleftharpoons 2NH_4OH$	0.510				2	40
		$NO_2^- + 7H^+ + 6e^- \rightleftharpoons NH_4OH + H_2O$	0.806				2	41
3	KCl satd.	$HNO_2 + H^+ + e^- \rightleftharpoons NO + H_2O$	0.983				3	42
		$HNO_3 + 3H^+ + 3e^- \rightleftharpoons NO + 2H_2O$	0.957				2	43
1) 4 2) 3	 KCl satd.	$NO_+ e^- \rightleftharpoons NO$	1.45	± 50			4 3	44
		$NO_2 + 2H^+ + 2e^- \rightleftharpoons NO + H_2O$	1.049				2	45
		$NO_2^- + 2H^+ + e^- \rightleftharpoons NO + H_2O$	1.202				2	46
		$NO_3^- + 4H^+ + 3e^- \rightleftharpoons NO + 2H_2O$	0.957		28		2	47
		$N_2O_4 + 4H^+ + 4e^- \rightleftharpoons 2NO + 2H_2O$	1.035		-11		2	48
4		$N_2O_4 + 4H^+ + 2e^- \rightleftharpoons 2NO^+ + 2H_2O$	0.77	± 80			4	49
		$2HNO_2 + 4H^+ + 4e^- \rightleftharpoons N_2O + 3H_2O$	1.297		-327		2	50
		$2HNO_3 + 8H^+ + 8e^- \rightleftharpoons N_2O + 5H_2O$	1.116				2	51
		$2NO + 2H^+ + 2e^- \rightleftharpoons N_2O + H_2O$	1.591				2	52
		$2NO_2 + 6H^+ + 6e^- \rightleftharpoons N_2O + 3H_2O$	1.229				2	53
		$2NO_2^- + 6H^+ + 4e^- \rightleftharpoons N_2O + 3H_2O$	1.396				2	54
		$2NO_3^- + 10H^+ + 8e^- \rightleftharpoons N_2O + 5H_2O$	1.116				2	55
		$HNO_3 + H^+ + e^- \rightleftharpoons NO_2 + H_2O$	0.775				2	56
		$NO_3^- + 2H^+ + e^- \rightleftharpoons NO_2 + H_2O$	0.775				2	57
		$NO_2 + e^- \rightleftharpoons NO_2^-$	0.895				2	58
		$NO_3^- + 2H^+ + 2e^- \rightleftharpoons NO_2^- + H_2O$	0.835				2	59
		$2HNO_3 + 2H^+ + 2e^- \rightleftharpoons N_2O_4 + 2H_2O$	0.803				2	60
		$2NO_3^- + 4H^+ + 2e^- \rightleftharpoons N_2O_4 + 2H_2O$	0.803		107		2	61
		$N_2O_4 + 2e^- \rightleftharpoons 2NO_2^-$	0.867				2	62

Nitrogen 1

Standard value V	Solvent	Electrode Reference Number	Standard value V	Solvent	Electrode Reference Number	Standard value V	Solvent	Electrode Reference Number
-3.40	1	5	0.957	1	47			
-3.09	1	6	0.983	1	42			
-0.970	1	1	1.035	1	48			
-0.330	1	2	1.049	1	45			
-0.23	1	35	1.065	1	10			
0.057	1	23	1.093	1	8			
0.092	1	38	1.116	1	51			
0.275	1	28	1.116	1	55			
0.373	1	4	1.202	1	46			
0.387	1	37	1.229	1	53			
0.510	1	40	1.246	1	15			
0.647	1	32	1.246	1	20			
0.695	1	26	1.275	1	29			
0.712	1	12	1.297	1	50			
0.727	1	39	1.35	1	30			
0.73	1	34	1.357	1	21			
0.755	1	22	1.363	1	18			
0.77	1	3	1.396	1	54			
0.77	1	49	1.42	1	36			
0.775	1	56	1.45	1	44			
0.775	1	57	1.454	1	13			
0.789	1	24	1.520	1	19			
0.803	1	60	1.591	1	52			
0.803	1	61	1.678	1	16			
0.806	1	41	1.766	1	17			
0.835	1	59	1.96	1	25			
0.836	1	31	2.65	1	14			
0.86	1	11						
0.864	1	27						
0.867	1	62						
0.895	1	58						
0.897	1	33						
0.934	1	7						
0.934	1	9						
0.957	1	43						

BIBLIOGRAPHY : Nitrogen

1) Data reported by A.J. De Bethune, N.A. Swendeman-Loud "Standard Aqueous Electrode Potentials and Temperature Coefficients at 25 °C" (Clifford A. Hampel, Ill.)

2) Data reported by M. Pourbaix "Atlas d'Equilibres Electrochimiques à 25 °C" (Gauthier-Villars Editeur, Paris, 1963)

3) G. Schimdt, U.Neumann, Z. Physik. Chem. (Frankfurt), 154 (1967), 150 - 165

4) T. Mussini, Chim. Ind., 50 (1968), 783

Electronically conducting phase	Intermediate species	Composition of the solution	Solvent	Temperature °C	Pressure Torr	Measuring method
a	b	c	d		e	d
Ind.			1	25	760	IV
Ind.			1	25	760	IV
Ind.	H_3PO_2		1	25	760	IV
Ind.	H_3PO_2, H_3PO_3		1	25	760	IV
Ind.			1	25	760	IV
Ind.			1	25	760	IV
Ind.			1	25	760	IV
Ind.			1	25	760	IV
Ind.			1	25	760	IV
Ind.			1	25	760	IV
Ind.			1	25	760	IV
Ind.	H_3PO_4		1	25	760	IV
Ind.			1	25	760	IV
Ind.			1	25	760	IV
Ind.	$H_4P_2O_6$		1	25	760	IV
Ind.	H_3PO_3, H_3PO_4		1	25	760	IV
Ind.	H_3PO_3, $H_4P_2O_6$		1	25	760	IV
Ind.			1	25	760	IV
Ind.			1	25	760	IV
Ind.			1	25	760	IV
Ind.			1	25	760	IV
Ind.	$H_4P_2O_6$		1	25	760	IV
Ind.	$H_4P_2O_6$, H_3PO_4		1	25	760	IV
P (red)			1	25	760	IV
P(white)			1	25	760	IV
P (red)	H_3PO_2		1	25	760	IV
P(white)	H_3PO_2		1	25	760	IV
P (red)			1	25	760	IV
P(white)			1	25	760	IV
P (red)			1	25	760	IV
P(white)			1	25	760	IV
P (red)	H_3PO_3		1	25	760	IV
P(white)	H_3PO_3		1	25	760	IV

Comparison electrode	Liquid junction	Electrode reaction	Standard value	Uncertainty	Temperature Coefficient	Notes	Reference	Electrode Reference Number
			V	mV	µV/°C			
d			f		g	h		
		$HPO_3^{2-} + 3H^+ + 2e^- \rightleftharpoons H_2PO_2^- + H_2O$	-0.323				1	1
		$H_2PO_3^- + 2H^+ + 2e^- \rightleftharpoons H_2PO_2^- + H_2O$	-0.504				1	2
		$H_2PO_3^- + 3H^+ + 2e^- \rightleftharpoons H_3PO_2 + H_2O$	-0.446				1	3
		$H_3PO_3 + 2H^+ + 2e^- \rightleftharpoons H_3PO_2 + H_2O$	-0.499		-360		1, 2	4
		$HPO_4^{2-} + 2H^+ + 2e^- \rightleftharpoons HPO_3^{2-} + H_2O$	-0.234				1	5
		$H_2PO_4^- + H^+ + 2e^- \rightleftharpoons HPO_3^{2-} + H_2O$	-0.447				1	6
		$HP_2O_6^{3-} + H^+ + 2e^- \rightleftharpoons 2HPO_3^{2-}$	0.275				1	7
		$H_2P_2O_6^{2-} + 2e^- \rightleftharpoons 2HPO_3^{2-}$	0.061				1	8
		$PO_4^{3-} + 3H^+ + 2e^- \rightleftharpoons HPO_3^{2-} + H_2O$	0.121		-490		1,2	9
		$P_2O_6^{4-} + 2H^+ + 2e^- \rightleftharpoons 2HPO_3^{2-}$	0.570				1	10
		$H_2PO_4^- + 2H^+ + 2e^- \rightleftharpoons H_2PO_3^- + H_2O$	-0.260				1	11
		$H_3PO_4 + H^+ + 2e^- \rightleftharpoons H_2PO_3^- + H_2O$	-0.329				1	12
		$H_2P_2O_6^{2-} + 2H^+ + 2e^- \rightleftharpoons 2H_2PO_3^-$	0.423				1	13
		$H_3P_2O_6^- + H^+ + 2e^- \rightleftharpoons 2H_2PO_3^-$	0.340				1	14
		$H_4P_2O_6 + 2e^- \rightleftharpoons 2H_2PO_3^-$	0.275				1	15
		$H_3PO_4 + 2H^+ + 2e^- \rightleftharpoons H_3PO_3 + H_2O$	-0.276		-360		1,2	16
		$H_4P_2O_6 + 2H^+ + 2e^- \rightleftharpoons 2H_3PO_3$	0.380				1	17
		$2HPO_4^{2-} + 3H^+ + 2e^- \rightleftharpoons HP_2O_6^{3-} + 2H_2O$	-0.744				1	18
		$2HPO_4^{2-} + 4H^+ + 2e^- \rightleftharpoons H_2P_2O_6^{2-} + 2H_2O$	-0.551				1	19
		$2H_2PO_4^- + 2H^+ + 2e^- \rightleftharpoons H_2P_2O_6^{2-} + 2H_2O$	-0.955				1	20
		$2H_2PO_4^- + 3H^+ + 2e^- \rightleftharpoons H_3P_2O_6^- + 2H_2O$	-0.872				1	21
		$2H_2PO_4^- + 4H^+ + 2e^- \rightleftharpoons H_4P_2O_6 + 2H_2O$	-0.807				1	22
		$2H_3PO_4 + 2H^+ + 2e^- \rightleftharpoons H_4P_2O_6 + 2H_2O$	-0.933				1	23
		$H_2PO_2^- + 2H^+ + e^- \rightleftharpoons P + 2H_2O$	-0.248				1	24
		$H_2PO_2^- + 2H^+ + e^- \rightleftharpoons P + 2H_2O$	-0.391				1	25
		$H_3PO_2 + H^+ + e^- \rightleftharpoons P + 2H_2O$	-0.365				1	26
		$H_3PO_2 + H^+ + e^- \rightleftharpoons P + 2H_2O$	-0.508		-420		1, 2	27
		$HPO_3^{2-} + 5H^+ + 3e^- \rightleftharpoons P + 3H_2O$	-0.298				1	28
		$HPO_3^{2-} + 5H^+ + 3e^- \rightleftharpoons P + 3H_2O$	-0.346				1	29
		$H_2PO_3^- + 4H^+ + 3e^- \rightleftharpoons P + 3H_2O$	-0.419				1	30
		$H_2PO_3^- + 4H^+ + 3e^- \rightleftharpoons P + 3H_2O$	-0.467				1	31
		$H_3PO_3 + 3H^+ + 3e^- \rightleftharpoons P + 3H_2O$	-0.454				1	32
		$H_3PO_3 + 3H^+ + 3e^- \rightleftharpoons P + 3H_2O$	-0.502				1	33

Electronically conducting phase	Intermediate species	Composition of the solution	Solvent	Temperature °C	Pressure Torr	Measuring method
a	b	c	d		e	d
P(red)			1	25	760	IV
P(white)			1	25	760	IV
P (red)			1	25	760	IV
P(white)			1	25	760	IV
P(red)	H_3PO_4		1	25	760	IV
P(white)	H_3PO_4		1	25	760	IV
P(red)			1	25	760	IV
P(white)			1	25	760	IV
Ind.	P_4H_2 (s)		1	25	760	IV
Ind.	P_4H_2 (s), H_3PO_2		1	25	760	IV
Ind.	P_4H_2 (s)		1	25	760	IV
Ind.	P_4H_2 (s)		1	25	760	IV
Ind.	P_4H_2 (s), H_3PO_3		1	25	760	IV
Ind.	P_4H_2 (s)		1	25	760	IV
Ind.	P_4H_2 (s)		1	25	760	IV
Ind.	P_4H_2 (s) , H_3PO_4		1	25	760	IV
P(red)	P_4H_2 (s)		1	25	760	IV
P(white)	P_4H_2 (s)		1	25	760	IV
Ind.	P_4H_2 (s)		1	25	760	IV
Ind.	PH_3 (gas)		1	25	760	IV
Ind.	PH_3 (gas), H_3PO_2		1	25	760	IV
Ind.	PH_3 (gas)		1	25	760	IV
Ind.	PH_3 (gas)		1	25	760	IV
Ind.	PH_3 (gas), H_3PO_3		1	25	760	IV
Ind.	PH_3 (gas)		1	25	760	IV
Ind.	PH_3 (gas)		1	25	760	IV
Ind.	PH_3 (gas), H_3PO_4		1	25	760	IV
P(red)	PH_3 (gas)		1	25	760	IV
P(white)	PH_3 (gas)		1	25	760	IV
Ind.	PH_3 (gas), P_4H_2(s)		1	25	760	IV
Ind.	PH_3 (gas), P_2H_4 (l)		1	25	760	IV
Ind.	PH_3 (gas)		1	25	760	IV
P(red)	P_2H_4 (l)		1	25	760	IV

Comparison electrode	Liquid junction	Electrode reaction	Standard value V	Uncertainty mV	Temperature Coefficient µV/°C	Notes	Reference	Electrode Reference Number
d			f		g	h		
		$HPO_4^{2-} + 7H^+ + 5e^- \rightleftharpoons P + 4H_2O$	-0.288				1	34
		$HPO_4^{2-} + 7H^+ + 5e^- \rightleftharpoons P + 4H_2O$	-0.316				1	35
		$H_2PO_4^- + 6H^+ + 5e^- \rightleftharpoons P + 4H_2O$	-0.358				1	36
		$H_2PO_4^- + 6H^+ + 5e^- \rightleftharpoons P + 4H_2O$	-0.386				1	37
		$H_3PO_4 + 5H^+ + 5e^- \rightleftharpoons P + 4H_2O$	-0.383				1	38
		$H_3PO_4 + 5H^+ + 5e^- \rightleftharpoons P + 4H_2O$	-0.411				1	39
		$PO_4^{3-} + 8H^+ + 5e^- \rightleftharpoons P + 4H_2O$	-0.128				1	40
		$PO_4^{3-} + 8H^+ + 5e^- \rightleftharpoons P + 4H_2O$	-0.156				1	41
		$4H_2PO_2^- + 10H^+ + 6e^- \rightleftharpoons P_4H_2 + 8H_2O$	-0.376				1	42
		$4H_3PO_2 + 6H^+ + 6e^- \rightleftharpoons P_4H_2 + 8H_2O$	-0.455				1	43
		$4HPO_3^{2-} + 22H^+ + 14e^- \rightleftharpoons P_4H_2 + 12H_2O$	-0.346				1	44
		$4H_2PO_3^- + 18H^+ + 14e^- \rightleftharpoons P_4H_2 + 12H_2O$	-0.450				1	45
		$4H_3PO_3 + 14H^+ + 14e^- \rightleftharpoons P_4H_2 + 12H_2O$	-0.480				1	46
		$4HPO_4^{2-} + 30H^+ + 22e^- \rightleftharpoons P_4H_2 + 16H_2O$	-0.305				1	47
		$4H_2PO_4^- + 26H^+ + 22e^- \rightleftharpoons P_4H_2 + 16H_2O$	-0.383				1	48
		$4H_3PO_4 + 22H^+ + 22e^- \rightleftharpoons P_4H_2 + 16H_2O$	-0.406				1	49
		$4P + 2H^+ + 2e^- \rightleftharpoons P_4H_2$	-0.633				1	50
		$4P + 2H^+ + 2e^- \rightleftharpoons P_4H_2$	-0.347				1	51
		$4PO_4^{3-} + 34H^+ + 22e^- \rightleftharpoons P_4H_2 + 16H_2O$	-0.176				1	52
		$H_2PO_2^- + 5H^+ + 4e^- \rightleftharpoons PH_3 + 2H_2O$	-0.145				1	53
		$H_3PO_2 + 4H^+ + 4e^- \rightleftharpoons PH_3 + 2H_2O$	-0.174				1	54
		$HPO_3^{2-} + 8H^+ + 6e^- \rightleftharpoons PH_3 + 3H_2O$	-0.205				1	55
		$H_2PO_3^- + 7H^+ + 6e^- \rightleftharpoons PH_3 + 3H_2O$	-0.265				1	56
		$H_3PO_3 + 6H^+ + 6e^- \rightleftharpoons PH_3 + 3H_2O$	-0.282				1	57
		$HPO_4^{2-} + 10H^+ + 8e^- \rightleftharpoons PH_3 + 4H_2O$	-0.212				1	58
		$H_2PO_4^- + 9H^+ + 8e^- \rightleftharpoons PH_3 + 4H_2O$	-0.265				1	59
		$H_3PO_4 + 8H^+ + 8e^- \rightleftharpoons PH_3 + 4H_2O$	-0.281				1	60
		$P + 3H^+ + 3e^- \rightleftharpoons PH_3$	-0.111				1	61
		$P + 3H^+ + 3e^- \rightleftharpoons PH_3$	-0.063		-104		1,2	62
		$P_4H_2 + 10H^+ + 10e^- \rightleftharpoons 4PH_3$	-0.006				1	63
		$P_2H_4 + 2H^+ + 2e^- \rightleftharpoons 2PH_3$	0.006				1	64
		$PO_4^{3-} + 11H^+ + 8e^- \rightleftharpoons PH_3 + 4H_2O$	-0.123				1	65
		$2P + 4H^+ + 4e^- \rightleftharpoons P_2H_4$	-0.169				1	66

Electronically conducting phase	Intermediate species	Composition of the solution	Solvent	Temperature °C	Pressure Torr	Measuring method
a	b	c	d		e	d
P(white)	P_2H_4 (l)		1	25	760	IV
Ind.	P_2H_4 (l), P_4H_2 (s)		1	25	760	IV
Ind.			1	25	760	IV
Ind.			1	25	760	IV

Comparison electrode	Liquid junction	Electrode reaction	Standard value V	Uncertainty mV	Temperature Coefficient μV/°C	Notes	Reference	Electrode Reference Number
d			f		g	h		
		$2P + 4H^+ + 4e^- \rightleftharpoons P_2H_4$	-0.100				1	67
		$P_4H_2 + 6H^+ + 6e^- \rightleftharpoons 2P_2H_4$	-0.014				1	68
		$2HPO_4^{2-} + 2H^+ + 2e^- \rightleftharpoons P_2O_6^{4-} + 2H_2O$	-1.039				1	69
		$2PO_4^{3-} + 4H^+ + 2e^- \rightleftharpoons P_2O_6^{4-} + 2H_2O$	-0.328				1	70

Standard value V	Solvent	Electrode Reference Number	Standard value V	Solvent	Electrode Reference Number	Standard value V	Solvent	Electrode Reference Number
-1.039	1	69	-0.316	1	35			
-0.955	1	20	-0.305	1	47			
-0.933	1	23	-0.298	1	28			
-0.872	1	21	-0.288	1	34			
-0.807	1	22	-0.282	1	57			
-0.744	1	18	-0.281	1	60			
-0.633	1	50	-0.276	1	16			
-0.551	1	19	-0.265	1	56			
-0.508	1	27	-0.265	1	59			
-0.504	1	2	-0.260	1	11			
-0.502	1	33	-0.248	1	24			
-0.499	1	4	-0.234	1	5			
-0.480	1	46	-0.212	1	58			
-0.467	1	31	-0.205	1	55			
-0.455	1	43	-0.176	1	52			
-0.454	1	32	-0.174	1	54			
-0.450	1	45	-0.169	1	66			
-0.447	1	6	-0.156	1	41			
-0.446	1	3	-0.145	1	53			
-0.419	1	30	-0.128	1	40			
-0.411	1	39	-0.123	1	65			
-0.406	1	49	-0.111	1	61			
-0.391	1	25	-0.100	1	67			
-0.386	1	37	-0.063	1	62			
-0.383	1	38	-0.014	1	68			
-0.383	1	48	-0.006	1	63			
-0.376	1	42	0.006	1	64			
-0.365	1	26	0.061	1	8			
-0.358	1	36	0.121	1	9			
-0.347	1	51	0.275	1	15			
-0.346	1	29	0.275	1	7			
-0.346	1	44	0.340	1	14			
-0.329	1	12	0.380	1	17			
-0.328	1	70	0.423	1	13			
-0.323	1	1	0.570	1	10			

BIBLIOGRAPHY : Phosphorus

1) Data reported by M. Pourbaix "Atlas d'Equilibres Electrochimiques à 25 °C " (Gauthier-Villars Editeur, Paris, 1962), 504 - 515

2) Data reported by A.J. De Bethune, N.A. Swendeman-Loud "Standard Aqueous Electrode Potentials and Temperature Coefficients at 25 °C" (Clifford A. Hampel, Ill.)

Electronically conducting phase	Intermediate species	Composition of the solution	Solvent	Temperature °C	Pressure Torr	Measuring method
a	b	c	d		e	d
As	$AsBr_3$	$AsBr_3$: $AlBr_3$ = 0.1 (M:M); $AlBr_3$: $C_6H_5CH_3$ = = 0.25 (M:M)	25	25		I
As	As_2O_3		1	25	760	IV
As			1	25	760	IV
As			1	25	760	IV
As			1	25	760	IV
As	As_2O_5		1	25	760	IV
As	As_4O_6	NaOH : 0.1 %; HNO_3 : 0.1 %; H_2O_2 : 3%	1	25		I
As	$HAsO_2$		1	25	760	IV
As	AsH_3 (gas)		1	25	760	IV
Ind.	H_3AsO_4		1	25	760	IV
Ind.			1	25	760	IV
Ind.			1	25	760	IV
Ind.	As_2O_3		1	25	760	IV
Ind.	As_2O_3 , As_2O_5		1	25	760	IV
Ind.	As_2O_3		1	25	760	IV
Ind.	As_2O_3		1	25	760	IV
Ind.	As_2O_3 , H_3AsO_4		1	25	760	IV
Ind.	$HAsO_2$		1	25	760	IV
Ind.	$HAsO_2$		1	25	760	IV
Ind.	$HAsO_2$, H_3AsO_4		1	25	760	IV

Comparison electrode	Liquid junction	Electrode reaction	Standard value	Uncertainty	Temperature Coefficient	Notes	Reference	Electrode Reference Number
			V	mV	μV/°C			
d			f		g	h		
6		$AsBr_3 + 3e^- \rightleftharpoons As + 3Br^-$	0.590			1	3	1
		$As_2O_3 + 6H^+ + 6e^- \rightleftharpoons 2As + 3H_2O$	0.234				1	2
		$AsO^+ + 2H^+ + 3e^- \rightleftharpoons As + H_2O$	0.254				1	3
		$AsO_2^- + 4H^+ + 3e^- \rightleftharpoons As + 2H_2O$	0.429				1	4
		$AsO_4^{3-} + 8H^+ + 5e^- \rightleftharpoons As + 4H_2O$	0.648				1	5
		$As_2O_5 + 10H^+ + 10e^- \rightleftharpoons 2As + 5H_2O$	0.429				1	6
3		$As_4O_6 + 10H^+ + 10e^- \rightleftharpoons 4As + 5H_2O + 1/2\,O_2$	0.285				4	7
		$HAsO_2 + 3H^+ + 3e^- \rightleftharpoons As + 2H_2O$	0.248		-510		1,2	8
		$As + 3H^+ + 3e^- \rightleftharpoons AsH_3$	-0.608		-50		1,2	9
		$H_3AsO_4 + 3H^+ + 2e^- \rightleftharpoons AsO^+ + 3H_2O$	0.550				1	10
		$AsO_4^{3-} + 4H^+ + 2e^- \rightleftharpoons AsO_2^- + 2H_2O$	0.977				1	11
		$HAsO_4^{2-} + 3H^+ + 2e^- \rightleftharpoons AsO_2^- + 2H_2O$	0.609				1	12
		$2AsO_4^{3-} + 10H^+ + 4e^- \rightleftharpoons As_2O_3 + 5H_2O$	1.270				1	13
		$As_2O_5 + 4H^+ + 4e^- \rightleftharpoons As_2O_3 + 2H_2O$	0.721				1	14
		$2HAsO_4^{2-} + 8H^+ + 4e^- \rightleftharpoons As_2O_3 + 5H_2O$	0.901				1	15
		$2H_2AsO_4^- + 6H^+ + 4e^- \rightleftharpoons As_2O_3 + 5H_2O$	0.687				1	16
		$2H_3AsO_4 + 4H^+ + 4e^- \rightleftharpoons As_2O_3 + 5H_2O$	0.580				1	17
		$HAsO_4^{2-} + 4H^+ + 2e^- \rightleftharpoons HAsO_2 + 2H_2O$	0.881				1	18
		$H_2AsO_4^- + 3H^+ + 2e^- \rightleftharpoons HAsO_2 + 2H_2O$	0.666				1	19
		$H_3AsO_4 + 2H^+ + 2e^- \rightleftharpoons HAsO_2 + 2H_2O$	0.560		-364		1	20

Arsenic 1

Standard value V	Solvent	Electrode Reference Number	Standard value V	Solvent	Electrode Reference Number	Standard value V	Solvent	Electrode Reference Number
-0.608	1	9						
0.234	1	2						
0.248	1	8						
0.254	1	3						
0.285	1	7						
0.429	1	4						
0.429	1	6						
0.550	1	10						
0.560	1	20						
0.580	1	17						
0.609	1	12						
0.648	1	5						
0.666	1	19						
0.687	1	16						
0.721	1	14						
0.881	1	18						
0.901	1	15						
0.977	1	11						
1.270	1	13						
0.590	25	1						

BIBLIOGRAPHY : Arsenic

1) Data reported by M. Pourbaix "Atlas d'Equilibres Electrochimiques à 25 °C" (Gauthier-Villars Editeur, Paris, 1963), pp. 516 - 523

2) Data reported by A.J. De Bethune, N.A. Swendeman-Loud "Standard Aqueous Electrode Potentials and Temperature Coefficients at 25 °C " (Clifford A. Hampel, Ill.)

3) V.S. Galinker, Zh. Obsch. Khim., 19 (1949), 2048 - 2050

4) N.S. Yagn, F.T. Bunazhnov, V.G. Sokolova, Zap. Leningr. Gorn. Inst., 42 (1961), 35 - 40

Electronically conducting phase	Intermediate species	Composition of the solution	Solvent	Temperature °C	Pressure Torr	Measuring method
a	b	c	d		e	d
Ind.	$HSbO_2$		1	25	760	IV
Sb	$HSbO_2$		1	25	760	IV
Sb		HCl : 1 and 2 m	1	25 ± 0.25		I
Sb		CH_3COOH-CH_3COONa and Clark and Lubs buffers, pH 4 up to 7,5	1	25		I
Sb			1	25	760	IV
Sb			1	25	760	IV
Sb	Sb_2O_3 (senarmontite)	CH_3COOH-CH_3COONa and Clark and Lubs buffers, pH 4 up to 7,5	1	25		I
Sb	Sb_2O_3 (valentinite)		1	25	760	IV
Sb	SbH_3 (gas)		1	25	760	IV
Ind.			1	25	760	IV
Ind.			1	25	760	IV
Ind.	Sb_2O_5		1	25	760	IV
Ind.			1	25	760	IV
Ind.	Sb_2O_3 (senarmontite)		1	25	760	IV
Ind.	Sb_2O_3 (valentinite)		1	25	760	IV
Ind.	Sb_2O_3 (senarmontite), Sb_2O_4		1	25	760	IV
Ind.	Sb_2O_3 (valentinite), Sb_2O_4		1	25	760	IV
Ind.	Sb_2O_3 (senarmontite), Sb_2O_5		1	25	760	IV
Ind.	Sb_2O_3 (valentinite), Sb_2O_5		1	25	760	IV
Ind.	Sb_2O_4, Sb_2O_5		1	25	760	IV

Comparison electrode	Liquid junction	Electrode reaction	Standard value V	Uncertainty mV	Temperature Coefficient μV/°C	Notes	Reference	Electrode Reference Number
d			f		g	h		
		$SbO_3^- + 3H^+ + 2e^- \rightleftharpoons HSbO_2 + H_2O$	0.678				1	1
		$HSbO_2 + 3H^+ + 3e^- \rightleftharpoons Sb + 2H_2O$	0.230				1	2
1		$Sb(Cl)_4^- + 3e^- \rightleftharpoons Sb + 4Cl^-$	0.17				2	3
3		$SbO^+ + 2H^+ + 3e^- \rightleftharpoons Sb + H_2O$	0.212				3	4
		$SbO_2^- + 4H^+ + 3e^- \rightleftharpoons Sb + 2H_2O$	0.446				1	5
		$SbO_2^- + 2H_2O + 3e^- \rightleftharpoons Sb + 4OH^-$	-0.66				4	6
3		$Sb_2O_3 + 6H^+ + 6e^- \rightleftharpoons 2Sb + 3H_2O$	0.152		-375		3,4	7
		$Sb_2O_3 + 6H^+ + 6e^- \rightleftharpoons 2Sb + 3H_2O$	0.167				1	8
		$Sb + 3H^+ + 3e^- \rightleftharpoons SbH_3$	-0.510		-60		1,4	9
		$SbO_2^+ + 2H^+ + 2e^- \rightleftharpoons SbO^+ + H_2O$	0.720				1	10
		$SbO_3^- + 4H^+ + 2e^- \rightleftharpoons SbO^+ + 2H_2O$	0.704				1	11
		$Sb_2O_5 + 6H^+ + 4e^- \rightleftharpoons 2SbO^+ + 3H_2O$	0.581				1	12
		$SbO_3^- + 2H^+ + 2e^- \rightleftharpoons SbO_2^- + H_2O$	0.353				1	13
		$2SbO_3^- + 6H^+ + 4e^- \rightleftharpoons Sb_2O_3 + 3H_2O$	0.794				1	14
		$2SbO_3^- + 6H^+ + 4e^- \rightleftharpoons Sb_2O_3 + 3H_2O$	0.772				1	15
		$Sb_2O_4 + 2H^+ + 2e^- \rightleftharpoons Sb_2O_3 + H_2O$	0.863				1	16
		$Sb_2O_4 + 2H^+ + 2e^- \rightleftharpoons Sb_2O_3 + H_2O$	0.819				1	17
		$Sb_2O_5 + 4H^+ + 4e^- \rightleftharpoons Sb_2O_3 + 2H_2O$	0.671				1	18
		$Sb_2O_5 + 4H^+ + 4e^- \rightleftharpoons Sb_2O_3 + 2H_2O$	0.649				1	19
		$Sb_2O_5 + 2H^+ + 2e^- \rightleftharpoons Sb_2O_4 + H_2O$	0.479				1	20

Antimony 1

Standard value V	Solvent	Electrode Reference Number	Standard value V	Solvent	Electrode Reference Number	Standard value V	Solvent	Electrode Reference Number
-0.66	1	6						
-0.510	1	9						
0.152	1	7						
0.167	1	8						
0.17	1	3						
0.212	1	4						
0.230	1	2						
0.353	1	13						
0.446	1	5						
0.479	1	20						
0.581	1	12						
0.649	1	19						
0.671	1	18						
0.678	1	1						
0.704	1	11						
0.720	1	10						
0.772	1	15						
0.794	1	14						
0.819	1	17						
0.863	1	16						

BIBLIOGRAPHY : Antimony

1) Data reported by M. Pourbaix "Atlas d'Equilibres Electrochimiques à 25 °C" (Gauthier-Villars Editeur, Paris, 1963), pp. 524 - 532

2) N.G. Nozakdzee, V.A. Kazakov, A.T. Vagramyan, Elektrokhim., 4 (1968), 1464 - 1466

3) A.R. Tourky, E.M. Khairy, J. Chem. Soc., 1952, 2626 - 2633

4) Data reported by A.J. De Bethune, N.A. Swendeman-Loud "Standard Aqueous Electrode Potentials and Temperature Coefficients at 25 °C" (Clifford A. Hampel, Ill.)

Electronically conducting phase	Intermediate species	Composition of the solution	Solvent	Temperature °C	Pressure Torr	Measuring method
a	b	c	d		e	d
Bi		HCl, BiOCl: satd.	1	25		I
Bi		LiCl : 0.01 M	7	25		I
Bi			1	25	760	IV
Bi			1	25	760	IV
Bi	Bi_2O_3		1	25	760	IV
Bi	Bi_2O_3		1	25	760	IV
Ind.	Bi_2O_4		1	25	760	IV
Ind.	Bi_2O_5		1	25	760	IV
Ind.	Bi_4O_7		1	25	760	IV
Bi	BiOCl	HCl, $HClO_4$: 0.9928 up to 3.014 m	1	25		I
Bi			1	25	760	IV
Bi	$Bi(OH)_3$		1	25	760	IV
Bi	BiH_3 (gas)		1	25	760	IV
Ind.	Bi_2O_4		1	25	760	IV
Ind.	Bi_2O_5		1	25	760	IV
Ind.	Bi_4O_7		1	25	760	IV
Ind.	Bi_2O_3, Bi_4O_7		1	25	760	IV
Ind.	Bi_2O_4, Bi_2O_5		1	25	760	IV
Ind.	Bi_4O_7		1	25	760	IV
Ind.	Bi_2O_4		1	25	760	IV
Ind.	Bi_2O_5		1	25	760	IV
Ind.	Bi_4O_7		1	25	760	IV
Ind.	$Bi(OH)_3$, Bi_4O_7		1	25	760	IV

Comparison electrode	Liquid junction	Electrode reaction	Standard value	Uncertainty	Temperature Coefficient	Notes	Reference	Electrode Reference Number
			V	mV	µV/°C			
d			f		g	h		
1		$Bi^{3+} + 3e^- \rightleftharpoons Bi$	0.200	±1			1	1
1		$Bi^{3+} + 3e^- \rightleftharpoons Bi$	-0.513				2	2
		$Bi(Cl)_4^- + 3e^- \rightleftharpoons Bi + 4Cl^-$	0.16				3	3
		$BiO^+ + 2H^+ + 3e^- \rightleftharpoons Bi + H_2O$	0.320				4	4
		$Bi_2O_3 + 6H^+ + 6e^- \rightleftharpoons 2Bi + 3H_2O$	0.371				4	5
		$Bi_2O_3 + 3H_2O + 6e^- \rightleftharpoons 2Bi + 6OH^-$	-0.46		-1214		5	6
		$Bi_2O_4 + 8H^+ + 2e^- \rightleftharpoons 2Bi^{3+} + 4H_2O$	1.910				4	7
		$Bi_2O_5 + 10H^+ + 4e^- \rightleftharpoons 2Bi^{3+} + 5H_2O$	1.759				4	8
		$Bi_4O_7 + 14H^+ + 2e^- \rightleftharpoons 4Bi^{3+} + 7H_2O$	2.279				4	9
1	$HClO_4$	$BiOCl + H_2 + e^- \rightleftharpoons Bi + Cl^- + H_2O$	0.1697	±0.4	-400		6	10
		$BiOH^{2+} + H^+ + 3e^- \rightleftharpoons Bi + H_2O$	0.254				4	11
		$Bi(OH)_3 + 3H^+ + 3e^- \rightleftharpoons Bi + 3H_2O$	0.478				4	12
		$Bi + 3H^+ + 3e^- \rightleftharpoons BiH_3$	-0.800				4	13
		$Bi_2O_4 + 4H^+ + 2e^- \rightleftharpoons 2BiO^+ + 2H_2O$	1.593				4	14
		$Bi_2O_5 + 6H^+ + 4e^- \rightleftharpoons 2BiO^+ + 3H_2O$	1.605				4	15
		$Bi_4O_7 + 6H^+ + 2e^- \rightleftharpoons 4BiO^+ + 3H_2O$	1.644				4	16
		$Bi_4O_7 + 2H^+ + 2e^- \rightleftharpoons 2Bi_2O_3 + H_2O$	1.338				4	17
		$Bi_2O_5 + 2H^+ + 2e^- \rightleftharpoons Bi_2O_4 + H_2O$	1.607				4	18
		$2Bi_2O_4 + 2H^+ + 2e^- \rightleftharpoons Bi_4O_7 + H_2O$	1.541				4	19
		$Bi_2O_4 + 6H^+ + 2e^- \rightleftharpoons 2BiOH^{2+} + 2H_2O$	1.792				4	20
		$Bi_2O_5 + 8H^+ + 4e^- \rightleftharpoons 2BiOH^{2+} + 3H_2O$	1.700				4	21
		$Bi_4O_7 + 10H^+ + 2e^- \rightleftharpoons 4BiOH^{2+} + 3H_2O$	2.042				4	22
		$Bi_4O_7 + 5H_2O + 2H^+ + 2e^- \rightleftharpoons 4Bi(OH)_3$	0.690				4	23

Bismuth 1

Standard value V	Solvent	Electrode Reference Number
-0.800	1	13
-0.46	1	6
0.16	1	3
0.1697	1	10
0.200	1	1
0.254	1	11
0.320	1	4
0.371	1	5
0.478	1	12
0.690	1	23
1.338	1	17
1.541	1	19
1.593	1	14
1.605	1	15
1.607	1	18
1.644	1	16
1.700	1	21
1.759	1	8
1.792	1	20
1.910	1	7
2.042	1	22
2.279	1	9
-0.513	7	2

Standard value V	Solvent	Electrode Reference Number

Standard value V	Solvent	Electrode Reference Number

BIBLIOGRAPHY: Bismuth

1) V. Cupr, Pubs. Faculte Sci. Univ. Masaryk, 296 (1947)

2) B. Jakuszewki, Z. Kozlowskii, Roczn. Chem., 38 (1964), 93

3) W.M. Latimer, "Oxidation Potentials" (Prentice Hall, N.J. 1952)

4) Data reported by M. Pourbaix "Atlas d'Equilibres Electrochimiques à 25 °C" (Gauthier-Villars Editeur, Paris, 1963), pp. 533 - 539

5) Data reported by A.J. De Bethune, N.A. Swendeman-Loud "Standard Aqueous Electrode Potentials and Temperature Coefficients at 25 °C" (Clifford A. Hampel, Ill.)

6) V.P. Vasil'ev, N.K. Grechina, Elektrokhim., 5 (1969), 426 - 429

Electronically conducting phase	Intermediate species	Composition of the solution	Solvent	Temperature °C	Pressure Torr	Measuring method
a	b	c	d		e	d
Ind.			1	25	760	IV
Ind.			1	25	760	IV
Ind.			1	25	760	IV
Ind.			1	25	760	IV
V			1	25	760	IV
V			1	25	760	IV
V			1	25	760	IV
V	V_2O_2		1	25	760	IV
V			1	25	760	IV
Ind.			1	25	760	IV
Pt		$V_2(SO_4)_3$ + VSO_4 : 0.00523 up to 0.1410 m; H_2SO_4 : 0.05 up to 1.0 m	1	25		I
Ind.			1	25	760	IV
V	V_2O_3		1	25	760	IV
Ind.			1	25	760	IV
Pt		$VOSO_4$: 0.00534 up to 0.0350 m; $V_2(SO_4)_3$: 0.0050 up to 0.03466; H_2SO_4 : 0.05 up to 1.0 m	1	25		I
Ind.			1	25	760	IV
Ind.			1	25	760	IV
Ind.			1	25	760	IV
Ind.	V_2O_4		1	25	760	IV
Ind.			1	25	760	IV
Ind.			1	25	760	IV
Glass		NH_4VO_3 : 0.025 m; $VOSO_4$: 0.025 m	1	25		I
Glass		NH_4VO_3: 0.025 m; $VOSO_4$: 0.025 m	1	25		I
Ind.	V_2O_5 (evolved)		1	25	760	IV
Ind.	V_2O_5 (not evolved)		1	25	760	IV
Ind.			1	25	760	IV
Ind.	V_2O_2 , V_2O_3		1	25	760	IV
Ind.	V_2O_3		1	25	760	IV
Ind.	V_2O_3		1	25	760	IV
Ind.	V_2O_3		1	25	760	IV
Ind.	V_2O_3 , V_2O_4		1	25	760	IV
Ind.	V_2O_4		1	25	760	IV

Comparison electrode	Liquid junction	Electrode reaction	Standard value V	Uncertainty mV	Temperature Coefficient µV/°C	Notes	Reference	Electrode Reference Number
d			f		g	h		
		$2HVO_4^{2-} + 5H^+ + 2e^- \rightleftharpoons HV_2O_5^- + 3H_2O$	1.281				1	1
		$2H_2VO_4^- + 3H^+ + 2e^- \rightleftharpoons HV_2O_5^- + 3H_2O$	0.719				1	2
		$H_3V_2O_7^- + 2H^+ + 2e^- \rightleftharpoons HV_2O_5^- + 2H_2O$	0.501				1	3
		$2VO_4^{3-} + 7H^+ + 2e^- \rightleftharpoons HV_2O_5^- + 3H_2O$	1.962				1	4
		$HV_6O_{17}^{3-} + 16H_2O + 30e^- \rightleftharpoons 6V + 33\ OH^-$	-1.154				2	5
		$V^{2+} + 2e^- \rightleftharpoons V$	-1.175				1	6
		$VO_2^+ + 4H^+ + 5e^- \rightleftharpoons V + 2H_2O$	-0.25				3	7
		$V_2O_2 + 4H^+ + 4e^- \rightleftharpoons 2V + 2H_2O$	-0.820				1	8
		$V(OH)_4^+ + 4H^+ + 5e^- \rightleftharpoons V + 4H_2O$	-0.254				2	9
		$HV_2O_5^- + 9H^+ + 4e^- \rightleftharpoons 2V^{2+} + 5H_2O$	0.338				1	10
19		$V^{3+} + e^- \rightleftharpoons V^{2+}$	-0.255	±1	400		4	11
		$VO^+ + 2H^+ + e^- \rightleftharpoons V^{2+} + H_2O$	0.126				1	12
		$V_2O_3 + 6H^+ + 2e^- \rightleftharpoons 2V^{2+} + 3H_2O$	0.161				1	13
		$VOH^{2+} + H^+ + e^- \rightleftharpoons V^{2+} + H_2O$	-0.082				1	14
19		$VO^{2+} + 2H^+ + e^- \rightleftharpoons V^{3+} + H_2O$	0.337	±1	-1000		4	15
		$HV_2O_5^- + 5H^+ + 2e^- \rightleftharpoons 2VO^+ + 3H_2O$	0.551				1	16
		$VO^{2+} + e^- \rightleftharpoons VO^+$	-0.044				1	17
		$VO_4^{3-} + 6H^+ + 2e^- \rightleftharpoons VO^+ + 3H_2O$	1.256				1	18
		$V_2O_4 + 4H^+ + 2e^- \rightleftharpoons 2VO^+ + 2H_2O$	0.246				1	19
		$H_2VO_4^- + 4H^+ + e^- \rightleftharpoons VO^{2+} + 3H_2O$	1.314				1	20
		$H_3V_2O_7^- + 7H^+ + 2e^- \rightleftharpoons 2VO^{2+} + 5H_2O$	1.096				1	21
3		$VO_2^+ + 2H^+ + e^- \rightleftharpoons VO^{2+} + H_2O$	0.991				5	22
3		$VO_3^- + 4H^+ + e^- \rightleftharpoons VO^{2+} + 2H_2O$	1.250				5	23
		$V_2O_5 + 6H^+ + 2e^- \rightleftharpoons 2VO^{2+} + 3H_2O$	0.958				1	24
		$V_2O_5 + 6H^+ + 2e^- \rightleftharpoons 2VO^{2+} + 3H_2O$	0.998				1	25
		$V(OH)_4^+ + 2H^+ + e^- \rightleftharpoons VO^{2+} + 3H_2O$	1.00				2	26
		$V_2O_3 + 2H^+ + 2e^- \rightleftharpoons V_2O_2 + H_2O$	-0.549				1	27
		$HV_2O_5^- + 3H^+ + 2e^- \rightleftharpoons V_2O_3 + 2H_2O$	0.515				1	28
		$2VO^{2+} + H_2O + 2e^- \rightleftharpoons V_2O_3 + 2H^+$	-0.080				1	29
		$2VO_4^{3-} + 10H^+ + 4e^- \rightleftharpoons V_2O_3 + 5H_2O$	1.238				1	30
		$V_2O_4 + 2H^+ + 2e^- \rightleftharpoons V_2O_3 + H_2O$	0.210				1	31
		$2HVO_4^{2-} + 6H^+ + 2e^- \rightleftharpoons V_2O_4 + 4H_2O$	1.586				1	32

Electronically conducting phase	Intermediate species	Composition of the solution	Solvent	Temperature °C	Pressure Torr	Measuring method
a	b	c	d		e	d
Ind.	V_2O_4		1	25	760	IV
Ind.	V_2O_4		1	25	760	IV
Ind.	V_2O_4 , V_2O_5 (evolved)		1	25	760	IV
Ind.	V_2O_4 , V_2O_5 (not evolved)		1	25	760	IV
Ind.			1	25	760	IV
Hg	V (acetylacetone)$_3$	KCl : 10 F	1	25		II

Comparison electrode	Liquid junction	Electrode reaction	Standard value V	Uncertainty mV	Temperature Coefficient μV/°C	Notes	Reference	Electrode Reference Number
d			f		g	h		
		$2H_2VO_4^- + 4H^+ + 2e^- \rightleftharpoons V_2O_4 + 4H_2O$	1.022				1	33
		$H_3V_2O_7^- + 3H^+ + 2e^- \rightleftharpoons V_2O_4 + 3H_2O$	0.806				1	34
		$V_2O_5 + 2H^+ + 2e^- \rightleftharpoons V_2O_4 + H_2O$	0.666				1	35
		$V_2O_5 + 2H^+ + 2e^- \rightleftharpoons V_2O_4 + H_2O$	0.708				1	36
		$VO^{2+} + H^+ + e^- \rightleftharpoons VOH^{2+}$	0.164				1	37
3		$V(\text{acetylacetone})_3 + e^- \rightleftharpoons V(\text{acetylacetone})_3^-$	1.0				6	38

Vanadium 1

Standard value V	Solvent	Electrode Reference Number
-1.175	1	6
-1.154	1	5
-0.820	1	8
-0.549	1	27
-0.255	1	11
-0.254	1	9
-0.25	1	7
-0.082	1	14
-0.080	1	29
-0.044	1	17
0.126	1	12
0.161	1	13
0.164	1	37
0.210	1	31
0.246	1	19
0.337	1	15
0.338	1	10
0.501	1	3
0.515	1	28
0.551	1	16
0.666	1	35
0.708	1	36
0.719	1	2
0.806	1	34
0.958	1	24
0.991	1	22
0.998	1	25
1.00	1	26
1.0	1	38
1.022	1	33
1.096	1	21
1.238	1	30
1.250	1	23
1.256	1	18
1.281	1	1
1.314	1	20
1.586	1	32
1.962	1	4

BIBLIOGRAPHY : Vanadium

1) Data reported by M. Pourbaix "Atlas d'Equilibres Electrochimiques à 25 °C" Gauthier-Villars Editeur, Paris, 1963), pp. 234 - 245

2) Data reported by A.J. De Bethune, N.A. Swendeman-Loud " Standard Aqueous Electrode Potentials and Temperature Coefficients at 25 °C" (Clifford A. Hampel, Ill.)

3) W.M. Latimer, "Oxidation Potentials" (Prentice Hall, N.J. 1952)

4) G. Jones, J.N. Colvin, J. Am. Chem. Soc., 66 (1944), 1563 - 1571

5) F.Z. Dzhabarov, S.V. Gorbachev, Zh. Neorgan. Khim., 9 (1964), 2399 - 2402

6) W.P. Schaefer, Inorg. Chem., 4 (1965), 624 - 648

Electronically conducting phase	Intermediate species	Composition of the solution	Solvent	Temperature °C	Pressure Torr	Measuring method
a	b	c	d		e	d
Nb			1	25	760	IV
Nb	NbO		1	25	760	IV
Nb	Nb_2O_5		1	25	760	IV
Ind.	NbO, NbO_2		1	25	760	IV
Ind.	NbO_2, Nb_2O_5		1	25	760	IV

Comparison electrode	Liquid junction	Electrode reaction	Standard value V	Uncertainty mV	Temperature Coefficient μV/°C	Notes	Reference	Electrode Reference Number
d			f		g	h		
		$Nb^{3+} + 3e^- \rightleftharpoons Nb$	-1.099				1	1
		$NbO + 2H^+ + 2e^- \rightleftharpoons Nb + H_2O$	-0.733				2	2
		$Nb_2O_5 + 10H^+ + 10e^- \rightleftharpoons 2Nb + 5H_2O$	-0.644		-390		1	3
		$NbO_2 + 2H^+ + 2e^- \rightleftharpoons NbO + H_2O$	-0.625				2	4
		$Nb_2O_5 + 2H^+ + 2e^- \rightleftharpoons 2NbO_2 + H_2O$	-0.289				2	5

Niobium 1

Standard value V	Solvent	Electrode Reference Number	Standard value V	Solvent	Electrode Reference Number	Standard value V	Solvent	Electrode Reference Number
-1.099	1	1						
-0.733	1	2						
-0.644	1	3						
-0.625	1	4						
-0.289	1	5						

BIBLIOGRAPHY: Niobium

1) Data reported by A.J. De Bethune, N.A. Swendeman-Loud "Standard Aqueous Electrode Potentials and Temperature Coefficients at 25 °C" (Clifford A. Hampel, Ill.)

2) Data reported by M. Pourbaix, "Atlas d'Equilibres Electrochimiques à 25 °C" (Gauthier-Villars Editeur, Paris, 1963), pp. 246 - 250

Electronically conducting phase	Intermediate species	Composition of the solution	Solvent	Temperature °C	Pressure Torr	Measuring method
a	b	c	d		e	d
Ta	Ta_2O_5		1	25	760	IV

Comparison electrode	Liquid junction	Electrode reaction	Standard value	Uncertainty	Temperature Coefficient	Notes	Reference	Electrode Reference Number
			V	mV	μV/°C			
d			f		g	h		
		$Ta_2O_5 + 10H^+ + 10e^- \rightleftharpoons 2Ta + 5H_2O$	-0.750		-377		1, 2	1

Tantalum 1

Standard value V	Solvent	Electrode Reference Number	Standard value V	Solvent	Electrode Reference Number	Standard value V	Solvent	Electrode Reference Number
-0.750	1	1						

BIBLIOGRAPHY: Tantalum

1) Data reported by M. Pourbaix "Atlas d'Equilibres Electrochimiques à 25 °C" (Gauthier-Villars Editeur, Paris, 1963), pp. 251 - 255

2) Data reported by A.J. De Bethune, N.A. Swendeman-Loud "Standard Aqueous Electrode Potentials and Temperature Coefficients at 25 °C" (Clifford A. Hampel, Ill.)

Group VI

Electronically conducting phase	Intermediate species	Composition of the solution	Solvent	Temperature °C	Pressure Torr	Measuring method
a	b	c	d		e	d
Ind.	O_2		1	25	760	IV
Ind.	O_2		1	25	760	IV
Ind.	H_2O_2		1	25	760	IV
Ind.	O		1	25	760	IV
Ind.	O_2, H_2O (gas)		1	25	760	IV
Pt	O_2	H_2SO_4 : 2 N	1	25		III
Ind.	O_3		1	25	760	IV
Ind.	OH (gas)		1	25	760	IV
Ind.	H_2O_2, HO_2 (in sol.)		1	25	760	IV
Hg (DME)	O_2, H_2O_2	O_2, H_2O_2: pH in the range of 6 - 13	1	25 ± 0.1		II
Ind.	O_2, O		1	25	760	IV
Ind.	O_2		1	25	760	IV
Ind.	O_3, O_2		1	25	760	IV
Ind.	O_2, O_3		1	25	760	IV
Ind.	OH (gas)		1	25	760	IV
Ind.	OH (in sol.)		1	25	760	IV
Ind.			1	25	760	IV
Ind.	OH, H_2O_2 (in sol.)		1	25	760	IV
Ind.			1	25	760	IV
Ind.	O_2		1	25	760	IV
Ind.	OH (gas)		1	25	760	IV

Comparison electrode	Liquid junction	Electrode reaction	Standard value V	Uncertainty mV	Temperature Coefficient μV/°C	Notes	Reference	Electrode Reference Number
d			f		g	h		
		$H_2O + O_2 + 2e^- \rightleftharpoons HO_2^- + OH^-$	-0.076				1	1
		$O_2 + H^+ + 2e^- \rightleftharpoons HO_2^-$	0.338				2	2
		$H_2O_2 + 2H^+ + 2e^- \rightleftharpoons 2H_2O$	1.776		-658		1	3
		$O + 2H^+ + 2e^- \rightleftharpoons H_2O$	2.421		-1148		2,1	4
		$O_2 + 4H^+ + 4e^- \rightleftharpoons 2H_2O$	1.185		-230		1	5
1		$O_2 + 4H^+ + 4e^- \rightleftharpoons 2H_2O$	1.229			1	3	6
		$O_3 + 6H^+ + 6e^- \rightleftharpoons 3H_2O$	1.511				2	7
		$OH + H^+ + e^- \rightleftharpoons H_2O$	2.85		-1855		1	8
		$HO_2 + H^+ + e^- \rightleftharpoons H_2O_2$	1.495				1	9
3	KNO_3 satd.	$O_2 + 2H^+ + 2e^- \rightleftharpoons H_2O_2$	0.695	± 5	-1033		4,1	10
		$O_2 + 2H^+ + 2e^- \rightleftharpoons O + H_2O$	0.037				2	11
		$O_2 + e^- \rightleftharpoons O_2^-$	-0.563				1	12
		$O_3 + 2H^+ + 2e^- \rightleftharpoons O_2 + H_2O$	2.076		-483		2,1	13
		$O_3 + H_2O + 2e^- \rightleftharpoons O_2 + 2OH^-$	1.24		-1318		1	14
		$HO_2^- + H_2O + e^- \rightleftharpoons OH + 2OH^-$	-0.262				1	15
		$HO_2^- + H_2O + e^- \rightleftharpoons OH + 2OH^-$	-0.245				1	16
		$HO_2^- + H_2O + 2e^- \rightleftharpoons 3OH^-$	0.878				1	17
		$H_2O_2 + H^+ + e^- \rightleftharpoons OH + H_2O$	0.71		540		1	18
		$O_2^- + H_2O + e^- \rightleftharpoons OH^- + HO_2^-$	0.413				1	19
		$O_2 + 2H_2O + 4e^- \rightleftharpoons 4OH^-$	0.401		-1680		1	20
		$OH + e^- \rightleftharpoons OH^-$	2.02		-2689		1	21

Oxygen 1

Standard value V	Solvent	Electrode Reference Number	Standard value V	Solvent	Electrode Reference Number	Standard value V	Solvent	Electrode Reference Number
-0.563	1	12						
-0.262	1	15						
-0.245	1	16						
-0.076	1	1						
0.037	1	11						
0.338	1	2						
0.401	1	20						
0.413	1	19						
0.695	1	10						
0.71	1	18						
0.878	1	17						
1.185	1	5						
1.229	1	6						
1.24	1	14						
1.495	1	9						
1.511	1	7						
1.776	1	3						
2.02	1	21						
2.076	1	13						
2.421	1	4						
2.85	1	8						

BIBLIOGRAPHY: Oxygen

1) Data reported by A.J. De Bethune, N.A. Swendeman-Loud "Standard Aqueous Electrode Potentials and Temperature Coefficients at 25 ^{o}C" (Clifford A. Hampel, Ill.)

2) Data reported by M. Pourbaix "Atlas d'Equilibres Electrochimiques à 25 ^{o}C" (Gauthier-Villars Editeur, Paris, 1963), pp. 540 - 544

3) J.P. Hoare, Electrochim. Acta, 11 (1966), 203 - 210

4) D.M.H. Kern, J. Am. Chem. Soc., 76 (1954), 4208 - 4214

Electronically conducting phase	Intermediate species	Composition of the solution	Solvent	Temperature °C	Pressure Torr	Measuring method
a	b	c	d		e	d
Ag	Ag_2S	H_2S : 1 atm.; HCl : 0.1 m	1	25± 0.05		I
S	Bi_2S_3		1	25	760	IV
S	CdS		1	25	760	IV
S	Ce_2S_3		1	25	760	IV
Ind.	$(CH_3)_2SO$ (in sol.) $(CH_3)_2SO_2$ (in sol.)		1	25	760	IV
S	CoS (α)		1	25	760	IV
S	CuS		1	25	760	IV
S	Cu_2S		1	25	760	IV
S	FeS		1	25	760	IV
Hg	HgS	H_2S : 1 Atm.; HCl : 0.1 m	1	25		I
S			1	25	760	IV
Ind.			1	25	760	IV
Ind.			1	25	760	IV
Ind.			1	25	760	IV
Ind.			1	25	760	IV
Ind.			1	25	760	IV
Ind.			1	25	760	IV
Ind.	H_2S (in sol.)		1	25	760	IV
S	H_2S (gas)		1	25	760	IV
S	H_2S (in sol.)		1	25	760	IV
Ind.	H_2S (gas)		1	25	760	IV
Ind.	H_2S (gas)		1	25	760	IV
Ind.	H_2S (in sol.)		1	25	760	IV
Ind.			1	25	760	IV
Ind.	H_2SO_3 (in sol.)		1	25	760	IV
Ind.	H_2SO_3 (in sol.)		1	25	760	IV
Ind.			1	25	760	IV
Ind.			1	25	760	IV
Ind.	H_2SO_3 (in sol.)		1	25	760	IV
Ind.	H_2SO_3 (in sol.)		1	25	760	IV
S	La_2S_3		1	25	760	IV
S	MnS		1	25	760	IV

Comparison electrode	Liquid junction	Electrode reaction	Standard value	Uncertainty	Temperature Coefficient	Notes	Reference	Electrode Reference Number
			V	mV	µV/°C			
d			f		g	h		
1		$2Ag^{+} + S + 2e^{-} \rightleftharpoons Ag_2S$	-1.99				1	1
		$2Bi^{3+} + 3S + 6e^{-} \rightleftharpoons Bi_2S_3$	-1.425				2	2
		$Cd^{2+} + S + 2e^{-} \rightleftharpoons CdS$	-1.305			1	2,3	3
		$2Ce^{3+} + 3S + 6e^{-} \rightleftharpoons Ce_2S_3$	-0.575				2	4
		$(CH_3)_2SO_2 + 2H^{+} + 2e^{-} \rightleftharpoons (CH_3)_2SO + H_2O$	0.23				4	5
		$Co^{2+} + S + 2e^{-} \rightleftharpoons CoS$	-1.091				2	6
		$Cu^{2+} + S + 2e^{-} \rightleftharpoons CuS$	-1.52				2	7
		$2Cu^{+} + S + 2e^{-} \rightleftharpoons Cu_2S$	-1.88				2	8
		$Fe^{2+} + S + 2e^{-} \rightleftharpoons FeS$	-0.992				2	9
1		$Hg^{2+} + S + 2e^{-} \rightleftharpoons HgS$	-2.00				5	10
		$S + H^{+} + 2e^{-} \rightleftharpoons HS^{-}$	-0.065				6	11
		$S_2^{2-} + 2H^{+} + 2e^{-} \rightleftharpoons 2HS^{-}$	0 298				6	12
		$S_3^{2-} + 3H^{+} + 4e^{-} \rightleftharpoons 3HS^{-}$	0.097				6	13
		$S_4^{2-} + 4H^{+} + 6e^{-} \rightleftharpoons 4HS^{-}$	0.033				6	14
		$S_5^{2-} + 5H^{+} + 8e^{-} \rightleftharpoons 5HS^{-}$	0.003				6	15
		$S_2O_3^{2-} + 8H^{+} + 8e^{-} \rightleftharpoons 2HS^{-} + 3H_2O$	0.200				6	16
		$SO_4^{2-} + 9H^{+} + 8e^{-} \rightleftharpoons HS^{-} + 4H_2O$	0.252				6	17
		$HSO_4^{-} + 9H^{+} + 8e^{-} \rightleftharpoons H_2S + 4H_2O$	0.289				6	18
		$S + 2H^{+} + 2e^{-} \rightleftharpoons H_2S$	0.171				6	19
		$S + 2H^{+} + 2e^{-} \rightleftharpoons H_2S$	0.142		-209		4,6	20
		$S_5^{2-} + 10H^{+} + 8e^{-} \rightleftharpoons 5H_2S$	0.299				6	21
		$SO_4^{2-} + 10H^{+} + 8e^{-} \rightleftharpoons H_2S + 4H_2O$	0.311				6	22
		$SO_4^{2-} + 10H^{+} + 8e^{-} \rightleftharpoons H_2S + 4H_2O$	0.303				6	23
		$S_2O_6^{2-} + 2H^{+} + 2e^{-} \rightleftharpoons 2HSO_3^{-}$	0.455				6	24
		$SO_4^{2-} + 4H^{+} + 2e^{-} \rightleftharpoons H_2SO_3 + H_2O$	0.172		810		4	25
		$S_2O_6^{2-} + 4H^{+} + 2e^{-} \rightleftharpoons 2H_2SO_3$	0.564		1100		4,6	26
		$S_2O_8^{2-} + 2H^{+} + 2e^{-} \rightleftharpoons 2HSO_4^{-}$	2.123				6	27
		$2HSO_3^{-} + 3H^{+} + 2e^{-} \rightleftharpoons HS_2O_4^{-} + 2H_2O$	0.060				6	28
		$2H_2SO_3 + H^{+} + 2e^{-} \rightleftharpoons HS_2O_4^{-} + 2H_2O$	-0.056				6	29
		$2H_2SO_3 + H^{+} + 2e^{-} \rightleftharpoons HS_2O_4^{-} + 2H_2O$	-0.082				4	30
		$2La^{3+} + 3S + 6e^{-} \rightleftharpoons La_2S_3$	-0.601				2	31
		$Mn^{2+} + S + 2e^{-} \rightleftharpoons MnS$	-0.801				2	32

Electronically conducting phase	Intermediate species	Composition of the solution	Solvent	Temperature °C	Pressure Torr	Measuring method
a	b	c	d		e	d
S	NiS (α)		1	25	760	IV
S	NiS (β)		1	25	760	IV
S	NiS (γ)		1	25	760	IV
S	PbS		1	25	760	IV, II
S	H_2SO_3 (in sol.)		1	25	760	IV
S			1	25	760	IV
S	S_2Cl_2 (in sol.)		1	25	760	IV
S	SO		1	25	760	IV
S	SO_2		1	25	760	IV
S			1	25	760	IV
Ag			1	25		I
S	BaS_2O_3 (in sol.)		1	25		V
Pt	BaS_2O_3 (s)	BaS_2O_3 : (0.21808 up to 0.64710) · 10^{-3} m	1	25		III
S	CaS_2O_3 (in sol.)	1) $CaCl_2$: (13.28 up to 25.25) · 10^{-3} m; BaS_2O_3 : (14.62 up to 16.69) · 10^{-3} m; 2) $CaCl_2$: (14.8 up to 29.6) · 10^{-3} m; $Na_2S_2O_3$: (2.012 and 2.018) · 10^{-3} m	1	25		V
S	CdS_2O_3 (in sol.)	1) $CdCl_2$: (4.68 up to 11.61) · 10^{-3} m; BaS_2O_3 : (16.52 up to 25.42) · 10^{-3} m; 2) $CdCl_2$: (1.393 up to 34.93) · 10^{-3} m; $Na_2S_2O_3$: (2.013 and 2.020) · 10^{-3} m	1	25		V
S	CoS_2O_3 (in sol.)	$CoCl_2$: (10.36 up to 22.83) · 10^{-3} m; BaS_2O_3 : (14.09 up to 16.47) · 10^{-3} m	1	25		V
Hg		$Na_2S_2O_3$: 3.94 · 10^{-3} up to 0.626 m; $HgNO_3$: 10^{-3} m and 2 · 10^{-3} m	1	25 ± 0.05		I
Hg		$Na_2S_2O_3$: 3.94 · 10^{-3} up to 0.626 m; $HgNO_3$: 10^{-3} m and 2 · 10^{-3} m	1	25 ± 0.05		I
Hg		$Na_2S_2O_3$: 3.94 · 10^{-3} up to 0.626 m; $HgNO_3$: 10^{-3} m and 2 · 10^{-3} m	1	25 ± 0.05		I
S		KCl : (30.23 up to 56.08) · 10^{-3} m; BaS_2O_3 : (14.01 up to 15.42) · 10^{-3} m	1	25		V
S	MgS_2O_3 (in sol.)	1) $MgCl_2$: (13.45 up to 20.54) · 10^{-3} m; BaS_2O_3 : (14.49 up to 15.66) · 10^{-3} m; 2) $MgCl_2$: (14.8 up to 29.6) · 10^{-3} m; $Na_2S_2O_3$: (2.012 and 2.018) · 10^{-3} m	1	25		V
S	MnS_2O_3 (in sol.)	$MnCl_2$: (10.64 up to 21.02) · 10^{-3} m; BaS_2O_3 : (14.03 up to 15.98) · 10^{-3} m	1	25		V
S		NaCl : (45.76 up to 72.25) · 10^{-3} m ; BaS_2O_3 : (14.60 up to 15.99) · 10^{-3} m	1	25		V

Comparison electrode	Liquid junction	Electrode reaction	Standard value V	Uncertainty mV	Temperature Coefficient μV/°C	Notes	Reference	Electrode Reference Number
d			f		g	h		
		$Ni^{2+} + S + 2e^- \rightleftharpoons NiS$	-1.025				7	33
		$Ni^{2+} + S + 2e^- \rightleftharpoons NiS$	-1.186				7	34
		$Ni^{2+} + S + 2e^- \rightleftharpoons NiS$	-1.296				7	35
		$Pb^{2+} + S + 2e^- \rightleftharpoons PbS$	-1.296			1	2,3	36
		$H_2SO_3 + 4H^+ + 4e^- \rightleftharpoons S + 3H_2O$	0.449		-660		4,6	37
		$HSO_4^- + 7H^+ + 6e^- \rightleftharpoons S + 4H_2O$	0.339				6	38
		$S_2Cl_2 + 2e^- \rightleftharpoons 2S + 2Cl^-$	1.23		-640		4	39
		$SO + 2H^+ + 2e^- \rightleftharpoons S + H_2O$	1.507				6	40
		$SO_2 + 4H^+ + 4e^- \rightleftharpoons S + 2H_2O$	0.451				6	41
		$S_2O_3^{2-} + 6H^+ + 4e^- \rightleftharpoons 2S + 3H_2O$	0.465				6	42
3		$Ag(S_2O_3)_2^{3-} + 12H^+ + 8e^- \rightleftharpoons Ag^+ + 4S + 6H_2O$	0.365				8	43
		$BaS_2O_3 + 6H^+ + 4e^- \rightleftharpoons Ba^{2+} + 2S + 3H_2O$	0.432		-400	2	9	44
		$BaS_2O_3 + 6H^+ + 4e^- \rightleftharpoons Ba^{2+} + 2S + 3H_2O$	0.394			3	10	45
		$CaS_2O_3 + 6H^+ + 4e^- \rightleftharpoons Ca^{2+} + 2S + 3H_2O$	0.436			2 4 1	10 11 9	46
		$CdS_2O_3 + 6H^+ + 4e^- \rightleftharpoons Cd^{2+} + 2S + 3H_2O$	0.407			2 4 1	10 11	47
		$CoS_2O_3 + 6H^+ + 4e^- \rightleftharpoons Co^{2+} + 2S + 3H_2O$	0.435			2	10	48
3		$Hg(S_2O_3)_2^{2-} + 12H^+ + 8e^- \rightleftharpoons Hg^{2+} + 4S + 6H_2O$	0.245			5	12	49
3		$Hg(S_2O_3)_3^{4-} + 18H^+ + 12e^- \rightleftharpoons Hg^{2+} + 6S + 9H_2O$	0.306			5	12	50
3		$Hg(S_2O_3)_4^{6-} + 24H^+ + 16e^- \rightleftharpoons Hg^{2+} + 8S + 12H_2O$	0.341			5	12	51
		$KS_2O_3 + 6H^+ + 4e^- \rightleftharpoons K^+ + 2S + 3H_2O$	0.452			2	10	52
		$MgS_2O_3 + 6H^+ + 4e^- \rightleftharpoons Mg^{2+} + 2S + 3H_2O$	0.438			2 4 1	10 11	53
		$MnS_2O_3 + 6H^+ + 4e^- \rightleftharpoons Mn^{2+} + 2S + 3H_2O$	0.438			2	10	54
		$NaS_2O_3^- + 6H^+ + 4e^- \rightleftharpoons Na^+ + 2S + 3H_2O$	0.455			2	10	55

Electronically conducting phase	Intermediate species	Composition of the solution	Solvent	Temperature °C	Pressure Torr	Measuring method
a	b	c	d		e	d
S	NiS_2O_3 (in sol.)	$NiCl_2$: (10.48 up to 23.59) · 10^{-3} m; BaS_2O_3 : (14.12 up to 16.67) · 10^{-3} m	1	25		V
S	SrS_2O_3 (in sol.)	$SrCl_2$: (8.68 up to 17.40) · 10^{-3} m; BaS_2O_3 : (13.72 up to 15.53) · 10^{-3} m	1	25		V
S	ZnS_2O_3 (in sol.)	1) $ZnCl_2$: (7.53 up to 14.92) · 10^{-3} m; BaS_2O_3 : (13.83 up to 15.88) · 10^{-3} m 2) $ZnCl_2$: (17.14 up to 47.60) · 10^{-3} m; $Na_2S_2O_3$: 2.514 · 10^{-3} m	1	25		V
S			1	25	760	IV
Pt		$Ce_2(SO_4)_3$: up to 0.5302 N	1	25 ± 0.02		III
Pt		$Er_2(SO_4)_3$: up to 1.4536 N	1	25 ± 0.02		III
Pt		$Gd_2(SO\ )_3$: up to 0.3272 N	1	25± 0.02		III
S	Hg_2SO_4 (s)	Hg_2SO_4 : (0.9091 up to 1.1120) · 10^{-3} m	1	25 ± 0.02		V
Pt		$Ho_2(SO_4)_3$: up to 0.6509 N	1	25 ± 0.02		III
Ind.		K_2SO_4 : (0.11338 up to 1.1309) · 10^{-3} m	1	25		III
Pt		1) $La_2(SO_4)_3$: (0.39428 up to 1.8202) · 10^{-3} N 2) $La_2(SO_4)_3$: up to 0.1021 N	1	25 ± 0.02		III
1)Glass electrode 2)Pt, H_2	$MgSO_4$ (in sol.)	1) $MgSO_4$: (0.16196 up to 1.6759) · 10^{-3} m 2) $MgSO_4$: (3.795 up to 6.993) · 10^{-3} m; HCl : (3.294 up to 7.666) · 10^{-3} m	1	25		III, I
Ind.	$MnSO_4$ (in sol.)		1	25		III
Ind.		Na_2SO_4 : (0.10384 up to 1.1901) · 10^{-3} N	1	25		III
Pt		$Nd_2(SO_4)_3$: up to 0.4807	1	25± 0.02		III
Pt		$Pr_2(SO_4)_3$: up to 1.1477 N	1	25 ± 0.02		III
Pt		$Sm_2(SO_4)_3$: up to 0.2575 N	1	25 ± 0.02		III
Ind.	UO_2SO_4 (in sol.)	$UO_2(ClO_4)_2$: (2.916 up to 12.73) · 10^{-3} m; $HClO_4$: (6.08 up to 18.35) · 10^{-3}m; Li_2SO_4 : (1.028 up to 5.006) · 10^{-3} m	1	25		III
Pt		$Y_2(SO_4)_3$: up to 1.2351 N	1	25 ± 0.02		III
Pt		$Yb_2(SO_4)_3$: up to 3.3345 N	1	25 ± 0.02		III
S			1	25	760	IV
S			1	25	760	IV
Pt	S	H_2S : 0.093 - 0.477 M	1	25		I
Pt		H_2S : 0.093 - 0.477 M	1	25		I
Ind.			1	25	760	IV

Comparison electrode	Liquid junction	Electrode reaction	Standard value V	Uncertainty mV	Temperature Coefficient μV/°C	Notes	Reference	Electrode Reference Number
d			f		g	h		
		$NiS_2O_3 + 6H^+ + 4e^- \rightleftharpoons Ni^{2+} + 2S + 3H_2O$	0.434			2	10	56
		$SrS_2O_3 + 6H^+ + 4e^- \rightleftharpoons Sr^{2+} + 2S + 3H_2O$	0.435			2	10	57
		$ZnS_2O_3 + 6H^+ + 4e^- \rightleftharpoons Zn^{2+} + 2S + 3H_2O$	0.430			2	10	58
						4		
						1	11	
		$SO_4^{2-} + 8H^+ + 6e^- \rightleftharpoons S + 4H_2O$	0.3572		-168		4	59
		$CeSO_4^+ + 8H^+ + 6e^- \rightleftharpoons Ce^{3+} + S + 4H_2O$	0.322			3	13	60
		$ErSO_4^+ + 8H^+ + 6e^- \rightleftharpoons Er^{3+} + S + 4H_2O$	0.322			3	13	61
		$GdSO_4^+ + 8H^+ + 6e^- \rightleftharpoons Gd^{3+} + 4H_2O$	0.321			3	13	62
		$Hg_2SO_4 + 8H^+ + 6e^- \rightleftharpoons Hg_2^{2+} + S + 4H_2O$	0.297			2	14	63
		$HoSO_4^+ + 8H^+ + 6e^- \rightleftharpoons Ho^{3+} + S + 4H_2O$	0.322			3	13	64
		$KSO_4^- + 8H^+ + 6e^- \rightleftharpoons K^+ + S + 4H_2O$	0.348			3	15	65
		$LaSO_4^+ + 8H^+ + 6e^- \rightleftharpoons La^{3+} + S + 4H_2O$	0.322			3	15	66
						1	18	
3		$MgSO_4 + 8H^+ + 6e^- \rightleftharpoons Mg^{2+} + S + 4H_2O$	0.338		-800	1	16	67
4						3	17	
						5		
		$MnSO_4 + 8H^+ + 6e^- \rightleftharpoons Mn^{2+} + S + 4H_2O$	0.346			3	18	68
		$NaSO_4^- + 8H^+ + 6e^- \rightleftharpoons Na^+ + S + 4H_2O$	0.350			3	19	69
		$NdSO_4^+ + 8H^+ + 6e^- \rightleftharpoons Nd^{3+} + S + 4H_2O$	0.321			3	13	70
		$PrSO_4^+ + 8H^+ + 6e^- \rightleftharpoons Pr^{3+} + S + 4H_2O$	0.322			3	13	71
		$SmSO_4^+ + 8H^+ + 6e^- \rightleftharpoons Sm^{3+} + S + 4H_2O$	0.321			3	13	72
		$UO_2SO_4 + 8H^+ + 6e^- \rightleftharpoons UO_2^{2+} + S + 4H_2O$	0.328			4	19	73
		$YSO_4^+ + 8H^+ + 6e^- \rightleftharpoons Y^{3+} + S + 4H_2O$	0.323			3	13	74
		$YbSO_4^+ + 8H^+ + 6e^- \rightleftharpoons Yb^{3+} + S + 4H_2O$	0.322			3	13	75
		$S_4O_6^{2-} + 12H^+ + 10e^- \rightleftharpoons 4S + 6H_2O$	0.416				6	76
		$S_5O_6^{2-} + 12H^+ + 10e^- \rightleftharpoons 5S + 6H_2O$	0.484				6	77
3	KCl satd.	$S + 2e^- \rightleftharpoons S^{2-}$	-0.47627		-980		4, 20	78
3	KCl satd.	$S_2^{2-} + 2e^- \rightleftharpoons 2S^{2-}$	-0.5242		-1080		20	79
		$SO_3^{2-} + 6H^+ + 6e^- \rightleftharpoons S^{2-} + 3H_2O$	0.231				6	80

Electronically conducting phase	Intermediate species	Composition of the solution	Solvent	Temperature °C	Pressure Torr	Measuring method
a	b	c	d		e	d
Ind.			1	25	760	IV
Ind.			1	25	760	IV
Pt	S	H_2S : 0.093 - 0.477 M	1	25		I
Pt		H_2S : 0.093 - 0.477 M	1	25		I
Pt	S	H_2S : 0.093 - 0.477 M	1	25		I
Pt		H_2S : 0.093 - 0.477 M	1	25		I
Pt	S	H_2S : 0.093 - 0.477 M	1	25		I
Pt		H_2S : 0.093 - 0.477 M	1	25		I
Pt	S	H_2S : 0.093 - 0.477 M	1	25		I
Ind.			1	25	760	IV
S	Sb_2S_3		1	25	760	IV
Ind.			1	25	760	IV
Ind.			1	25	760	IV
Ind.			1	25	760	IV
Ind.	H_2SO_3 (in sol.)		1	25	760	IV
Ind.			1	25	760	IV
Ind.			1	25	760	IV
Ind.			1	25	760	IV
Ind.			1	25	760	IV
Ind.			1	25	760	IV
Ind.			1	25	760	IV
Ind.			1	25	760	IV
Ind.			1	25	760	IV
Ind.			1	25	760	IV
Ind.	H_2SO_3 (in sol.)		1	25	760	IV
Ind.	SO_2		1	25	760	IV
S	Tl_2S		1	25	760	IV
S	ZnS (α)		1	25	760	IV

Comparison electrode	Liquid junction	Electrode reaction	Standard value V	Uncertainty mV	Temperature Coefficient μV/°C	Notes	Reference	Electrode Reference Number
d			f		g	h		
		$S_2O_3^{2-} + 6H^+ + 8e^- \rightleftharpoons 2S^{2-} + 3H_2O$	-0.006				6	81
		$SO_4^{2-} + 8H^+ + 8e^- \rightleftharpoons S^{2-} + 4H_2O$	0.149				6	82
3	KCl satd.	$2S + 2e^- \rightleftharpoons S_2^{2-}$	-0.42836		-888		20	83
3	KCl satd.	$2S_3^{2-} + 2e^- \rightleftharpoons 3S_2^{2-}$	-0.5056		-1040		20	84
3	KCl satd.	$3S + 2e^- \rightleftharpoons S_3^{2-}$	-0.38973		-811		20	85
3	KCl satd.	$3S_4^{2-} + 2e^- \rightleftharpoons 4S_3^{2-}$	-0.4778		-1000		20	86
3	KCl satd.	$4S + 2e^- \rightleftharpoons S_4^{2-}$	-0.36039		-750		20	87
3	KCl satd.	$4S_5^{2-} + 2e^- \rightleftharpoons 5S_4^{2-}$	-0.4407		-920		20	88
3	KCl satd.	$5S + 2e^- \rightleftharpoons S_5^{2-}$	-0.34030		-700			89
		$5S_2O_3^{2-} + 30H^+ + 24e^- \rightleftharpoons 2S_5^{2-} + 15H_2O$	0.331				6	90
		$2Sb^{3+} + 3S + 6e^- \rightleftharpoons Sb_2S_3$	-1.391				21	91
		$SO_4^{2-} + H_2O + 2e^- \rightleftharpoons SO_3^{2-} + 2OH^-$	-0.93		-1389		4	92
		$S_2O_6^{2-} + 2e^- \rightleftharpoons 2SO_3^{2-}$	0.026				6	93
		$2HSO_3^- + 4H^+ + 4e^- \rightleftharpoons S_2O_3^{2-} + 3H_2O$	0.491				6	94
		$2H_2SO_3 + 2H^+ + 4e^- \rightleftharpoons S_2O_3^{2-} + 3H_2O$	0.400		-1260		4	95
		$2SO_3^{2-} + 6H^+ + 4e^- \rightleftharpoons S_2O_3^{2-} + 3H_2O$	0.705				6	96
		$2SO_3^{2-} + 3H_2O + 4e^- \rightleftharpoons S_2O_3^{2-} + 6OH^-$	-0.571		-1146		4	97
		$S_4O_6^{2-} + 2e^- \rightleftharpoons 2S_2O_3^{2-}$	0.08		-1110		4	98
		$S_2O_8^{2-} + 2e^- \rightleftharpoons 2SO_4^{2-}$	2.010		-1260		4,6	99
		$2HSO_3^- + 2H^+ + 2e^- \rightleftharpoons S_2O_4^{2-} + 2H_2O$	-0.013				6	100
		$2SO_3^{2-} + 2H_2O + 2e^- \rightleftharpoons S_2O_4^{2-} + 4OH^-$	-1.12		-710		4	101
		$2SO_3^{2-} + 4H^+ + 2e^- \rightleftharpoons S_2O_4^{2-} + 2H_2O$	0.416				6	102
		$2SO_4^{2-} + 4H^+ + 2e^- \rightleftharpoons S_2O_6^{2-} + 2H_2O$	-0.22		520		4	103
		$4HSO_3^- + 8H^+ + 6e^- \rightleftharpoons S_4O_6^{2-} + 6H_2O$	0.581				6	104
		$4H_2SO_3 + 4H^+ + 6e^- \rightleftharpoons S_4O_6^{2-} + 6H_2O$	0.509		-1310		4,6	105
		$4SO_2 + 4H^+ + 6e^- \rightleftharpoons S_4O_6^{2-} + 2H_2O$	0.510				6	106
		$2Tl^+ + S + 2e^- \rightleftharpoons Tl_2S$	-1.042				2	107
		$Zn^{2+} + S + 2e^- \rightleftharpoons ZnS$	-1.191				2	108

Sulfur 1

Standard value V	Solvent	Electrode Reference Number
-2.00	1	10
-1.99	1	1
-1.88	1	8
-1.52	1	7
-1.425	1	2
-1.391	1	91
-1.305	1	3
-1.296	1	35
-1.296	1	36
-1.191	1	108
-1.186	1	34
-1.12	1	101
-1.091	1	6
-1.042	1	107
-1.025	1	33
-0.992	1	9
-0.93	1	92
-0.801	1	32
-0.601	1	31
-0.575	1	4
-0.571	1	97
-0.5242	1	79
-0.5056	1	84
-0.4778	1	86
-0.47627	1	78
-0.4407	1	88
-0.42836	1	83
-0.38973	1	85
-0.36039	1	87
-0.34030	1	89
-0.22	1	103
-0.082	1	30
-0.065	1	11
-0.056	1	29
-0.013	1	100
-0.006	1	81
0.003	1	15
0.026	1	93
0.033	1	14
0.060	1	28
0.08	1	98
0.097	1	13
0.142	1	20
0.149	1	82
0.171	1	19
0.172	1	25
0.200	1	16
0.23	1	5
0.231	1	80
0.245	1	49
0.252	1	17
0.289	1	18
0.297	1	63
0.298	1	12
0.299	1	21
0.303	1	23
0.306	1	50
0.311	1	27
0.321	1	70
0.321	1	72
0.321	1	62
0.322	1	66
0.322	1	60
0.322	1	71
0.322	1	64
0.322	1	61
0.322	1	75
0.323	1	74
0.328	1	73
0.331	1	90
0.338	1	67
0.339	1	38
0.341	1	51
0.346	1	68
0.348	1	65
0.350	1	69
0.3572	1	59
0.365	1	43
0.394	1	45
0.400	1	95
0.407	1	47
0.416	1	76
0.416	1	102
0.430	1	58
0.432	1	44
0.434	1	56
0.435	1	57
0.435	1	48
0.436	1	46
0.438	1	53
0.438	1	54
0.449	1	37
0.451	1	41
0.452	1	52
0.455	1	24
0.455	1	55
0.465	1	42
0.484	1	77
0.491	1	94
0.509	1	105
0.510	1	106
0.564	1	26
0.581	1	104
0.705	1	96
1.23	1	39

Sulfur 2

Standard value V	Solvent	Electrode Reference Number	Standard value V	Solvent	Electrode Reference Number	Standard value V	Solvent	Electrode Reference Number
1.507	1	40						
2.010	1	99						
2.123	1	27						

NOTES: Sulfur

1) The value is an average of those given by several authors, they all are in the same range of uncertainty

2) Solubility data

3) Conductometric data

4) Spectroscopic data

5) Potentiometric data

BIBLIOGRAPHY: Sulfur

1) J.R. Goates, A.G. Cole, E.L. Gray, N.D. Faux, J. Am. Chem. Soc., 73 (1951), 707 - 708

2) F.D. Rossini, Nat. Bur. Stand., Circular 500, 1952

3) P. Kivalo, A. Ringbom, Suomen Kem., 29 B (1956), 109

4) Data reported by A.J. De Bethune, N.A. Swendeman-Loud "Standard Aqueous Electrode Potentials and Temperature Coefficients at 25 °C" (Clifford A. Hampel, Ill.)

5) J.R. Goates, A.G. Cole, E.L. Gray, J. Am. Chem. Soc., 73 (1951), 3596 - 3597

6) Data reported by M. Pourbaix "Atlas d'Equilibres Electrochimiques à 25 °C" (Gauthier-Villars Editeur, Paris, 1963), pp. 545 - 553

7) A. Ringbom, "Solubilities of Sulfides", Report to Analytical Section, I.U.P.A.C. July 1953

8) H. Chateau, B. Hervier, J. Pouradier, J. Phys. Chem., 61 (1957), 250 - 251; J. Chim. Phys., 54 (1957), 246

9) C.W. Davies, P.A.H. Wyatt, Trans. Faraday Soc., 45 (1949), 770

10) T.O. Denney, C.B. Monk, Trans. Faraday Soc., 47 (1951), 992 - 998

11) F.G.R. Gimblett, C.B. Monk, Trans. Faraday Soc., 51 (1955), 793 - 802

12) V.F. Toropova, Zh. Obshch. Khim., 24 (1954), 423 - 427

13) F.H. Spedding, S. Jaffe, J. Am. Chem. Soc., 76 (1954), 882 - 884

14) S.A. Brown, J.E. Land, J. Am. Chem. Soc., 79 (1957), 3015 - 3016

15) I.L. Jenkins, C.B. Monk, J. Am. Chem. Soc., 72 (1950), 2695 - 2698

16) H.S. Dunsmore, J.C. James, J. Chem. Soc., 1951, 2925 - 2930

17) H.W. Jones, C.B. Monk, Trans. Faraday Soc., 48 (1952), 929 - 933

18) J.C. James, Thesis, London, 1947

19) E.W. Davies, C.B. Monk, Trans. Faraday Soc., 53 (1957), 442 - 449

20) G. Maronny, G. Valensi, Proc. Intern. Comm. Electrochem. Thermodynam. and Kinet., 9th Meeting, 1957, 155 - 165

21) R. Akeret, Diss. Eidg. Techn. Hochschule, Zürich, 1953

Electronically conducting phase	Intermediate species	Composition of the solution	Solvent	Temperature °C	Pressure Torr	Measuring method
a	b	c	d		e	d
Ind.			1	25	760	IV
Se			1	25	760	IV
Ind.			1	25	760	IV
Ind.	H_2Se (in sol.)		1	25	760	IV
Ind.	H_2Se (in sol.), H_2SeO_3		1	25	760	IV
Se	H_2Se (gas)		1	25	760	IV
Se	H_2Se (in sol.)		1	25	760	IV
Ind.			1	25	760	IV
Ind.	H_2SeO_3		1	25	760	IV
Ind.	H_2SeO_3		1	25	760	IV
Se			1	25	760	IV
Se-Pt	H_2SeO_3	H_2SeO_3 : 0.003 up to 0.56 m	1	25 ± 1		I
Se			1	25	760	IV
Se			1	25	760	IV
Se			1	25	760	IV
Ind.			1	25	760	IV
Ind.			1	25	760	IV
Ind.			1	25	760	IV

Comparison electrode	Liquid junction	Electrode reaction	Standard value V	Uncertainty mV	Temperature Coefficient μV/°C	Notes	Reference	Electrode Reference Number
d			f		g	h		
		$HSeO_3^- + 6H^+ + 6e^- \rightleftharpoons HSe^- + 3H_2O$	0.349				1	1
		$Se + H^+ + 2e^- \rightleftharpoons HSe^-$	-0.510				1	2
		$SeO_3^{2-} + 7H^+ + 6e^- \rightleftharpoons HSe^- + 3H_2O$	0.414				1	3
		$HSeO_3^- + 7H^+ + 6e^- \rightleftharpoons H_2Se + 3H_2O$	0.386				1	4
		$H_2SeO_3 + 6H^+ + 6e^- \rightleftharpoons H_2Se + 3H_2O$	0.360				1	5
		$Se + 2H^+ + 2e^- \rightleftharpoons H_2Se$	-0.369				1	6
		$Se + 2H^+ + 2e^- \rightleftharpoons H_2Se$	-0.399		-28		3,1	7
		$SeO_4^{2-} + 3H^+ + 2e^- \rightleftharpoons HSeO_3^- + H_2O$	1.075				1	8
		$HSeO_4^- + 3H^+ + 2e^- \rightleftharpoons H_2SeO_3 + H_2O$	1.090				1	9
		$SeO_4^{2-} + 4H^+ + 2e^- \rightleftharpoons H_2SeO_3 + H_2O$	1.151		553		3,1	10
		$HSeO_3^- + 5H^+ + 4e^- \rightleftharpoons Se + 3H_2O$	0.778				1	11
3	KCl satd.	$H_2SeO_3 + 4H^+ + 4e^- \rightleftharpoons Se + 3H_2O$	-0.74		-520		3,2	12
		$Se + 2e^- \rightleftharpoons Se^{2-}$	-0.924		-890		3,1	13
		$SeO_3^{2-} + 6H^+ + 4e^- \rightleftharpoons Se + 3H_2O$	0.875				1	14
		$SeO_3^{2-} + 3H_2O + 4e^- \rightleftharpoons Se + 3OH^-$	-0.366		-1318		3	15
		$SeO_3^{2-} + 6H^+ + 6e^- \rightleftharpoons Se^{2-} + 3H_2O$	0.276				1	16
		$SeO_4^{2-} + 2H^+ + 2e^- \rightleftharpoons SeO_3^{2-} + H_2O$	0.880				1	17
		$SeO_4^{2-} + H_2O + 2e^- \rightleftharpoons SeO_3^{2-} + 2OH^-$	0.05		-1187		3	18

Selenium 1

Standard value V	Solvent	Electrode Reference Number	Standard value V	Solvent	Electrode Reference Number	Standard value V	Solvent	Electrode Reference Number
-0.924	1	13						
-0.74	1	12						
-0.510	1	2						
-0.399	1	7						
-0.369	1	6						
-0.366	1	15						
0.05	1	18						
0.276	1	16						
0.349	1	1						
0.360	1	5						
0.386	1	4						
0.414	1	3						
0.778	1	11						
0.875	1	14						
0.880	1	17						
1.075	1	8						
1.090	1	9						
1.151	1	10						

BIBLIOGRAPHY: Selenium

1) Data reported by M. Pourbaix "Atlas d'Equilibres Electrochimiques à 25 °C" (Gauthier-Villars Editeur, Paris, 1963), pp. 554 - 559

2) Sh. D. Osman-Zade, A.T. Vagramyan, Elektrokhim., 3 (1966), 85 - 87

3) Data reported by A.J. De Bethune, N.A. Swendeman-Loud "Standard Aqueous Electrode Potentials and Temperature Coefficients at 25 °C" (Clifford A. Hampel, Ill.)

Electronically conducting phase	Intermediate species	Composition of the solution	Solvent	Temperature °C	Pressure Torr	Measuring method
a	b	c	d		e	d
Ind.			1	25	760	IV
Te	H_2Te (gas)		1	25	760	IV
Te	H_2Te (gas)	HCl : 0.003 up to 0.02 N; $LaCl_3$ excess	1	30 ± 0.5		III
Ind.	H_2Te (gas)		1	25	760	IV
Ind.	H_2Te (in sol.)		1	25	760	IV
Te	H_2Te (in sol.)		1	25	760	IV
Ind.	H_2Te (in sol.)		1	25	760	IV
Ind.	H_2Te (in sol.)		1	25	760	IV
Te	H_2Te_2 (gas)	HCl: 0.003 up to 0.02 N; $LaCl_3$ excess	1	30 ± 0.5		III
Ind.	H_2TeO_4					
Ind.			1	25	760	IV
Ind.	H_2TeO_4		1	25	760	IV
Te			1	25	760	IV
Te			1	25	760	IV
Te			1	25	760	IV
Te	TeO_2		1	25	760	IV
Te	$TeO_2 \cdot H_2O$	Citric acid + Na_2HPO_4 buffers; pH: 2.2 up to 8.0	1	25± 0.1		I
Te			1	25	760	IV
Te			1	25	760	IV
Te			1	25	760	IV
Te		Na_2Te_2 : 0.0152 up to 0.240 M; NaOH : 0.15, 0.75, 1.5, 3 and 5 N	1	25 ± 0.5		III
Ind.			1	25	760	IV
Ind.			1	25	760	IV
Ind.			1	25	760	IV
Te		pH in the range of 9 - 13	1	25		I
Ind.			1	25	760	IV
Ind.			1	25	760	IV
Ind.	H_2TeO_4		1	25	760	IV
Ind.	TeO_2		1	25	760	IV
Ind.	$TeO_2 \cdot H_2O$		1	25	760	IV
Ind.	TeO_2, H_2TeO_4		1	25	760	IV
Ind.	$TeO_2 \cdot H_2O$, H_2TeO_4		1	25	760	IV

Comparison electrode	Liquid junction	Electrode reaction	Standard value	Uncertainty	Temperature Coefficient	Notes	Reference	Electrode Reference Number
			V	mV	μV/°C			
d			f		g	h		
		$Te_2^{2-} + 2H^+ + 2e^- \rightleftharpoons 2HTe^-$	-0.795				1	1
		$Te + 2H^+ + 2e^- \rightleftharpoons H_2Te$	-0.717		280		1,4	2
3	KCl	$Te_2 + 4H^+ + 4e^- \rightleftharpoons 2H_2Te$	-0.50				2	3
		$Te_2^{2-} + 4H^+ + 2e^- \rightleftharpoons 2H_2Te$	-0.595				1	4
		$HTeO_2^+ + 5H^+ + 6e^- \rightleftharpoons H_2Te + 2H_2O$	0.121				1	5
		$Te + 2H^+ + 2e^- \rightleftharpoons H_2Te$	-0.739				1	6
		$Te_2^{2-} + 4H^+ + 2e^- \rightleftharpoons 2H_2Te$	-0.638				1	7
		$Te^{4+} + 2H^+ + 6e^- \rightleftharpoons H_2Te$	0.132				1	8
3	KCl	$Te_2 + 2H^+ + 2e^- \rightleftharpoons H_2Te_2$	-0.365				2	9
		$H_2TeO_4 + 3H^+ + 2e^- \rightleftharpoons HTeO_2^+ + 2H_2O$	0.953				1	10
		$HTeO_4^- + 2H^+ + 2e^- \rightleftharpoons HTeO_3^- + H_2O$	0.813				1	11
		$H_2TeO_4 + H^+ + 2e^- \rightleftharpoons HTeO_3^- + H_2O$	0.631				1	12
		$HTeO_2^+ + 3H^+ + 4e^- \rightleftharpoons Te + 2H_2O$	0.551				1	13
		$HTeO_3^- + 5H^+ + 4e^- \rightleftharpoons Te + 3H_2O$	0.713				1	14
		$Te^{4+} + 4e^- \rightleftharpoons Te$	0.568				1	15
		$TeO_2 + 4H^+ + 4e^- \rightleftharpoons Te + 2H_2O$	0.521		-370		1,4	16
3		$TeO_2 \cdot H_2O + 4H^+ + 4e^- \rightleftharpoons Te + 3H_2O$	0.593				3	17
		$TeO_3^{2-} + 6H^+ + 4e^- \rightleftharpoons Te + 3H_2O$	0.827				1	18
		$TeO_3^{2-} + 3H_2O + 4e^- \rightleftharpoons Te + 6OH^-$	-0.57		-1230		4	19
		$Te + 2e^- \rightleftharpoons Te^{2-}$	-1.143				4	20
		$2Te + 2e^- \rightleftharpoons Te_2^{2-}$	-0.790				6	21
		$Te_2^{2-} + 2e^- \rightleftharpoons 2Te^{2-}$	-1.445				1	22
		$2HTeO_2^+ + 6H^+ + 10e^- \rightleftharpoons Te_2^{2-} + 4H_2O$	0.273				1	23
		$2HTeO_3^- + 10H^+ + 10e^- \rightleftharpoons Te_2^{2-} + 6H_2O$	0.402				1	24
3	KCl satd.	$2Te + 2e^- \rightleftharpoons Te_2^{2-}$	-0.845				5	25
		$2Te^{4+} + 10e^- \rightleftharpoons Te_2^{2-}$	0.286				1	26
		$2TeO_3^{2-} + 12H^+ + 10e^- \rightleftharpoons Te_2^{2-} + 6H_2O$	0.493				1	27
		$H_2TeO_4 + 6H^+ + 2e^- \rightleftharpoons Te^{4+} + 4H_2O$	0.920				1	28
		$HTeO_4^- + 3H^+ + 2e^- \rightleftharpoons TeO_2 + 2H_2O$	1.202				1	29
		$HTeO_4^- + 3H^+ + 2e^- \rightleftharpoons TeO_2 \cdot H_2O + H_2O$	1.036				1	30
		$H_2TeO_4 + 2H^+ + 2e^- \rightleftharpoons TeO_2 + 2H_2O$	1.020		130		1,4	31
		$H_2TeO_4 + 2H^+ + 2e^- \rightleftharpoons TeO_2 \cdot H_2O + H_2O$	0.854				1	32

Tellurium 2

Electronically conducting phase	Intermediate species	Composition of the solution	Solvent	Temperature °C	Pressure Torr	Measuring method
a	b	c	d		e	d
Ind.	TeO_2, TeO_3		1	25	760	IV
Ind.	$TeO_2 \cdot H_2O$, TeO_3		1	25	760	IV
Ind.	TeO_2		1	25	760	IV
Ind.	$TeO_2 \cdot H_2O$		1	25	760	IV
Ind.			1	25	760	IV
Ind.			1	25	760	IV
Ind.			1	25	760	IV

Comparison electrode	Liquid junction	Electrode reaction	Standard value V	Uncertainty mV	Temperature Coefficient μV/°C	Notes	Reference	Electrode Reference Number
d			f		g	h		
		$TeO_3 + 2H^+ + 2e^- \rightleftharpoons TeO_2 + H_2O$	1.020				1	33
		$TeO_3 + 2H^+ + 2e^- \rightleftharpoons TeO_2 \cdot H_2O$	0.850				1	34
		$TeO_4^{2-} + 4H^+ + 2e^- \rightleftharpoons TeO_2 + 2H_2O$	1.509				1	35
		$TeO_4^{2-} + 4H^+ + 2e^- \rightleftharpoons TeO_2 \cdot H_2O + H_2O$	1.343				1	36
		$HTeO_4^- + H^+ + 2e^- \rightleftharpoons TeO_3^{2-} + H_2O$	0.584				1	37
		$TeO_4^{2-} + 2H^+ + 2e^- \rightleftharpoons TeO_3^{2-} + H_2O$	0.892				1	38
		$TeO_4^{2-} + H_2O + 2e^- \rightleftharpoons TeO_3^{2-} + 2OH^-$	0.4			1	4	39

Tellurium 1

Standard value V	Solvent	Electrode Reference Number	Standard value V	Solvent	Electrode Reference Number	Standard value V	Solvent	Electrode Reference Number
-1.445	1	22	1.036	1	30			
-1.143	1	20	1.202	1	29			
-0.845	1	25	1.343	1	36			
-0.795	1	1	1.509	1	35			
-0.790	1	21						
-0.739	1	6						
-0.717	1	2						
-0.638	1	7						
-0.595	1	4						
-0.57	1	19						
-0.50	1	3						
-0.365	1	9						
0.121	1	5						
0.132	1	8						
0.273	1	23						
0.286	1	26						
0.4	1	39						
0.402	1	24						
0.493	1	27						
0.521	1	16						
0.551	1	13						
0.568	1	15						
0.584	1	37						
0.593	1	17						
0.631	1	12						
0.713	1	14						
0.813	1	11						
0.827	1	18						
0.850	1	34						
0.854	1	32						
0.892	1	38						
0.920	1	28						
0.953	1	10						
1.020	1	31						
1.020	1	33						

NOTES: Tellurium

1) U^o value is not exactly known.

BIBLIOGRAPHY: Tellurium

1) Data reported by M. Pourbaix "Atlas d'Equilibres Electrochimiques à 25 °C" (Gauthier-Villars Editeur, Paris, 1963), pp. 560 - 571

2) S.A. Awad, J. Phys. Chem., 66 (1962), 890 - 894

3) J.A. Ricketts, L.W. Tresselt, J. Am. Chem. Soc., 81 (1959), 2305 - 2307

4) Data reported by A.J. De Bethune, N.A. Swendeman-Loud "Standard Aqueous Electrode Potentials and Temperature Coefficients at 25 °C" (Clifford A. Hampel, Ill.)

5) A.J. Panson, J. Phys. Chem., 68 (1964), 1721 - 1724

6) V.M. Komandenko, Elektrokhim., 7 (1971), 1255 - 1257

Electronically conducting phase	Intermediate species	Composition of the solution	Solvent	Temperature °C	Pressure Torr	Measuring method
a	b	c	d		e	d
Po			1	25	760	IV
Po			1	25	760	IV
Po-Au		HNO_3 : 1 N	1	22		I
Po-Au		HCl: 1 N	1	22		I
Po	PoO_2		1	25	760	IV
Po			1	25	760	IV
Po			1	25	760	IV
Ind.	PoO_2		1	25	760	IV
Ind.	PoO_3		1	25	760	IV
Ind.			1	25	760	IV
Po-Au		HCl: 1 N	1	22		I
Po	PoH_2 (gas)		1	25	760	IV
Ind.	PoO_2, PoO_3		1	25	760	IV
Ind.	PoO_3		1	25	760	IV

Comparison electrode	Liquid junction	Electrode reaction	Standard value V	Uncertainty mV	Temperature Coefficient μV/°C	Notes	Reference	Electrode Reference Number
d			f		g	h		
		$Po^{2+} + 2e^- \rightleftharpoons Po$	0.651		-430		1,2	1
		$Po^{3+} + 3e^- \rightleftharpoons Po$	0.56				3	2
3	KCl satd.	$Po^{4+} + 4e^- \rightleftharpoons Po$	0.76				4	3
3	KCl satd.	$Po(Cl)_4^{2-} + 2e^- \rightleftharpoons Po + 4Cl^-$	0.38				4	4
		$PoO_2 + 4H^+ + 4e^- \rightleftharpoons Po + 2H_2O$	0.724				1	5
		$PoO_3^{2-} + 6H^+ + 4e^- \rightleftharpoons Po + 3H_2O$	0.748				1	6
		$PoO_3^{2-} + 3H_2O + 4e^- \rightleftharpoons Po + 6OH^-$	-0.49				2	7
		$PoO_2 + 4H^+ + 2e^- \rightleftharpoons Po^{2+} + 2H_2O$	0.798		-260		1,2	8
		$PoO_3 + 6H^+ + 4e^- \rightleftharpoons Po^{2+} + 3H_2O$	1.161				1	9
		$PoO_3^{2-} + 6H^+ + 2e^- \rightleftharpoons Po^{2+} + 3H_2O$	0.847				1	10
3	KCl satd.	$Po(Cl)_6^{2-} + 2e^- \rightleftharpoons Po(Cl)_4^{2-} + 2Cl^-$	0.72				4	11
		$Po + 2H^+ + 2e^- \rightleftharpoons PoH_2$	-1.000			1	1,3,5	12
		$PoO_3 + 2H^+ + 2e^- \rightleftharpoons PoO_2 + H_2O$	1.524				1	13
		$PoO_3 + 2e^- \rightleftharpoons PoO_3^{2-}$	1.474				1	14

Polonium 1

Standard value V	Solvent	Electrode Reference Number	Standard value V	Solvent	Electrode Reference Number	Standard value V	Solvent	Electrode Reference Number
-1.000	1	12						
-0.49	1	7						
0.38	1	4						
0.56	1	2						
0.651	1	1						
0.72	1	11						
0.724	1	5						
0.748	1	6						
0.76	1	3						
0.798	1	8						
0.847	1	10						
1.161	1	9						
1.474	1	14						
1.524	1	13						

NOTES: Polonium

1) The value is an average of those given by several authors; they all are in the same range of uncertainty.

BIBLIOGRAPHY: Polonium

1) Data reported by M. Pourbaix "Atlas d'Equilibres Electrochimiques à 25 °C" (Gauthier-Villars, Editeur, Paris, 1963)

2) Data reported by A.J. De Bethune, N.A. Swendeman-Loud "Standard Aqueous Electrode Potentials and Temperature Coefficients at 25 °C" (Clifford A. Hampel, Ill.)

3) Sheng-Lan Hsia, Hua Hsueh T'ung Rao, 1 (1964), 38 - 41

4) K.W. Bagnall, J.H. Freeman, J. Chem. Soc., 1956, 2770 - 2774

5) A.C. Wahl, N.A. Bonner, Radioactivity Applied to Chemistry (Wiley, New York, 1951)

Electronically conducting phase	Intermediate species	Composition of the solution	Solvent	Temperature °C	Pressure Torr	Measuring method
a	b	c	d		e	d
Cr			1	25	760	IV
Cr			1	25	760	IV
Cr	CrO		1	25	760	IV
Cr	CrO_2		1	25	760	IV
Cr			1	25	760	IV
Cr			1	25	760	IV
Cr	Cr_2O_3		1	25	760	IV
Cr			1	25	760	IV
Cr			1	25	760	IV
Cr	$Cr(OH)_3$		1	25	760	IV
Cr	$Cr(OH)_3 \cdot nH_2O$		1	25	760	IV
Cr	$Cr(OH)_3$		1	25	760	IV
Cr	$Cr(OH)_3 \cdot nH_2O$		1	25	760	IV
Cr			1	25	760	IV
Cr	H_2CrO_4		1	25	760	IV
Ind.			1	25	760	IV
Ind.			1	25	760	IV
Ind.	Cr_2O_3		1	25	760	IV
Ind.			1	25	760	IV
Ind.			1	25	760	IV
Ind.	$Cr(OH)_3$		1	25	760	IV
Ind.	$Cr(OH)_3 \cdot nH_2O$		1	25	760	IV
Ind.	CrO_2		1	25	760	IV
Ind.			1	25	760	IV
Pt		$Cr_2(SO_4)_3$: 0.136 up to 0.544 m; H_2CrO_4 : 0.544 down to 0.136 m; H_2SO_4 : 4 up to 16 g per 100 cm^3	1	25		I
Cr			1	25		IV
Ind.			1	25	760	IV
Ind.	H_2CrO_4		1	25	760	IV
Hg (DME)		KCN : 1 m; $K_3[Cr(CN)_6]$: (0.198 up to 3.40) · · 10^{-3} m	1	25 ± 0.04		II
Ind.	CrO		1	25	760	IV
Ind.	CrO		1	25	760	IV
Ind.	CrO, Cr_2O_3		1	25	760	IV

Comparison electrode	Liquid junction	Electrode reaction	Standard value V	Uncertainty mV	Temperature Coefficient μV/°C	Notes	Reference	Electrode Reference Number
d			f		g	h		
		$Cr^{2+} + 2e^- \rightleftharpoons Cr$	-0.913				1	1
		$Cr^{3+} + 3e^- \rightleftharpoons Cr$	-0.744		468		1,2	2
		$CrO + 2H^+ + 2e^- \rightleftharpoons Cr + H_2O$	-0.588				1	3
		$CrO_2 + 2H_2O + 4e^- \rightleftharpoons Cr + 4OH^-$	-1.27				2	4
		$CrO_2^- + 4H^+ + 3e^- \rightleftharpoons Cr + 2H_2O$	-0.213				1	5
		$CrO_3^{3-} + 6H^+ + 3e^- \rightleftharpoons Cr + 3H_2O$	0.374				1	6
		$Cr_2O_3 + 6H^+ + 6e^- \rightleftharpoons 2Cr + 3H_2O$	-0.579				1	7
		$CrO_4^{2-} + 8H^+ + 6e^- \rightleftharpoons Cr + 4H_2O$	0.366				1	8
		$Cr_2O_7^{2-} + 14H^+ + 12e^- \rightleftharpoons 2Cr + 7H_2O$	0.294				1	9
		$Cr(OH)_3 + 3H^+ + 3e^- \rightleftharpoons Cr + 3H_2O$	-0.654				1	10
		$Cr(OH)_3 \cdot nH_2O + 3H^+ + 3e^- \rightleftharpoons Cr + (n+3)H_2O$	-0.512				1	11
		$Cr(OH)_3 + 3e^- \rightleftharpoons Cr + 3OH^-$	-1.48		-1304		2	12
		$Cr(OH)_3 \cdot nH_2O + 3e^- \rightleftharpoons Cr + 3OH^- + nH_2O$	-1.34		-990		2	13
		$HCrO_4^- + 7H^+ + 6e^- \rightleftharpoons Cr + 4H_2O$	0.303				1	14
		$H_2CrO_4 + 6H^+ + 6e^- \rightleftharpoons Cr + 4H_2O$	0.295				1	15
		$Cr^{3+} + e^- \rightleftharpoons Cr^{2+}$	-0.407				1	16
		$CrO_2^- + 4H^+ + e^- \rightleftharpoons Cr^{2+} + 2H_2O$	1.188				1	17
		$Cr_2O_3 + 6H^+ + 2e^- \rightleftharpoons 2Cr^{2+} + 3H_2O$	0.088				1	18
		$Cr(OH)^{2+} + H^+ + e^- \rightleftharpoons Cr^{2+} + H_2O$	-0.182				1	19
		$Cr(OH)_2^+ + 2H^+ + e^- \rightleftharpoons Cr^{2+} + 2H_2O$	0.185				1	20
		$Cr(OH)_3 + 3H^+ + e^- \rightleftharpoons Cr^{2+} + 3H_2O$	-0.136				1	21
		$Cr(OH)_3 \cdot nH_2O + 3H^+ + e^- \rightleftharpoons Cr^{2+} + (n+3)H_2O$	0.289				1	22
		$CrO_2 + 4H^+ + e^- \rightleftharpoons Cr^{3+} + 2H_2O$	1.556				1	23
		$CrO_4^{2-} + 8H^+ + 3e^- \rightleftharpoons Cr^{3+} + 4H_2O$	1.477				1	24
19	H_2SO_4 2N	$Cr_2O_7^{2-} + 14H^+ + 6e^- \rightleftharpoons 2Cr^{3+} + 7H_2O$	1.232		-1263		2,3	25
		$Cr(SCN)^{2+} + 3e^- \rightleftharpoons Cr + SCN^-$	-0.805				4	26
		$HCrO_4^- + 7H^+ + 3e^- \rightleftharpoons Cr^{3+} + 4H_2O$	1.350				1	27
		$H_2CrO_4 + 6H^+ + 3e^- \rightleftharpoons Cr^{3+} + 4H_2O$	1.335				1	28
3		$Cr(CN)_6^{3-} + e^- \rightleftharpoons Cr(CN)_6^{4-}$	-1.28				5	29
		$CrO_2^- + 2H^+ + e^- \rightleftharpoons CrO + H_2O$	0.538				1	30
		$CrO_3^{3-} + 4H^+ + e^- \rightleftharpoons CrO + 2H_2O$	0.297				1	31
		$Cr_2O_3 + 2H^+ + 2e^- \rightleftharpoons 2CrO + H_2O$	-0.561				1	32

Electronically conducting phase	Intermediate species	Composition of the solution	Solvent	Temperature °C	Pressure Torr	Measuring method
a	b	c	d		e	d
Ind.	CrO, $Cr(OH)_3$		1	25	760	IV
Ind.	CrO, $Cr(OH)_3 \cdot nH_2O$		1	25	760	IV
Ind.	CrO_2		1	25	760	IV
Ind.	CrO_2		1	25	760	IV
Ind.	CrO_2		1	25	760	IV
Ind.			1	25	760	IV
Ind.			1	25	760	IV
Ind.			1	25	760	IV
Ind.	Cr_2O_3, CrO_2		1	25	760	IV
Ind.	Cr_2O_3		1	25	760	IV
Ind.	Cr_2O_3		1	25	760	IV
Ind.	Cr_2O_3		1	25	760	IV
Ind.	CrO_2		1	25	760	IV
Ind.			1	25	760	IV
Ind.			1	25	760	IV
Ind.			1	25	760	IV
Ind.			1	25	760	IV
Ind.			1	25	760	IV
Ind.			1	25	760	IV
Ind.	$Cr(OH)_3$, CrO_2		1	25	760	IV
Ind.	$Cr(OH)_3$		1	25	760	IV
Ind.	$Cr(OH)_3$		1	25	760	IV
Ind.	$Cr(OH)_3$		1	25	760	IV
Ind.	$Cr(OH)_3 \cdot nH_2O$, CrO_2		1	25	760	IV
Ind.	$Cr(OH)_3 \cdot nH_2O$		1	25	760	IV
Ind.	$Cr(OH)_3 \cdot nH_2O$		1	25	760	IV
Ind.	$Cr(OH)_3 \cdot nH_2O$		1	25	760	IV
Ind.	$Cr(OH)_3 \cdot nH_2O$		1	25	760	IV
Hg (DME)		NaCl : 0.4 m; acetate buffer : 0.1 m; polyoxyethylene lauryl : $2 \cdot 10^{-6}$ m	1	25		II

Comparison electrode	Liquid junction	Electrode reaction	Standard value V	Uncertainty mV	Temperature Coefficient μV/°C	Notes	Reference	Electrode Reference Number
d			f		g	h		
		$Cr(OH)_3 + H^+ + e^- \rightleftharpoons CrO + 2H_2O$	-0.785				1	33
		$Cr(OH)_3 \cdot nH_2O + H^+ + e^- \rightleftharpoons CrO + (n+2)H_2O$	-0.360				1	34
		$CrO_4^{2-} + 4H^+ + 2e^- \rightleftharpoons CrO_2 + 2H_2O$	1.437				1	35
		$Cr_2O_7^{2-} + 6H^+ + 4e^- \rightleftharpoons 2CrO_2 + 3H_2O$	1.221				1	36
		$HCrO_4^- + 3H^+ + 2e^- \rightleftharpoons CrO_2 + 2H_2O$	1.246				1	37
		$CrO_4^{2-} + 4H^+ + 3e^- \rightleftharpoons CrO_2^- + 2H_2O$	0.945				1	38
		$Cr_2O_7^{2-} + 6H^+ + 6e^- \rightleftharpoons 2CrO_2^- + 3H_2O$	0.801				1	39
		$CrO_4^{2-} + 2H^+ + 3e^- \rightleftharpoons CrO_3^{3-} + H_2O$	0.359				1	40
		$2CrO_2 + 2H^+ + 2e^- \rightleftharpoons Cr_2O_3 + H_2O$	1.060				1	41
		$2CrO_4^{2-} + 10H^+ + 6e^- \rightleftharpoons Cr_2O_3 + 5H_2O$	1.311				1	42
		$Cr_2O_7^{2-} + 8H^+ + 6e^- \rightleftharpoons Cr_2O_3 + 4H_2O$	1.168				1	43
		$2HCrO_4^- + 8H^+ + 6e^- \rightleftharpoons Cr_2O_3 + 5H_2O$	1.184				1	44
		$CrO_2 + 3H^+ + e^- \rightleftharpoons Cr(OH)^{2+} + H_2O$	1.331				1	45
		$CrO_4^{2-} + 7H^+ + 3e^- \rightleftharpoons Cr(OH)^{2+} + 3H_2O$	1.402				1	46
		$Cr_2O_7^{2-} + 12H^+ + 6e^- \rightleftharpoons 2Cr(OH)^{2+} + 5H_2O$	1.258				1	47
		$HCrO_4^- + 6H^+ + 3e^- \rightleftharpoons Cr(OH)^{2+} + 3H_2O$	1.275				1	48
		$CrO_4^{2-} + 6H^+ + 3e^- \rightleftharpoons Cr(OH)_2^+ + 2H_2O$	1.279				1	49
		$Cr_2O_7^{2-} + 10H^+ + 6e^- \rightleftharpoons 2Cr(OH)_2^+ + 3H_2O$	1.135				1	50
		$HCrO_4^- + 5H^+ + 3e^- \rightleftharpoons Cr(OH)_2^+ + 2H_2O$	1.152				1	51
		$CrO_2 + H_2O + H^+ + e^- \rightleftharpoons Cr(OH)_3$	1.284				1	52
		$CrO_4^{2-} + 5H^+ + 3e^- \rightleftharpoons Cr(OH)_3 + H_2O$	1.386				1	53
		$Cr_2O_7^{2-} + 8H^+ + 6e^- \rightleftharpoons 2Cr(OH)_3 + H_2O$	1.242				1	54
		$HCrO_4^- + 4H^+ + 3e^- \rightleftharpoons Cr(OH)_3 + H_2O$	1.259				1	55
		$CrO_2 + (n+1)H_2O + H^+ + e^- \rightleftharpoons Cr(OH)_3 \cdot nH_2O$	0.859				1	56
		$CrO_4^{2-} + nH_2O + 5H^+ + 3e^- \rightleftharpoons Cr(OH)_3 \cdot nH_2O + H_2O$	1.244				1	57
		$CrO_4^{2-} + (n+4)H_2O + 3e^- \rightleftharpoons Cr(OH)_3 \cdot nH_2O + 5OH^-$	-0.13		-1675		1,2	58
		$Cr_2O_7^{2-} + nH_2O + 8H^+ + 6e^- \rightleftharpoons 2Cr(OH)_3 \cdot nH_2O + H_2O$	1.101				1	59
		$HCrO_4^- + nH_2O + 4H^+ + 3e^- \rightleftharpoons Cr(OH)_3 \cdot nH_2O + H_2O$	1.117				1	60
3		[Cr (III) - trans-1,2 - cyclohexanediaminetetracetate]$^-$ + e^- $\rightleftharpoons$ [Cr (II) - trans-1,2 - cyclohexanediaminetetracetate]$^{2-}$	1.198				6	61

Chromium 1

Standard value V	Solvent	Electrode Reference Number	Standard value V	Solvent	Electrode Reference Number	Standard value V	Solvent	Electrode Reference Number
-1.48	1	12	1.117	1	60			
-1.34	1	13	1.135	1	50			
-1.28	1	29	1.152	1	51			
-1.27	1	4	1.168	1	43			
-0.913	1	1	1.184	1	44			
-0.805	1	26	1.188	1	17			
-0.785	1	33	1.198	1	61			
-0.744	1	2	1.221	1	36			
-0.654	1	10	1.232	1	25			
-0.588	1	3	1.242	1	54			
-0.579	1	7	1.244	1	57			
-0.561	1	32	1.246	1	37			
-0.512	1	11	1.258	1	47			
-0.407	1	16	1.259	1	55			
-0.360	1	34	1.275	1	48			
-0.213	1	5	1.279	1	49			
-0.182	1	19	1.284	1	52			
-0.136	1	21	1.311	1	42			
-0.13	1	58	1.331	1	45			
0.088	1	18	1.335	1	28			
0.185	1	20	1.350	1	27			
0.289	1	22	1.386	1	53			
0.294	1	9	1.402	1	46			
0.295	1	15	1.437	1	35			
0.297	1	31	1.477	1	24			
0.303	1	14	1.556	1	23			
0.359	1	40						
0.366	1	8						
0.374	1	6						
0.538	1	30						
0.801	1	39						
0.859	1	56						
0.945	1	38						
1.060	1	41						
1.101	1	59						

BIBLIOGRAPHY: Chromium

1) Data reported by M. Pourbaix "Atlas d'Equilibres Electrochimiques à 25 °C" (Gauthier-Villars Editeur, Paris, 1963), pp. 256 - 271

2) Data reported by A.J. De Bethune, N.A. Swendeman-Loud "Standard Aqueous Electrode Potentials and Temperature Coefficients at 25 °C" (Clifford A. Hampel, Ill.)

3) S.G. Dighe, J. Ind. Chem. Soc., 23 (1946), 291 - 299

4) C. Postmus, E.L. King, J. Phys. Chem., 59 (1955), 1208

5) D.N. Hume, I.M. Kolthoff, J. Am. Chem. Soc., 65 (1943), 1897 - 1901

6) N. Tanaka, T. Tomita, A. Yamada, Bull. Chem. Soc. Japan, 43 (1970), 2042 - 2045

Electronically conducting phase	Intermediate species	Composition of the solution	Solvent	Temperature °C	Pressure Torr	Measuring method
a	b	c	d		e	d
Mo			1	25	760	IV
Mo	.MoO_2		1	25	760	IV
Mo			1	25	760	IV
Mo		Na_2MoO_4 : 0.025 up to 0.050 m; KOH : 0.531 m	1	25± 0.4		IV
Ind.			1	25	760	IV
Ind.	H_2MoO_4 (s)		1	25	760	IV
Ind.	MoO_2		1	25	760	IV
Ind.	MoO_3		1	25	760	IV
Ind.			1	25	760	IV
Pt		Molybdenum blue : 0.004 up to 0.01 m; HCl : 0.01 up to 8 N	1	30		I
Ind.	MoO_2		1	25	760	IV
Ind.	MoO_2, H_2MoO_4 (s)		1	25	760	IV
Ind.	MoO_2, MoO_3		1	25	760	IV
Ind.	MoO_2		1	25	760	IV

Comparison electrode	Liquid junction	Electrode reaction	Standard value V	Uncertainty mV	Temperature Coefficient μV/°C	Notes	Reference	Electrode Reference Number
d			f		g	h		
		$Mo^{3+} + 3e^- \rightleftharpoons Mo$	-0.200				1	1
		$MoO_2 + 4H^+ + 4e^- \rightleftharpoons Mo + 2H_2O$	-0.072				1	2
		$MoO_4^{2-} + 8H^+ + 6e^- \rightleftharpoons Mo + 4H_2O$	0.154				1	3
		$MoO_4^{2-} + 4H_2O + 6e^- \rightleftharpoons Mo + 8OH^-$	-0.92			1	2	4
		$HMoO_4^- + 7H^+ + 3e^- \rightleftharpoons Mo^{3+} + 4H_2O$	0.390				1	5
		$H_2MoO_4 + 6H^+ + 3e^- \rightleftharpoons Mo^{3+} + 4H_2O$	-0.623				1	6
		$MoO_2 + 4H^+ + e^- \rightleftharpoons Mo^{3+} + 2H_2O$	0.311				1	7
		$MoO_3 + 6H^+ + 3e^- \rightleftharpoons Mo^{3+} + 3H_2O$	0.317				1	8
		$MoO_4^{2-} + 8H^+ + 3e^- \rightleftharpoons Mo^{3+} + 4H_2O$	0.508				1	9
3 or 1	HCl	$MoO_2^{2+} + 2H^+ + e^- \rightleftharpoons MoO^{3+} + H_2O$	0.482				3	10
		$HMoO_4^- + 3H^+ + 2e^- \rightleftharpoons MoO_2 + 2H_2O$	0.429				1	11
		$H_2MoO_4 + 2H^+ + 2e^- \rightleftharpoons MoO_2 + 2H_2O$	-1.091				1	12
		$MoO_3 + 2H^+ + 2e^- \rightleftharpoons MoO_2 + H_2O$	0.320				1	13
		$MoO_4^{2-} + 4H^+ + 2e^- \rightleftharpoons MoO_2 + 2H_2O$	0.606				1	14

Molybdenum 1

Standard value V	Solvent	Electrode Reference Number	Standard value V	Solvent	Electrode Reference Number	Standard value V	Solvent	Electrode Reference Number
-1.091	1	12						
-0.92	1	4						
-0.623	1	6						
-0.200	1	1						
-0.072	1	2						
0.154	1	3						
0.311	1	7						
0.317	1	8						
0.320	1	13						
0.390	1	5						
0.429	1	11						
0.482	1	10						
0.508	1	9						
0.606	1	14						

NOTES: Molybdenum

1) Solution heat data

BIBLIOGRAPHY: Molybdenum

1) Data reported by M. Pourbaix "Atlas d'Equilibres Electrochimiques à 25 °C" (Gauthier-Villars Editeur, Paris, 1963), pp. 272 - 279

2) R.L. Graham, L.G. Hepler, J. Am. Chem. Soc., 78 (1956), 4846 - 4848

3) H.K. El Shamy, A.M. El Aggan, J. Am. Chem. Soc., 75 (1953), 1187 - 1190

Tungsten 1

Electronically conducting phase	Intermediate species	Composition of the solution	Solvent	Temperature °C	Pressure Torr	Measuring method
a	b	c	d		e	d
W	WO_2		1	25	760	IV
W	WO_3		1	25	760	IV
W			1	25	760	IV
W		Na_2WO_4 : 0.0005 up to 0.01 m; NaOH: 0.05 up to 0.1 m	1	25		I
Pt		$K_4[W(CN)_8]$: 10^{-5} up to 10^{-2} m; $K[W(CN)_8]$: 10^{-5} up to 10^{-2} m	1	25± 0.1		I
Ind.	WO_2		1	25	760	IV
Ind.	WO_2, W_2O_5		1	25	760	IV
Ind.	W_2O_5, WO_3		1	25	760	IV
Ind.	W_2O_5		1	25	760	IV

Comparison electrode	Liquid junction	Electrode reaction	Standard value V	Uncertainty mV	Temperature Coefficient μV/°C	Notes	Reference	Electrode Reference Number
d			f		g	h		
		$WO_2 + 4H^+ + 4e^- \rightleftharpoons W + 2H_2O$	-0.119				1	1
		$WO_3 + 6H^+ + 6e^- \rightleftharpoons W + 3H_2O$	-0.090		-400		2	2
		$WO_4^{2-} + 8H^+ + 6e^- \rightleftharpoons W + 4H_2O$	0.049				1	3
3	KCl satd.	$WO_4^{2-} + 4H_2O + 6e^- \rightleftharpoons W + 8OH^-$	-1.0069	± 0.4	-1360		2,3	4
32		$[W(CN)_8]^{3-} + e^- \rightleftharpoons [W(CN)_8]^{4-}$	0.457			1	4	5
		$WO_4^{2-} + 4H^+ + 2e^- \rightleftharpoons WO_2 + 2H_2O$	0.386				1	6
		$W_2O_5 + 2H^+ + 2e^- \rightleftharpoons 2WO_2 + H_2O$	-0.031				1	7
		$2WO_3 + 2H^+ + 2e^- \rightleftharpoons W_2O_5 + H_2O$	-0.029				1	8
		$2WO_4^{2-} + 6H^+ + 2e^- \rightleftharpoons W_2O_5 + 3H_2O$	0.801				1	9

Tungsten 1

Standard value V	Solvent	Electrode Reference Number	Standard value V	Solvent	Electrode Reference Number	Standard value V	Solvent	Electrode Reference Number
-1.0069	1	4						
-0.119	1	1						
-0.090	1	2						
-0.031	1	7						
-0.029	1	8						
0.049	1	3						
0.386	1	6						
0.457	1	5						
0.801	1	9						

NOTES: Tungsten

1) Agar-agar saturated with KCl

BIBLIOGRAPHY: Tungsten

1) Data reported by M. Pourbaix "Atlas d'Equilibres Electrochimiques à 25 °C" (Gauthier-Villars Editeur, Paris, 1963), pp. 280 - 285

2) Data reported by A.J. De Bethune, N.A. Swendeman-Loud "Standard Aqueous Electrode Potentials and Temperature Coefficients at 25 °C" (Clifford A. Hampel, Ill.)

3) S.E.S. El Wakkad, T.M. Salam, H.A. Risk, J.G. Ebaid, J. Chem. Soc., 1957, 3776 - 3779

4) H. Baadsgaard, W.D. Treadwell, Helv. Chim. Acta, 38 (1955), 1669 - 1679

Group VII

Electronically conducting phase	Intermediate species	Composition of the solution	Solvent	Temperature °C	Pressure Torr	Measuring method
a	b	c	d		e	d
Ind.	F_2		1	25	760	IV
Ind.	F		7	25	760	IV
Ind.	F_2O, F_2		1	25	760	IV
Ind.	F_2O		1	25	760	IV
Ind.	HF (gas), F_2		1	25	760	IV
Ni	F	1 % H_2O (wt)	28	25		I
Ind.	HF (in sol.), F_2		1	25	760	IV
Ind.	F_2		1	25	760	IV
Ind.	HF (gas), F_2O		1	25	760	IV
Ind.	F_2O, HF (in sol.)		1	25	760	IV
Ind.	F_2O		1	25	760	IV

Comparison electrode	Liquid junction	Electrode reaction	Standard value	Uncertainty	Temperature Coefficient	Notes	Reference	Electrode Reference Number
			V	mV	μV/°C			
d			f		g	h		
		$F_2 + 2e^- \rightleftharpoons 2F^-$	2.866		-1830		1	1
		$F + e^- \rightleftharpoons F^-$	2.43	± 10			2	2
		$F_2O + 2H^+ + 2e^- \rightleftharpoons F_2 + H_2O$	1.439				3	3
		$F_2O + 2H^+ + 4e^- \rightleftharpoons 2F^- + H_2O$	2.153		-1184		3,1	4
		$F_2 + 2H^+ + 2e^- \rightleftharpoons 2HF$	2.806				3	5
1		$F + H^+ + e^- \rightleftharpoons HF$	2.85	± 20			4	6
		$F_2 + 2H^+ + 2e^- \rightleftharpoons 2HF$	3.053		-600		3,1	7
		$F_2 + H^+ + 2e^- \rightleftharpoons HF_2^-$	2.979				3	8
		$F_2O + 4H^+ + 4e^- \rightleftharpoons 2HF + H_2O$	2.123				3	9
		$F_2O + 4H^+ + 4e^- \rightleftharpoons 2HF + H_2O$	2.246				3	10
		$F_2O + 3H^+ + 4e^- \rightleftharpoons HF_2^- + H_2O$	2.209				3	11

Fluorine 1

Standard value V	Solvent	Electrode Reference Number
1.439	1	3
2.123	1	9
2.153	1	4
2.209	1	11
2.246	1	10
2.806	1	5
2.866	1	1
2.979	1	8
3.053	1	7
2.43	7	2
2.85	28	6

BIBLIOGRAPHY: Fluorine

1) Data reported by A.J. De Bethune, N.A. Swendeman-Loud "Standard Aqueous Electrode Potentials and Temperature Coefficients at 25 °C" (Clifford A. Hampel, Ill.)

2) B. Jakuszewski, S. Taniewska-Osinska, Zesz. Nauk Univ. Łodz, Nauk Mat. Pezyrod, Ser. II, 1959, 97 - 100

3) Data reported by M. Pourbaix "Atlas d'Equilibres Electrochimiques à 25 °C" (Gauthier-Villars Editeur, Paris, 1963), pp. 579 - 589

4) N. Watanabe, T. Chang, K. Nakanishi, Denki Kagaku, 36 (1968), 600 - 604

Electronically conducting phase	Intermediate species	Composition of the solution	Solvent	Temperature °C	Pressure Torr	Measuring method
a	b	c	d		e	d
Ind.	Cl_2		1	25	760	IV
Ind.	Cl_2 (in sol.)		1	25	760	IV
Ind.	Cl_2O, Cl_2		1	25	760	IV
Ind.	Cl_2O, Cl_2 (in sol.)		1	25	760	IV
Ind.	ClO_2, Cl_2		1	25	760	IV
Ind.	ClO_2, Cl_2 (in sol.)		1	25	760	IV
Ind.	Cl_2		1	25	760	IV
Ind.	Cl_2 (in sol.)		1	25	760	IV
Ind.	Cl_2		1	25	760	IV
Ind.	Cl_2 (in sol.)		1	25	760	IV
Ind.	Cl_2		1	25	760	IV
Ind.	Cl_2 (in sol.)		1	25	760	IV
Pt or glass	Cl_2 (gas), HClO	NaClO : 0.1 N; HClO : 0.02 - 0.04 N	1	25		I
Ind.	HClO, Cl_2 (in sol.)		1	25	760	IV
Ind.	$HClO_2$, Cl_2		1	25	760	IV
Ind.	$HClO_2$, Cl_2 (in sol.)		1	25	760	IV
Pt	Cl_2	HCl : 1 m; P (Cl_2): 0.0544 up to 0.9758 Atms.	1	25± 0.02		I
	Cl_2		7	25		I
Pt	Cl_2	Et_4NClO_4 : 0.1 M	23	20± 0.5		I
Pt	Cl_2	Et_4NClO_4 : 0.1 M	5	20± 0.5		I
Ind.	Cl_2 (in sol.)		1	25	760	IV
Pt or glass		NaClO : 0.1 N; HClO : 0.02 - 0.04 N	1	25		I
Pt or glass		NaClO : 0.1 N; HClO : 0.02 - 0.04 N	1	25		I
Ind.	Cl_2O		1	25	760	IV
Ind.	ClO_2		1	25	760	IV
Ind.			1	25	760	IV
Ind.			1	25	760	IV
Ind.			1	25	760	IV
Pt or glass	HClO	NaClO : 0.1 N; HClO : 0.02 - 0.04 N	1	25		I
Ind.	$HClO_2$		1	25	760	IV

Comparison electrode	Liquid junction	Electrode reaction	Standard value	Uncertainty	Temperature Coefficient	Notes	Reference	Electrode Reference Number
			V	mV	μV/°C			
d			f		g	h		
		$2ClO^- + 4H^+ + 2e^- \rightleftharpoons Cl_2 + 2H_2O$	2.072				1	1
		$2ClO^- + 4H^+ + 2e^- \rightleftharpoons Cl_2 + 2H_2O$	2.036				1	2
		$Cl_2O + 2H^+ + 2e^- \rightleftharpoons Cl_2 + H_2O$	1.714				1	3
		$Cl_2O + 2H^+ + 2e^- \rightleftharpoons Cl_2 + H_2O$	1.679				1	4
		$2ClO_2 + 8H^+ + 8e^- \rightleftharpoons Cl_2 + 4H_2O$	1.549				1	5
		$2ClO_2 + 8H^+ + 8e^- \rightleftharpoons Cl_2 + 4H_2O$	1.540				1	6
		$2ClO_2^- + 8H^+ + 6e^- \rightleftharpoons Cl_2 + 4H_2O$	1.678				1	7
		$2ClO_2^- + 8H^+ + 6e^- \rightleftharpoons Cl_2 + 4H_2O$	1.666				1	8
		$2ClO_3^- + 12H^+ + 10e^- \rightleftharpoons Cl_2 + 6H_2O$	1.470				1	9
		$2ClO_3^- + 12H^+ + 10e^- \rightleftharpoons Cl_2 + 6H_2O$	1.463				1	10
		$2ClO_4^- + 16H^+ + 14e^- \rightleftharpoons Cl_2 + 8H_2O$	1.389				1	11
		$2ClO_4^- + 16H^+ + 14e^- \rightleftharpoons Cl_2 + 8H_2O$	1.385				1	12
1		$2HClO + 2H^+ + 2e^- \rightleftharpoons Cl_2 + 2H_2O$	1.611	± 3	-800	1	2	13
		$2HClO + 2H^+ + 2e^- \rightleftharpoons Cl_2 + 2H_2O$	1.594				1	14
		$2HClO_2 + 6H^+ + 6e^- \rightleftharpoons Cl_2 + 4H_2O$	1.640				1	15
		$2HClO_2 + 6H^+ + 6e^- \rightleftharpoons Cl_2 + 4H_2O$	1.628				1	16
1		$Cl_2 + 2e^- \rightleftharpoons 2Cl^-$	1.35827	± 0.02	-1260	2	3	17
1		$Cl_2 + 2e^- \rightleftharpoons 2Cl^-$	1.1272				4	18
33		$Cl_2 + 2e^- \rightleftharpoons 2Cl^-$	0.60	± 20			5	19
33		$Cl_2 + 2e^- \rightleftharpoons 2Cl^-$	1.06	± 20			7	20
		$Cl_2 + 2e^- \rightleftharpoons 2Cl^-$	1.395		-1260		1,6	21
1		$ClO^- + 2H^+ + 2e^- \rightleftharpoons Cl^- + H_2O$	1.698	± 3	-800	1	2	22
1		$ClO^- + H_2O + 2e^- \rightleftharpoons Cl^- + 2OH^-$	0.841	± 3	-1750	1	2	23
		$Cl_2O + 2H^+ + 2e^- \rightleftharpoons 2Cl^- + H_2O$	2.152				1	24
		$ClO_2 + 4H^+ + 5e^- \rightleftharpoons Cl^- + 2H_2O$	1.511				1	25
		$ClO_2^- + 4H^+ + 4e^- \rightleftharpoons Cl^- + 2H_2O$	1.599				1	26
		$ClO_3^- + 6H^+ + 6e^- \rightleftharpoons Cl^- + 3H_2O$	1.451				1	27
		$ClO_4^- + 8H^+ + 8e^- \rightleftharpoons Cl^- + 4H_2O$	1.389				1	28
1		$HClO + H^+ + 2e^- \rightleftharpoons Cl^- + H_2O$	1.482	± 3	-1090	1	2	29
		$HClO_2 + 3H^+ + 4e^- \rightleftharpoons Cl^- + 2H_2O$	1.570				1	30

Electronically conducting phase	Intermediate species	Composition of the solution	Solvent	Temperature °C	Pressure Torr	Measuring method
a	b	c	d		e	d
Ind.	ClO_2		1	25	760	IV
Ind.			1	25	760	IV
Ind.			1	25	760	IV
Ind.	$HClO_2$		1	25	760	IV
Ind.	Cl_2O, ClO_2		1	25	760	IV
Ind.	Cl_2O		1	25	760	IV
Ind.	Cl_2O		1	25	760	IV
Ind.	Cl_2O		1	25	760	IV
Ind.	Cl_2O, $HClO_2$		1	25	760	IV
Ind.	ClO_2		1	25	760	IV
Ind.	ClO_2		1	25	760	IV
Pt	ClO_2	$NaClO_2$: 0.001 up to 0.1 m; or $AgClO_2$ satd.; or $Pb(ClO_2)_2$ satd.	1	25		I
Pt	ClO_2 (in sol.)	ClO_2 : 0.00143 up to 0.0090 M; $NaClO_2$: 0.02842 up to 0.05933 M	1	25	Atm.	I
Ind.			1	25	760	IV
Ind.			1	25	760	IV
Ind.			1	25	760	IV
Ind.			1	25	760	IV
Ind.			1	25	760	IV
Ind.	HCl (gas), Cl_2		1	25	760	IV
Ind.	HCl (gas), Cl_2O		1	25	760	IV
Ind.	HCl (gas), ClO_2		1	25	760	IV
Ind.	HClO, ClO_2		1	25	760	IV
Ind.	HClO		1	25	760	IV
Ind.	HClO, $HClO_2$		1	25	760	IV
Ind.	$HClO_2$, ClO_2		1	25	760	IV
Ind.	$HClO_2$		1	25	760	IV
Ind.	$HClO_2$		1	25	760	IV

Comparison electrode	Liquid junction	Electrode reaction	Standard value V	Uncertainty mV	Temperature Coefficient µV/°C	Notes	Reference	Electrode Reference Number
d			f		g	h		
		$ClO_2 + 2H^+ + 3e^- \rightleftharpoons ClO^- + H_2O$	1.374				1	31
		$ClO_2^- + 2H^+ + 2e^- \rightleftharpoons ClO^- + H_2O$	1.474				1	32
		$ClO_2^- + H_2O + 2e^- \rightleftharpoons ClO^- + 2OH^-$	0.66		-1454		6	33
		$HClO_2 + H^+ + 2e^- \rightleftharpoons ClO^- + H_2O$	1.423				1	34
		$2ClO_2 + 6H^+ + 6e^- \rightleftharpoons Cl_2O + 3H_2O$	1.494				1	35
		$2ClO_2^- + 6H^+ + 4e^- \rightleftharpoons Cl_2O + 3H_2O$	1.661				1	36
		$2ClO_3^- + 10H^+ + 8e^- \rightleftharpoons Cl_2O + 5H_2O$	1.408				1	37
		$2ClO_4^- + 14H^+ + 12e^- \rightleftharpoons Cl_2O + 7H_2O$	1.336				1	38
		$2HClO_2 + 4H^+ + 4e^- \rightleftharpoons Cl_2O + 3H_2O$	1.603				1	39
		$ClO_3^- + 2H^+ + e^- \rightleftharpoons ClO_2 + H_2O$	1.152				1	40
		$ClO_4^- + 4H^+ + 3e^- \rightleftharpoons ClO_2 + 2H_2O$	1.177				1	41
3		$ClO_2 + e^- \rightleftharpoons ClO_2^-$	0.954		-2220		6,7	42
3		$ClO_2 + e^- \rightleftharpoons ClO_2^-$	0.936	±3	-1600		8	43
		$ClO_3^- + 2H^+ + 2e^- \rightleftharpoons ClO_2^- + H_2O$	1.155				1	44
		$ClO_3^- + H_2O + 2e^- \rightleftharpoons ClO_2^- + 2OH^-$	0.33		-1470		6	45
		$ClO_4^- + 4H^+ + 4e^- \rightleftharpoons ClO_2^- + 2H_2O$	1.173				1	46
		$ClO_4^- + 2H^+ + 2e^- \rightleftharpoons ClO_3^- + H_2O$	1.189		-410		1,6	47
		$ClO_4^- + H_2O + 2e^- \rightleftharpoons ClO_3^- + 2OH^-$	0.36		-1240		6	48
		$Cl_2 + 2H^+ + 2e^- \rightleftharpoons 2HCl$	0.987				1	49
		$Cl_2O + 4H^+ + 4e^- \rightleftharpoons 2HCl + H_2O$	1.351				1	50
		$ClO_2 + 5H^+ + 5e^- \rightleftharpoons HCl + 2H_2O$	1.436				1	51
		$ClO_2 + 3H^+ + 3e^- \rightleftharpoons HClO + H_2O$	1.522				1	52
		$ClO_2^- + 3H^+ + 2e^- \rightleftharpoons HClO + H_2O$	1.703				1	53
		$HClO_2 + 2H^+ + 2e^- \rightleftharpoons HClO + H_2O$	1.645		-550		1,6	54
		$ClO_2 + H^+ + e^- \rightleftharpoons HClO_2$	1.277		-1440		1,6	55
		$ClO_3^- + 3H^+ + 2e^- \rightleftharpoons HClO_2 + H_2O$	1.214		-250		1,6	56
		$ClO_4^- + 5H^+ + 4e^- \rightleftharpoons HClO_2 + 2H_2O$	1.201				1	57

Chlorine 1

Standard value V	Solvent	Electrode Reference Number
0.33	1	45
0.36	1	48
0.66	1	33
0.841	1	23
0.936	1	43
0.954	1	42
0.987	1	49
1.152	1	40
1.155	1	44
1.173	1	46
1.177	1	41
1.189	1	47
1.201	1	57
1.214	1	56
1.277	1	55
1.336	1	38
1.351	1	50
1.35827	1	17
1.374	1	31
1.385	1	12
1.389	1	11
1.389	1	28
1.395	1	21
1.408	1	37
1.423	1	34
1.436	1	51
1.451	1	27
1.463	1	10
1.470	1	9
1.474	1	32
1.482	1	29
1.494	1	35
1.511	1	25
1.522	1	52
1.540	1	6
1.549	1	5
1.570	1	30
1.594	1	14
1.599	1	26
1.603	1	39
1.611	1	13
1.628	1	16
1.640	1	15
1.645	1	54
1.661	1	35
1.666	1	8
1.678	1	7
1.679	1	4
1.698	1	22
1.703	1	53
1.714	1	3
2.036	1	2
2.072	1	1
2.152	1	24
1.1272	7	18
0.60	23	19
1.06	5	20

Standard value V	Solvent*	Electrode Reference Number

NOTES: Chlorine

1) T ranges from 10 to 50 ^{o}C

2) T range from 25 to 80 ^{o}C

BIBLIOGRAPHY: Chlorine

1) Data reported by M. Pourbaix "Atlas d'Equilibres Electrochimiqeus à 25 ^{o}C" (Gauthier-Villars Editeur, Paris, 1963), pp. 590 - 603

2) I.E. Flis, K.P. Mishchenko, N.V. Troitskaya, Zh. Fiz. Khim., 33 (1959), 1744 - 1749

3) A. Cerquetti, P. Longhi, T. Mussini, G. Natta, J. Electroanal. Chem., 20 (1969), 411 - 418

4) B. Jakuszewski, S. Taniewska-Osinska, Acta Chim. Soc. Sci. Łodz , 4 (1959), 17

5) J.C. Marchon, Compt. Rend., 267 C (1968), 1123 - 1126

6) Data reported by A.J. De Bethune, N.A. Swendeman-Loud "Standard Aqueous Electrode Potentials and Temperature Coefficients at 25 ^{o}C" (Clifford A. Hampel, Ill.)

7) T. Naito, Kogyo Kagaku Zasshi, 65 (1962), 749 - 752

8) N.V. Troitskaya, K.P. Mishchenko, I.E. Flis, Zh. Fiz. Khim., 33 (1959), 1614 - 1617

Electronically conducting phase	Intermediate species	Composition of the solution	Solvent	Temperature °C	Pressure Torr	Measuring method
a	b	c	d		e	d
Ind.	Br_2 (in sol.)		1	25	760	IV
Ind.	Br_2 (gas)		1	25	760	IV
Ind.	Br_2 (l)		1	25	760	IV
Ind.	Br_2 (in sol.)		1	25	760	IV
Ind.	HBrO, Br_2 (gas)		1	25	760	IV
Ind.	HBrO, Br_2 (l)		1	25	760	IV
Ind.	HBrO, Br_2 (in sol.)		1	25	760	IV
Ind.	$HBrO_3$, Br_2 (gas)		1	25	760	IV
Ind.	$HBrO_3$, Br_2 (l)		1	25	760	IV
Ind.	$HBrO_3$, Br_2 (in sol.)		1	25	760	IV
Ind.	Br_2 (l)		1	25	760	IV
Pt	Br_2 (in sol.)	HBr : 0.100 and 1.050 M; Br_2 : 0.000132 up to 0.0141 M	1	25		I
	Br_2 (in sol.)		7	25		I
Ind.			1	25	760	IV
Pt		$(C_2H_5)_4$ $NClO_4$: 0.1 m	23	20 ± 0.5		I
Pt		$(C_2H_5)_4$ $NClO_4$: 0.1 m	5	20 ± 0.5		I
Ind.			1	25	760	IV
Ind.			1	25	760	IV
Ind.			1	25	760	IV
Ind.			1	25	760	IV
Ind.			1	25	760	IV
Ind.	HBrO		1	25	760	IV
Ind.	$HBrO_3$		1	25	760	IV
Ind.	Br_2 (gas)		1	25	760	IV
Ind.	Br_2 (l)		1	25	760	IV
Ind.	Br_2 (in sol.)		1	25	760	IV
Pt	Br_2 (in sol.)	$(C_2H_5)_4$ $NClO_4$: 0.1 m	23	20 ± 0.5		I
Pt	Br (in sol.)	$(C_2H_5)_4$ $NClO_4$: 0.1 m	5	20 ± 0.5		I
Ind.			1	25	760	IV
Ind.			1	25	760	IV
Ind.			1	25	760	IV
Ind.	HBrO		1	25	760	IV

Comparison electrode	Liquid junction	Electrode reaction	Standard value V	Uncertainty mV	Temperature Coefficient μV/°C	Notes	Reference	Electrode Reference Number
d			f		g	h		
		$2BrO^- + 4H^+ + 2e^- \rightleftharpoons Br_2 + 2H_2O$	2.090				1	1
		$2BrO_3^- + 12H^+ + 10e^- \rightleftharpoons Br_2 + 6H_2O$	1.492				1	2
		$2BrO_3^- + 12H^+ + 10e^- \rightleftharpoons Br_2 + 6H_2O$	1.495		-418		1,2	3
		$2BrO_3^- + 12H^+ + 10e^- \rightleftharpoons Br_2 + 6H_2O$	1.491				1	4
		$2HBrO + 2H^+ + 2e^- \rightleftharpoons Br_2 + 2H_2O$	1.579				1	5
		$2HBrO + 2H^+ + 2e^- \rightleftharpoons Br_2 + 2H_2O$	1.596				1	6
		$2HBrO + 2H^+ + 2e^- \rightleftharpoons Br_2 + 2H_2O$	1.574				1	7
		$2HBrO_3 + 10H^+ + 10e^- \rightleftharpoons Br_2 + 6H_2O$	1.484				1	8
		$2HBrO_3 + 10H^+ + 10e^- \rightleftharpoons Br_2 + 6H_2O$	1.487				1	9
		$2HBrO_3 + 10H^+ + 10e^- \rightleftharpoons Br_2 + 6H_2O$	1.482				1	10
		$Br_2 + 2e^- \rightleftharpoons 2Br^-$	1.066		-629		1,2	11
5		$Br_2 + 2e^- \rightleftharpoons 2Br^-$	1.0873	± 0.1	-541		2,3	12
		$Br_2 + 2e^- \rightleftharpoons 2Br^-$	0.8617				4	13
		$Br_3^- + 2e^- \rightleftharpoons 3Br^-$	1.051					14
33		$Br_3^- + 2e^- \rightleftharpoons 3Br^-$	0.22	± 20			5	15
33		$Br_3^- + 2e^- \rightleftharpoons 3Br^-$	0.63	± 20			5	16
		$Br_5^- + 4e^- \rightleftharpoons 5Br^-$	1.068				1	17
		$BrO^- + 2H^+ + 2e^- \rightleftharpoons Br^- + H_2O$	1.589				1	18
		$BrO^- + H_2O + 2e^- \rightleftharpoons Br^- + 2OH^-$	0.761				2	19
		$BrO_3^- + 6H^+ + 6e^- \rightleftharpoons Br^- + 3H_2O$	1.423				1	20
		$BrO_3^- + 3H_2O + 6e^- \rightleftharpoons Br^- + 6OH^-$	0.61				2	21
		$HBrO + H^+ + 2e^- \rightleftharpoons Br^- + H_2O$	1.331				1	22
		$HBrO_3 + 5H^+ + 6e^- \rightleftharpoons Br^- + 3H_2O$	1.417				1	23
		$3Br_2 + 2e^- \rightleftharpoons 2Br_3^-$	1.145				1	24
		$3Br_2 + 2e^- \rightleftharpoons 2Br_3^-$	1.096				1	25
		$3Br_2 + 2e^- \rightleftharpoons 2Br_3^-$	1.159				1	26
33		$3Br_2 + 2e^- \rightleftharpoons 2Br_3^-$	0.83	± 10			5	27
33		$3Br_2 + 2e^- \rightleftharpoons 2Br_3^-$	1.24	± 10			5	28
		$3Br_5^- + 2e^- \rightleftharpoons 5Br_3^-$	1.156				1	29
		$3BrO^- + 6H^+ + 4e^- \rightleftharpoons Br_3^- + 3H_2O$	1.856				1	30
		$3BrO_3^- + 18H^+ + 16e^- \rightleftharpoons Br_3^- + 9H_2O$	1.472				1	31
		$3HBrO + 3H^+ + 4e^- \rightleftharpoons Br_3^- + 3H_2O$	1.470				1	32

Electronically conducting phase	Intermediate species	Composition of the solution	Solvent	Temperature °C	Pressure Torr	Measuring method
a	b	c	d		e	d
Ind.	$HBrO_3$		1	25	760	IV
Ind.	Br_2 (l)		1	25	760	IV
Ind.	Br_2 (in sol.)		1	25	760	IV
Ind.			1	25	760	IV
Ind.	HBrO		1	25	760	IV
Ind.			1	25	760	IV
	$HBrO_3$		1	25	760	IV
Ind.	HBrO		1	25	760	IV
Ind.	$HBrO_3$, HBrO		1	25	760	IV

Comparison electrode	Liquid junction	Electrode reaction	Standard value	Uncertainty	Temperature Coefficient	Notes	Reference	Electrode Reference Number
			V	mV	µV/°C			
d			f		g	h		
		$3HBrO_3 + 15H^+ + 16e^- \rightleftharpoons Br_3^- + 9H_2O$	1.462				1	33
		$5Br_2 + 2e^- \rightleftharpoons 2Br_5^-$	1.056				1	34
		$5Br_2 + 2e^- \rightleftharpoons 2Br_5^-$	1.161				1	35
		$5BrO^- + 10H^+ + 6e^- \rightleftharpoons Br_5^- + 5H_2O$	1.935				1	36
		$5HBrO + 5H^+ + 6e^- \rightleftharpoons Br_5^- + 5H_2O$	1.505				1	37
		$BrO_3^- + 4H^+ + 4e^- \rightleftharpoons BrO^- + 2H_2O$	1.341				1	38
		$HBrO_3 + 3H^+ + 4e^- \rightleftharpoons BrO^- + 2H_2O$	1.330				1	39
		$BrO_3^- + 5H^+ + 4e^- \rightleftharpoons HBrO + 2H_2O$	1.470				1	40
		$HBrO_3 + 4H^+ + 4e^- \rightleftharpoons HBrO + 2H_2O$	1.460				1	41

Bromine 1

Standard value V	Solvent	Electrode Reference Number	Standard value V	Solvent	Electrode Reference Number	Standard value V	Solvent	Electrode Reference Number
0.61	1	21	2.090	1	1			
0.761	1	19	0.8617	7	13			
1.051	1	14	0.22	23	15			
1.056	1	34	0.83	23	27			
1.066	1	11	0.63	5	16			
1.068	1	17	1.24	5	28			
1.0873	1	12						
1.096	1	25						
1.145	1	24						
1.156	1	29						
1.159	1	26						
1.161	1	35						
1.330	1	39						
1.331	1	22						
1.341	1	38						
1.417	1	23						
1.423	1	20						
1.460	1	41						
1.462	1	33						
1.470	1	32						
1.470	1	40						
1.472	1	31						
1.482	1	10						
1.484	1	8						
1.487	1	9						
1.491	1	4						
1.492	1	2						
1.495	1	3						
1.505	1	37						
1.574	1	7						
1.579	1	5						
1.589	1	18						
1.596	1	6						
1.856	1	30						
1.935	1	36						

BIBLIOGRAPHY : Bromine

1) Data reported by A.J. De Bethune, N.A. Swendeman-Loud "Standard Aqueous Electrode Potentials and Temperature Coefficients at 25 °C" (Clifford A. Hampel, Ill.)

2) Data reported by M. Pourbaix 'Atlas d'Equilibres Electrochimiques à 25 °C" (Gauthier-Villars Editeur, Paris, 1963), pp. 604 - 613

3) T. Mussine, G. Faita, Ric. Sci., 36 (1966), 175 - 182

4) B. Jakuszewski, S. Taniewska-Osinka, Acta Chim. Soc. Sci. Łodz , 4 (1959), 17

5) J.C. Marchon, Compt. Rend., 267 C (1968), 1123 - 1126

Electronically conducting phase	Intermediate species	Composition of the solution	Solvent	Temperature °C	Pressure Torr	Measuring method
a	b	c	d		e	d
Ind.	HI (gas), I_2 (gas)		1	25	760	IV
Ind.	HIO, HIO_4		1	25	760	IV
Ind.	HIO		1	25	760	IV
Ind.	HIO		1	25	760	IV
Ind.	HIO		1	25	760	IV
Ind.	HIO_3 , HIO_4		1	25	760	IV
Ind.	HIO, I_2		1	25	760	IV
Ind.	HIO, I_2 (in sol.)		1	25	760	IV
Ind.	HIO_3 , I_2 (gas)		1	25	760	IV
Ind.	HIO_3 , I_2		1	25	760	IV
Ind.	HIO_3 , I_2 (in sol.)		1	25	760	IV
Ind.	HIO_4 , I_2 (in sol.)		1	25	760	IV
Ind.	I_2 (in sol.)		1	25	760	IV
Ind.	I_2 (gas)		1	25	760	IV
Ind.	I_2		1	25	760	IV
Ind.	I_2 (in sol.)		1	25	760	IV
Ind.	IBr (in sol.), I_2		1	25	760	IV
Ind.	I_2		1	25	760	IV
Ind.	ICl (in sol.), I_2		1	25	760	IV
Ind.	I_2 (in sol.)		1	25	760	IV
Ind.	ICl_3 (s), I_2 (s)		1	25	760	IV
Pt	I_2 , ICN (in sol.)	$HClO_4$: 0.2 up to 0.4 M; NaCN : 0.6 m; KIO_3 : 0.002 and 0.1 m; I_2 : 1.0 m	1	25 ± 0.2		I
Ind.	I_2		1	25	760	IV
Ind.	I_2 (in sol.)		1	25	760	IV
Ind.	I_2 (gas)		1	25	760	IV
Ind.	I_2		1	25	760	IV
Ind.	I_2 (in sol.)		1	25	760	IV
Ind.	I_2 (in sol.)		1	25	760	IV
Ind.	I_2 (in sol.)		1	25	760	IV
Ind.	HIO_3		1	25	760	IV
Ind.	HIO_4		1	25	760	IV
Ind.			1	25	760	IV

Comparison electrode	Liquid junction	Electrode reaction	Standard value	Uncertainty	Temperature Coefficient	Notes	Reference	Electrode Reference Number
			V	mV	µV/°C			
d			f		g	h		
		$I_2 + 2H^+ + 2e^- \rightleftharpoons 2HI$	0.087				1	1
		$HIO_4 + 6H^+ + 6e^- \rightleftharpoons HIO + 3H_2O$	1.290				1	2
		$HIO_5^{2-} + 8H^+ + 6e^- \rightleftharpoons HIO + 4H_2O$	1.389				1	3
		$IO_3^- + 5H^+ + 4e^- \rightleftharpoons HIO + 2H_2O$	1.134				1	4
		$IO_4^- + 7H^+ + 6e^- \rightleftharpoons HIO + 3H_2O$	1.235				1	5
		$HIO_4 + 2H^+ + 2e^- \rightleftharpoons HIO_3 + H_2O$	1.626				1	6
		$2HIO + 2H^+ + 2e^- \rightleftharpoons I_2 + 2H_2O$	1.439		420		1,2	7
		$2HIO + 2H^+ + 2e^- \rightleftharpoons I_2 + 2H_2O$	1.354				1	8
		$2HIO_3 + 10H^+ + 10e^- \rightleftharpoons I_2 + 6H_2O$	1.166				1	9
		$2HIO_3 + 10H^+ + 10e^- \rightleftharpoons I_2 + 6H_2O$	1.186				1	10
		$2HIO_3 + 10H^+ + 10e^- \rightleftharpoons I_2 + 6H_2O$	1.169				1	11
		$2HIO_4 + 14H^+ + 14e^- \rightleftharpoons I_2 + 8H_2O$	1.300				1	12
		$2HIO_5^{2-} + 18H^+ + 14e^- \rightleftharpoons I_2 + 10H_2O$	1.384				1	13
		$2I^+ + 2e^- \rightleftharpoons I_2$	1.256				1	14
		$2I^+ + 2e^- \rightleftharpoons I_2$	1.357				1	15
		$2I^+ + 2e^- \rightleftharpoons I_2$	1.272				1	16
		$2IBr + 2e^- \rightleftharpoons I_2 + 2Br^-$	1.02				3	17
		$2I(Br)_2^- + 2e^- \rightleftharpoons I_2 + 4Br^-$	0.87				3	18
		$2ICl + 2e^- \rightleftharpoons I_2 + 2Cl^-$	1.19				3	19
		$ICl_2^- + e^- \rightleftharpoons 2Cl^- + I_2$	1.056				2	20
		$2ICl_3 + 6e^- \rightleftharpoons I_2 + 6Cl^-$	1.28				3	21
3		$2ICN + 2H^+ + 2e^- \rightleftharpoons I_2 + 2HCN$	0.625	±3			4	22
		$2IO^- + 4H^+ + 2e^- \rightleftharpoons I_2 + 2H_2O$	2.090				1	23
		$2IO^- + 4H^+ + 2e^- \rightleftharpoons I_2 + 2H_2O$	2.005				1	24
		$2IO_3^- + 12H^+ + 10e^- \rightleftharpoons I_2 + 6H_2O$	1.175				1	25
		$2IO_3^- + 12H^+ + 10e^- \rightleftharpoons I_2 + 6H_2O$	1.195				1	26
		$2IO_3^- + 12H^+ + 10e^- \rightleftharpoons I_2 + 6H_2O$	1.178		-364		1,2	27
		$2IO_4^- + 16H^+ + 14e^- \rightleftharpoons I_2 + 8H_2O$	1.314				1	28
		$2IO_5^{3-} + 20H^+ + 14e^- \rightleftharpoons I_2 + 10H_2O$	1.477				1	29
		$HIO_3 + 5H^+ + 4e^- \rightleftharpoons I^+ + 3H_2O$	1.143				1	30
		$HIO_4 + 7H^+ + 6e^- \rightleftharpoons I^+ + 4H_2O$	1.304				1	31
		$IO_3^- + 6H^+ + 4e^- \rightleftharpoons I^+ + 3H_2O$	1.155				1	32

Electronically conducting phase	Intermediate species	Composition of the solution	Solvent	Temperature °C	Pressure Torr	Measuring method
a	b	c	d		e	d
Ind.	HIO		1	25	760	IV
Ind.	HIO_3		1	25	760	IV
Ind.	HIO_4		1	25	760	IV
Ind.			1	25	760	IV
Ind.	I_2 (gas)		1	25	760	IV
Ind.	I_2		1	25	760	IV
Pt	I_2 (in sol.)	I_2 : 10^{-5} up to 10^{-2} M; KI : not reported; KCl: 0.1M	1	25		I
	I_2		7	25		I
Ind.			1	25	760	IV
Ind.			1	25	760	IV
Pt		$LiClO_4$: 0.1 m	23	25		I
Pt		$(C_2H_5)_4NClO_4$: 0.1 m	5	20 ± 0.5		I
Ind.			1	25	760	IV
Ind.			1	25	760	IV
Ind.			1	25	760	IV
Ind.			1	25	760	IV
Ind.			1	25	760	IV
Ind.			1	25	760	IV
Ind.	HIO		1	25	760	IV
Ind.	HIO_3		1	25	760	IV
Ind.	HIO_4		1	25	760	IV
Ind.			1	25	760	IV
Ind.	I_2		1	25	760	IV
Ind.	I_2 (in sol.)		1	25	760	IV
Pt	I_2 (in sol.)	$LiClO_4$: 0.1 m	23	25		I
Pt	I_2 (in sol.)	$(C_2H_5)_4NClO_4$: 0.1 m	5	20± 0.5		I
Ind.			1	25	760	IV
Ind.			1	25	760	IV
Ind.			1	25	760	IV
Ind.			1	25	760	IV
Ind.			1	25	760	IV
Ind.			1	25	760	IV
Ind.			1	25	760	IV

Comparison electrode	Liquid junction	Electrode reaction	Standard value V	Uncertainty mV	Temperature Coefficient µV/°C	Notes	Reference	Electrode Reference Number
d			f		g	h		
		$HIO + H^+ + 2e^- \rightleftharpoons I^- + H_2O$	0.987				1	33
		$HIO_3 + 5H^+ + 6e^- \rightleftharpoons I^- + 3H_2O$	1.077				1	34
		$HIO_4 + 7H^+ + 8e^- \rightleftharpoons I^- + 4H_2O$	1.215				1	35
		$HIO_5^{2-} + 9H^+ + 8e^- \rightleftharpoons I^- + 5H_2O$	1.288				1	36
		$I_2 + 2e^- \rightleftharpoons 2I^-$	0.636				1	37
		$I_2 + 2e^- \rightleftharpoons 2I^-$	0.5355		-148		1,2	38
34	no	$I_2 + 2e^- \rightleftharpoons 2I^-$	0.615	±1			8	39
		$I_2 + 2e^- \rightleftharpoons 2I^-$	0.357				5	40
		$I^+ + 2e^- \rightleftharpoons I^-$	0.946				1	41
		$I_3^- + 2e^- \rightleftharpoons 3I^-$	0.536		-214		1,2	42
7		$I_3^- + 2e^- \rightleftharpoons 3I^-$	-0.124				6	43
33		$I_3^- + 2e^- \rightleftharpoons 3I^-$	0.28	±10			7	44
		$IO^- + 2H^+ + 2e^- \rightleftharpoons I^- + H_2O$	1.313				1	45
		$IO^- + H_2O + 2e^- \rightleftharpoons I^- + 2OH^-$	0.485				2	46
		$IO_3^- + 6H^+ + 6e^- \rightleftharpoons I^- + 3H_2O$	1.085				1	47
		$IO_3^- + 3H_2O + 6e^- \rightleftharpoons I^- + 6OH^-$	0.26		-1162		2	48
		$IO_4^- + 8H^+ + 8e^- \rightleftharpoons I^- + 4H_2O$	1.227				1	49
		$IO_5^{3-} + 10H^+ + 8e^- \rightleftharpoons I^- + 5H_2O$	1.370				1	50
		$3HIO + 3H^+ + 4e^- \rightleftharpoons I_3^- + 3H_2O$	1.213				1	51
		$3HIO_3 + 15H^+ + 16e^- \rightleftharpoons I_3^- + 9H_2O$	1.145				1	52
		$3HIO_4 + 21H^+ + 22e^- \rightleftharpoons I_3^- + 12H_2O$	1.276				1	53
		$3HIO_5^{2-} + 27H^+ + 22e^- \rightleftharpoons I_3^- + 15H_2O$	1.357				1	54
		$3I_2 + 2e^- \rightleftharpoons 2I_3^-$	0.534				1	55
		$3I_2 + 2e^- \rightleftharpoons 2I_3^-$	0.789				1	56
7		$3I_2 + 2e^- \rightleftharpoons 2I_3^-$	0.520				6	57
33		$3I_2 + 2e^- \rightleftharpoons 2I_3^-$	0.92	±10			7	58
		$3I^+ + 4e^- \rightleftharpoons I_3^-$	1.151				1	59
		$3IO^- + 6H^+ + 4e^- \rightleftharpoons I_3^- + 3H_2O$	1.701				1	60
		$3IO_3^- + 18H^+ + 16e^- \rightleftharpoons I_3^- + 9H_2O$	1.154				1	61
		$3IO_5^{3-} + 3OH^+ + 22e^- \rightleftharpoons I_3^- + 15H_2O$	1.445				1	62
		$3IO_4^- + 24H^+ + 22e^- \rightleftharpoons I_3^- + 12H_2O$	1.290				1	63
		$IO_3^- + 4H^+ + 4e^- \rightleftharpoons IO^- + 2H_2O$	0.972				1	64
		$IO_5^{3-} + 8H^+ + 6e^- \rightleftharpoons IO^- + 4H_2O$	1.389				1	65

Electronically conducting phase	Intermediate species	Composition of the solution	Solvent	Temperature °C	Pressure Torr	Measuring method
a	b	c	d		e	d
Ind.	HIO_4		1	25	760	IV
Ind.			1	25	760	IV
Ind.			1	25	760	IV
Ind.	H_5IO_6		1	25	760	IV
Ind.			1	25	760	IV
Ind.			1	25	760	IV

Comparison electrode	Liquid junction	Electrode reaction	Standard value V	Uncertainty mV	Temperature Coefficient μV/°C	Notes	Reference	Electrode Reference Number
d			f		g	h		
		$HIO_4 + H^+ + 2e^- \rightleftharpoons IO_3^- + H_2O$	1.603				1	66
		$HIO_5^{2-} + 3H^+ + 2e^- \rightleftharpoons IO_3^- + 2H_2O$	1.898				1	67
		$H_3IO_6^{2-} + 2e^- \rightleftharpoons IO_3^- + 3OH^-$	0.7				2	68
		$H_5IO_6 + H^+ + 2e^- \rightleftharpoons IO_3^- + 3H_2O$	1.601				2	69
		$IO_4^- + 2H^+ + 2e^- \rightleftharpoons IO_3^- + H_2O$	1.653				1	70
		$IO_5^{3-} + 4H^+ + 2e^- \rightleftharpoons IO_3^- + 2H_2O$	2.223				1	71

Iodine 1

Standard value V	Solvent	Electrode Reference Number
0.087	1	1
0.26	1	48
0.485	1	46
0.534	1	55
0.5355	1	38
0.536	1	42
0.615	1	39
0.625	1	22
0.636	1	37
0.7	1	68
0.789	1	56
1.02	1	17
0.946	1	41
0.972	1	64
0.987	1	33
0.87	1	18
1.056	1	20
1.077	1	34
1.085	1	47
1.134	1	4
1.143	1	30
1.145	1	52
1.151	1	59
1.154	1	61
1.155	1	32
1.166	1	9
1.169	1	11
1.175	1	25
1.178	1	27
1.186	1	10
1.19	1	19
1.195	1	26
1.213	1	51
1.215	1	35
1.227	1	49
1.235	1	5
1.256	1	14
1.272	1	16
1.276	1	53
1.28	1	21
1.288	1	36
1.290	1	2
1.290	1	63
1.300	1	12
1.304	1	31
1.313	1	45
1.314	1	28
1.354	1	8
1.357	1	15
1.357	1	54
1.370	1	50
1.384	1	13
1.389	1	3
1.389	1	65
1.439	1	7
1.445	1	62
1.447	1	29
1.601	1	69
1.603	1	66
1.626	1	6
1.653	1	70
1.701	1	60
1.898	1	67
2.005	1	24
2.090	1	23
2.223	1	71
0.357	7	40
-0.124	23	43
0.520	23	57
0.28	5	44
0.92	5	58

BIBLIOGRAPHY: Iodine

1) Data reported by M. Pourbaix, "Atlas d'Equilibres Electrochimiques à 25 °C" (Gauthier-Villars Editeur, Paris, 1963), pp. 614 - 626

2) Data reported by A.J. De Bethune, N.A. Swendeman-Loud, "Standard Aqueous Electrode Potentials and Temperature Coefficients at 25 °C" (Clifford A. Hampel, Ill.)

3) W.M. Latimer, Oxidation Potentials, 2nd Ed. Prentice-Hall, New York, 1952

4) D.F. Bowersox, E.A. Butler, E.H. Swift, Anal. Chem. 28 (1956), 221 - 224

5) B. Jakuszewski, S. Taniewska-Osinka, Acta Chim. Soc. Sci. Łodz, 4 (1959), 17

6) J. Desbarres, Bull. Soc. Chim. France, 1961, 502 - 506

7) J.C. Marchon, Compt. Rend., 267 C (1968), 1123 - 1126

8) L.G. Lavrenova, T.V. Zegzhda, V.M. Shul'man, Elektrokhim., 7 (1971), 83 - 86

Electronically conducting phase	Intermediate species	Composition of the solution	Solvent	Temperature °C	Pressure Torr	Measuring method
a	b	c	d		e	d
Ind.	At_2		1	25	760	IV
Ind.	At_2 , HAtO		1	25	760	IV
Ind.	At_2		1	25	760	IV
Ind.			1	25	760	IV
Ind.	$HAtO_3$, $HAtO_3$		1	25	760	IV

Comparison electrode	Liquid junction	Electrode reaction	Standard value	Uncertainty	Temperature Coefficient	Notes	Reference	Electrode Reference Number
			V	mV	μV/°C			
d			f		g	h		
		$2AtO^- + 2H_2O + 2e^- \rightleftharpoons At_2 + 2OH^-$	0.0				1	1
		$2HAtO + 2H^+ + 2e^- \rightleftharpoons At_2 + 2H_2O$	0.7				1	2
		$At_2 + 2e^- \rightleftharpoons 2At^-$	0.2				1	3
		$AtO_3^- + 2H_2O + 4e^- \rightleftharpoons AtO^- + 4OH^-$	0.5				1	4
		$HAtO_3 + 4H^+ + 4e^- \rightleftharpoons HAtO + 2H_2O$	1.4				1	5

Astatine 1

Standard value V	Solvent	Electrode Reference Number	Standard value V	Solvent	Electrode Reference Number	Standard value V	Solvent	Electrode Reference Number
0.0	1	1						
0.2	1	3						
0.5	1	4						
0.7	1	2						
1.4	1	5						

BIBLIOGRAPHY : Astatine

1) Data reported by A.J. De Bethune, N.A. Swendeman-Loud "Standard Aqueous Electrode Potentials and Temperature Coefficients at 25 °C" (Clifford A. Hampel, Ill.)

Electronically conducting phase	Intermediate species	Composition of the solution	Solvent	Temperature °C	Pressure Torr	Measuring method
a	b	c	d		e	d
Ind.	Mn_2O_3		1	25	760	IV
Ind.			1	25	760	IV
Ind.	Mn_3O_4		1	25	760	IV
Ind.	Mn $(OH)_3$		1	25	760	IV
Mn			1	25	760	IV
Mn			1	25	760	IV
Mn			1	25	760	IV
Mn			1	25	760	IV
Mn	$MnCO_3$ (crystals)		1	25	760	IV
Mn	$MnCO_3$ (precipitate)		1	25	760	IV
Mn	MnO		1	25	760	IV
Mn	MnO_2		1	25	760	IV
Mn	Mn_2O_3		1	25	760	IV
Mn	Mn_3O_4		1	25	760	IV
Mn			1	25	760	IV
Mn	Mn $(OH)_2$		1	25	760	IV
Mn	Mn $(OH)_2$		1	25	760	IV
Mn	Mn $(OH)_3$		1	25	760	IV
Mn	Mn $(OH)_4$		1	25	760	IV
Mn	MnOOH		1	25	760	IV
Mn	MnO $(OH)_2$		1	25	760	IV
Mn			1	25		III
Mn		$MnCl_2$: (0.3225 up to 1.77524) · 10^{-3} M; $Na_4P_4O_{12}$: (0.83728 up to 1.44920) · 10^{-3} M	1	25		III
Mn	Mn (S_2O_3)	$MnCl_2$: (10.64 up to 21.02); BaS_2O_3 satd.	1	25± 0.03		V
	Mn (SO_4)		1	25		III
Au		Mn $(ClO_4)_2$: 0.01 up to 0.1 M; $HClO_4$: 3 M	1	25± 0.05		I
Ind.			1	25	760	IV
Pt	MnO_2	1) Not reported 2) $HClO_4$: 0.0250 up to 0.5000 M; Mn $(ClO_4)_2$: 0.002 up to 0.5000 M 3) $MnCl_2$: 0.001 up to 0.1 m; HCl 4) $MnCl_2$, HCl, CH_3COOH, Mn $(CH_3COO)_2$ pH : 4.32 up to 4.57; or $MnClO_4$ + $HClO_4$	1	25		I
Ind.	Mn_2O_3		1	25	760	IV

Comparison electrode	Liquid junction	Electrode reaction	Standard value V	Uncertainty mV	Temperature Coefficient μV/°C	Notes	Reference	Electrode Reference Number
d			f		g	h		
		$Mn_2O_3 + H_2O + 2e^- \rightleftharpoons 2HMnO_2^-$	-0.590				1	1
		$MnO_4^{2-} + 5H^+ + 4e^- \rightleftharpoons HMnO_2^- + 2H_2O$	1.234				1	2
		$Mn_3O_4 + 2H_2O + 2e^- \rightleftharpoons 3HMnO_2^- + H^+$	1.228				1	3
		$Mn(OH)_3 + e^- \rightleftharpoons HMnO_2^- + H_2O$	-0.148				1	4
		$HMnO_2^- + 3H^+ + 2e^- \rightleftharpoons Mn + 2H_2O$	-0.163			1	1,2	5
		$Mn^{2+} + 2e^- \rightleftharpoons Mn$	-1.185	± 5	-80	1	2,3,4	6
		$Mn^{3+} + 3e^- \rightleftharpoons Mn$	-0.2830				2	7
		$Mn^{4+} + 4e^- \rightleftharpoons Mn$	0.195				2	8
		$MnCO_3 + 2e^- \rightleftharpoons Mn + CO_3^{2-}$	-1.50		-1232		3	9
		$MnCO_3 + 2e^- \rightleftharpoons Mn + CO_3^{2-}$	-1.48		-1304		3	10
		$MnO + 2H^+ + 2e^- \rightleftharpoons Mn + H_2O$	-0.652			1	1,2	11
		$MnO_2 + 4H^+ + 4e^- \rightleftharpoons Mn + 2H_2O$	0.024				2	12
		$Mn_2O_3 + 6H^+ + 6e^- \rightleftharpoons 2Mn + 3H_2O$	-0.305				2	13
		$Mn_3O_4 + 8H^+ + 8e^- \rightleftharpoons 3Mn + 4H_2O$	-0.436				2	14
		$MnOH^+ + H^+ + 2e^- \rightleftharpoons Mn + H_2O$	-0.873				2	15
		$Mn(OH)_2 + 2H^+ + 2e^- \rightleftharpoons Mn + 2H_2O$	-0.727			1	1,2	16
		$Mn(OH)_2 + 2e^- \rightleftharpoons Mn + 2OH^-$	-1.56		-1079	1	3,5	17
		$Mn(OH)_3 + 3H^+ + 3e^- \rightleftharpoons Mn + 3H_2O$	-0.157				2	18
		$Mn(OH)_4 + 4H^+ + 4e^- \rightleftharpoons Mn + 4H_2O$	0.214				2	19
		$MnOOH + 3H^+ + 3e^- \rightleftharpoons Mn + 2H_2O$	-0.335				2	20
		$MnO(OH)_2 + 4H^+ + 4e^- \rightleftharpoons Mn + 3H_2O$	0.116				2	21
		$Mn(P_3O_9)^- + 2e^- \rightleftharpoons Mn + P_3O_9^{3-}$	-1.291			2	6	22
		$Mn(P_4O_{12})^{2-} + 2e^- \rightleftharpoons Mn + P_4O_{12}^{4-}$	-1.355			2	7	23
1		$Mn(S_2O_3) + 2e^- \rightleftharpoons Mn + S_2O_3^{2-}$	-1.243			3	8	24
		$Mn(SO_4) + 2e^- \rightleftharpoons Mn + SO_4^{2-}$	-1.252			2	9	25
35		$Mn^{3+} + e^- \rightleftharpoons Mn^{2+}$	1.5415	± 0.3	1230		3,10	26
		$Mn^{4+} + 2e^- \rightleftharpoons Mn^{2+}$	1.569				2	27
1) 3		$MnO_2 + 4H^+ + 2e^- \rightleftharpoons Mn^{2+} + 2H_2O$	1.224	± 13	-661		11	28
2) 1							12,13	
3) 4							14	
4) 1 or 4							3	
		$Mn_2O_3 + 6H^+ + 2e^- \rightleftharpoons 2Mn^{2+} + 3H_2O$	1.443			1	1,2	29

Electronically conducting phase	Intermediate species	Composition of the solution	Solvent	Temperature °C	Pressure Torr	Measuring method
a	b	c	d		e	d
Ind.			1	25	760	IV
Ind.			1	25	760	IV
Ind.	Mn_3O_4		1	25	760	IV
Ind.	$Mn(OH)_3$		1	25	760	IV
Ind.	$Mn(OH)_4$		1	25	760	IV
Ind.	MnOOH		1	25	760	IV
Ind.	$MnO(OH)_2$		1	25	760	IV
Ind.	MnO_2		1	25	760	IV
Ind.			1	25	760	IV
Pt	$[Mn(CO)_5]_2$	$Na[Mn(CO)_5]$: 0.01 M	1	20		I
Ind.	MnO, MnO_2		1	25	760	IV
Ind.	MnO, Mn_2O_3		1	25	760	IV
Ind.	MnO, Mn_3O_4		1	25	760	IV
Ind.	MnO, $Mn(OH)_3$		1	25	760	IV
Pt	MnO_2		1	25		I
Ind.	MnO_2		1	25	760	IV
Ind.	MnO_2		1	25	760	IV
Ind.	MnO_2		1	25	760	IV
Ind.	Mn_2O_3, MnO_2		1	25	760	IV
Pt		KOH : 0.02 up to 4 N; K_2MnO_4 : (0.069 up to 0.281) · 10^{-3} M; $KMnO_4$: (0.079 up to 0.371) · 10^{-3} M	1	25± 0.5		I
Pt (R.D.E.)		$KMnO_4$: 3 · 10^{-3} M and 6 · 10^{-4} M; NaOH : 0.03 up to 15 M; $NaClO_4$: 0 up to 0.97 M	1	25		III
Ind.	Mn_3O_4, Mn_2O_3		1	25	760	IV
Ind.	Mn_3O_4, $Mn(OH)_3$		1	25	760	IV
Ind.	Mn_3O_4, MnOOH		1	25	760	IV
Ind.	$Mn(OH)_2$, MnO_2		1	25	760	IV
Ind.	$Mn(OH)_2$, MnO_2		1	25	760	IV
Ind.	$Mn(OH)_2$, Mn_2O_3		1	25	760	IV
Ind.	$Mn(OH)_3$, Mn_3O_4		1	25	760	IV
Ind.	$Mn(OH)_2$, $Mn(OH)_3$		1	25	760	IV
Ind.	$Mn(OH)_2$, $Mn(OH)_3$		1	25	760	IV
Ind.	$Mn(OH)_2$, MnOOH		1	25	760	IV

Comparison electrode	Liquid junction	Electrode reaction	Standard value V	Uncertainty mV	Temperature Coefficient μV/°C	Notes	Reference	Electrode Reference Number
d			f		g	h		
		$MnO_4^- + 8H^+ + 5e^- \rightleftharpoons Mn^{2+} + 4H_2O$	1.507		-660		1,3	30
		$MnO_4^{2-} + 8H^+ + 4e^- \rightleftharpoons Mn^{2+} + 4H_2O$	1.742				1	31
		$Mn_3O_4 + 8H^+ + 2e^- \rightleftharpoons 3Mn^{2+} + 4H_2O$	1.824			1	1,2	32
		$Mn(OH)_3 + 3H^+ + e^- \rightleftharpoons Mn^{2+} + 3H_2O$	1.900				1	33
		$Mn(OH)_4 + 4H^+ + 2e^- \rightleftharpoons Mn^{2+} + 4H_2O$	1.608				2	34
		$MnOOH + 3H^+ + e^- \rightleftharpoons Mn^{2+} + 2H_2O$	1.352				2	35
		$MnO(OH)_2 + 4H^+ + 2e^- \rightleftharpoons Mn^{2+} + 3H_2O$	1.42				2	36
		$MnO_2 + 4H^+ + e^- \rightleftharpoons Mn^{3+} + 2H_2O$	0.948				1	37
		$MnO_4^- + 8H^+ + 4e^- \rightleftharpoons Mn^{3+} + 4H_2O$	1.506				1	38
3		$[Mn(CO)_5]_2 + 2e^- \rightleftharpoons 2[Mn(CO)_5]^-$	-0.68				15	39
		$MnO_2 + 2H^+ + 2e^- \rightleftharpoons MnO + H_2O$	0.7025				2	40
		$Mn_2O_3 + 2H^+ + 2e^- \rightleftharpoons 2MnO + H_2O$	0.840				2	41
		$Mn_3O_4 + 2H^+ + 2e^- \rightleftharpoons 3MnO + H_2O$	0.240			1	1,2	42
		$Mn(OH)_3 + H^+ + e^- \rightleftharpoons MnO + 2H_2O$	0.835				2	43
3		$MnO_4^- + 4H^+ + 3e^- \rightleftharpoons MnO_2 + 2H_2O$	1.679	±4	-666		3,11	44
		$MnO_4^- + 2H_2O + 2e^- \rightleftharpoons MnO_2 + 4OH^-$	0.595		-1778	1	2,3,5	45
		$MnO_4^{2-} + 4H^+ + 2e^- \rightleftharpoons MnO_2 + 2H_2O$	2.257				1	46
		$MnO_4^{2-} + 2H_2O + 2e^- \rightleftharpoons MnO_2 + 4OH^-$	0.60				3	47
		$2MnO_2 + 2H^+ + 2e^- \rightleftharpoons Mn_2O_3 + H_2O$	1.014			1	1,2	48
3	KCl satd.	$MnO_4^- + e^- \rightleftharpoons MnO_4^{2-}$	0.558	±2			16	49
22		$MnO_4^{2-} + e^- \rightleftharpoons MnO_4^{3-}$	0.274	±10			17	50
		$3Mn_2O_3 + 2H^+ + 2e^- \rightleftharpoons 2Mn_3O_4 + H_2O$	0.689				1	51
		$3Mn(OH)_3 + H^+ + e^- \rightleftharpoons Mn_3O_4 + 5H_2O$	2.014				1	52
		$3MnOOH + H^+ + e^- \rightleftharpoons Mn_3O_4 + 2H_2O$	0.417				2	53
		$MnO_2 + 2H^+ + 2e^- \rightleftharpoons Mn(OH)_2$	0.776				2	54
		$MnO_2 + 2H_2O + 2e^- \rightleftharpoons Mn(OH)_2 + 2OH^-$	-0.05		-1329		3	55
		$Mn_2O_3 + H_2O + 2H^+ + 2e^- \rightleftharpoons 2Mn(OH)_2$	0.538				2	56
		$Mn_3O_4 + 2H^+ + 2H_2O + 2e^- \rightleftharpoons 3Mn(OH)_2$	0.463			1	1,2	57
		$Mn(OH)_3 + H^+ + e^- \rightleftharpoons Mn(OH)_2 + H_2O$	0.98				2	58
		$Mn(OH)_3 + e^- \rightleftharpoons Mn(OH)_2 + OH^-$	0.15		-903		3	59
		$MnOOH + H^+ + e^- \rightleftharpoons Mn(OH)_2$	0.447				2	60

Electronically conducting phase	Intermediate species	Composition of the solution	Solvent	Temperature °C	Pressure Torr	Measuring method
a	b	c	d		e	d
Ind.	Mn $(OH)_2$, MnO$(OH)_2$		1	25	760	IV
Ind.	Mn $(OH)_3$, MnO_2		1	25	760	IV
Glass		$MnCl_2$: (8.27 up to 60.10)· 10^{-3} M; alanine :(18.67 up to 21.29) · 10^{-3} M; NaOH : (0.53 up to 0.65) · 0.65) · 10^{-3} M	1	25 ± 0.02		I
Pt		Mn $(ClO_4)_2$, K[Mn (CyDTA)], Na_2H_2CyDTA, $NaClO_4$, CH_3 COONa 0.05 M	1	25 ± 0.05		I
Pt		1) Mn $(ClO_4)_2$, Na_2H_2EDTA, K[Mn (EDTA)], $NaClO_4$, CH_3 COONa: 0.05 M; 2) K[Mn (EDTA)]: 10^{-3} M,HCl, CH_3 COOH + CH_3 COONa buffer, pH = 4.63	1	25		I,III
Glass		$MnCl_2$: (5.66 up to 60.08) · 10^{-3} M; glycyne: (22.62 up to 26.94) · 10^{-3} M; NaOH : (0.59 up to 2.10) · 10^{-3} M	1	25 ± 0.02		I
Glass		$MnCl_2$: (6.07 up to 57.96) · 10^{-3} M; glycylglycine : (12.30 up to 15.07) · 10^{-3} M; NaOH : (0.63 up to 2.72) · 10^{-3} M	1	25 ± 0.02		I
Pt, H_2		$MnCl_2$: (40.53 up to 48.72) · 10^{-3} M; glycylglycylglycine : (16.35 up to 21.50)· 10^{-3} M; NaOH : (5.85 up to 41.57) · 10^{-3} M	1	25 ± 0.03		I
Pt		Mn $(ClO_4)_2$, K[Mn (HEDTA)], Na_2H_2HEDTA, $NaClO_4$, CH_3 COONa:0.05 M	1	25± 0.05		I
Glass	Mn (8-hydroxyquinoline-5-sulfonate)	8-Hydroxyquinoline-5-sulfonate : 2.954 · 10^{-3} M; NaOH : 1.4 · 10^{-3} M; $MnCl_2$	1	25	743	I
Mn	Mn (malonate)		1	25		V

Comparison electrode	Liquid junction	Electrode reaction	Standard value	Uncertainty	Temperature Coefficient	Notes	Reference	Electrode Reference Number
			V	mV	μV/°C			
d			f		g	h		
		$MnO(OH)_2 + 2H^+ + 2e^- \rightleftharpoons Mn(OH)_2 + H_2O$	0.9670				2	61
		$MnO_2 + H_2O + H^+ + e^- \rightleftharpoons Mn(OH)_3$	0.573				1	62
3		$Mn(alanine)^{2+} + 2e^- \rightleftharpoons Mn + alanine$	-1.274			1	18	63
3	KNO_3 agar bridge	$[Mn(CyDTA)]^- + e^- \rightleftharpoons [Mn(CyDTA)]^{2-}$	0.814			4	19	64
3	KNO_3 agar bridge	$[Mn(EDTA)]^- + e^- \rightleftharpoons [Mn(EDTA)]^{2-}$	0.825			5	11,19	65
3		$[Mn(glycyne)]^{2+} + 2e^- \rightleftharpoons Mn + glycyne$	-1.286			6	18	66
3		$[Mn(glycylglycine)]^{2+} + 2e^- \rightleftharpoons Mn + glycylglycine$	-1.248			6	18	67
4		$[Mn(glycylglycylglycine)]^{2+} + 2e^- \rightleftharpoons$ Mn + glycylglycylglycine	-1.227			6	20	68
3	KNO agar bridge	$[Mn(HEDTA)]^- + e^- \rightleftharpoons [Mn(HEDTA)]^{2-}$	0.782			7	19	69
4		[Mn (8-hydroxyquinoline-5-sulfonate] + $2e^- \rightleftharpoons$ Mn + (8-hydroxyquinoline-5-sulfonate)$^{2-}$	-1.390			6	21	70
		[Mn (malonate)] + $2e^- \rightleftharpoons$ Mn + malonate^{2-}	-1.282			8	22	71

Manganese 1

Standard value V	Solvent	Electrode Reference Number
-1.56	1	17
-1.50	1	9
-1.48	1	10
-1.390	1	70
-1.355	1	23
-1.291	1	22
-1.286	1	66
-1.282	1	71
-1.274	1	63
-1.252	1	25
-1.248	1	67
-1.243	1	24
-1.227	1	68
-1.185	1	6
-0.873	1	15
-0.727	1	16
-0.68	1	39
-0.652	1	11
-0.590	1	1
-0.436	1	14
-0.335	1	20
-0.305	1	13
-0.2830	1	7
-0.163	1	5
-0.157	1	18
-0.148	1	4
-0.05	1	55
0.024	1	12
0.116	1	21
0.15	1	59
0.195	1	8
0.214	1	19
0.240	1	42
0.274	1	50
0.417	1	53

Standard value V	Solvent	Electrode Reference Number
0.447	1	60
0.463	1	57
0.538	1	56
0.558	1	49
0.573	1	62
0.595	1	45
0.60	1	47
0.689	1	51
0.7025	1	40
0.776	1	54
0.782	1	69
0.814	1	64
0.825	1	65
0.835	1	43
0.840	1	41
0.948	1	37
0.9670	1	61
0.98	1	58
1.014	1	48
1.224	1	28
1.228	1	3
1.234	1	2
1.352	1	35
1.42	1	36
1.443	1	29
1.506	1	38
1.507	1	30
1.5415	1	26
1.569	1	27
1.608	1	34
1.679	1	44
1.742	1	31
1.824	1	32
1.900	1	33
2.014	1	52

Standard value V	Solvent	Electrode Reference Number
2.257	1	46

NOTES : Manganese

1) The value is an average of those given by several authors when they all are in the same range of uncertainty
2) Conductometric data
3) Solubility data
4) H_4 CyDTA = trans -1,2 diaminocyclohexane-tetracetic acid
5) H_4 EDTA = Ethylenediamine-tetracetic acid
6) Potentiometric pH measurements
7) H_4 HEDTA = Hydroxyethylenediamine-tetracetic acid
8) Colorimetric pH measurements

BIBLIOGRAPHY : Manganese

1) Data reported by M. Pourbaix "Atlas d'Equilibres Electrochimiques à 25 ^{o}C" (Gauthier-Villars Editeur, Paris, 1963), pp. 286 - 293
2) J.P. Brenet, Z. Pavlovic, R. Popovic, Werkst. Korr., 19 (1968), 393 - 405
3) Data reported by A.J. De Bethune, N.A. Swendeman-Loud "Standard Aqueous Electrode Potentials and Temperature Coefficients at 25 ^{o}C" (Clifford A; Hampel, Ill.)
4) A. Walkley, J. Electrochem. Soc., 94 (1948), 41
5) T.A. Zordan, L.G. Hepler, Chem. Rev., 68 (1968), 737 - 745
6) H.W. Jones, C.B. Monk, C.W. Davies, J. Chem. Soc., 1949, 2693
7) H.W. Jones, C.B. Monk, J. Chem. Soc., 1950, 3475 - 3478
8) T.O. Donney, C.B. Monk, Trans. Faraday Soc., 47 (1951), 992 - 998
9) J.C. James, Thesis, London, 1947
10) L. Ciavatta, M. Grimaldi, J. Inorg. Chem., 31 (1965), 3071 - 3082
11) N. Tanaka, T. Shirikashi, H. Ogino, Bull. Chem. Soc. Japan, 38 (1965), 1515 - 1517
12) A. Witt Hutchison, J. Am. Chem. Soc., 69 (1947), 3051 - 3054
13) K.H. Maxwell, H.R. Thirsk, Proc. 6th Meeting Intern. Comm. Electrochem. Thermodynam. Kinetics, 1955, 390 - 409; J. Chem. Soc., 1955, 4054 - 4057
14) A.K. Covington, T. Cressey, B.G. Lever, H.R. Thirsk, Trans. Faraday Soc., 58 (1962), 1975 - 1988
15) W. Hieber, G. Wagner, Z. Naturforsch., 13 b (1958), 339 - 347
16) A. Carrington, M.C.R. Symons, J. Chem. Soc., 1956, 3373 - 3380
17) H. Schurig, K.E. Heusler, Fresenius' Z. Anal. Chem., 224 (1967), 45 - 62
18) C.B. Monk, Trans. Faraday Soc., 47 (1951), 297 - 302
19) R.E. Hamm, M.A. Suwyn, Inorg. Chem., 6 (1967), 139 - 142
20) J.I. Evans, C.B. Monk, Trans. Faraday Soc., 51 (1955), 1244 - 1250
21) R. Näsänen, E. Uisitalo, Acta Chem. Scand., 8 (1954) 112 - 118
22) D.I. Stock, C.W. Davies, J. Chem. Soc., 1949, 1371

Electronically conducting phase	Intermediate species	Composition of the solution	Solvent	Temperature °C	Pressure Torr	Measuring method
a	b	c	d		e	d
Tc			1	25	760	IV
Tc	TcO_2		1	25	760	IV
Tc			1	25	760	IV
Ind.	$HTcO_4$ (in sol.)		1	25	760	IV
Ind.	TcO_2		1	25	760	IV
Ind.			1	25	760	IV
Ind.	TcO_2, $HTcO_4$ (in sol.)		1	25	760	IV
Ind.	TcO_2, TcO_3		1	25	760	IV
Pt	TcO_2	$HTcO_4$: (0.135 up to 1.35) · 10^{-3} M	1	25 ± 0.05		I
Pt	TcO_2	$NaTcO_4$: 0.0683 M; NaOH : 0.0683 M	1	25± 0.05		I
Ind.	TcO_3, $HTcO_4$ (in sol.)		1	25	760	IV
Ind.	TcO_3		1	25	760	IV

Comparison electrode	Liquid junction	Electrode reaction	Standard value	Uncertainty	Temperature Coefficient	Notes	Reference	Electrode Reference Number
			V	mV	μV/°C			
d			f		g	h		
		$Tc^{2+} + 2e^- \rightleftharpoons Tc$	0.400				1	1
		$TcO_2 + 4H^+ + 4e^- \rightleftharpoons Tc + 2H_2O$	0.272				1,2	2
		$TcO_4^- + 8H^+ + 7e^- \rightleftharpoons Tc + 4H_2O$	0.472			1	1,3	3
		$HTcO_4 + 7H^+ + 5e^- \rightleftharpoons Tc^{2+} + 4H_2O$	0.501				1	4
		$TcO_2 + 4H^+ + 2e^- \rightleftharpoons Tc^{2+} + 2H_2O$	0.144				1	5
		$TcO_4^- + 8H^+ + 5e^- \rightleftharpoons Tc^{2+} + 4H_2O$	0.500				1	6
		$HTcO_4 + 3H^+ + 3e^- \rightleftharpoons TcO_2 + 2H_2O$	0.740				1	7
		$TcO_3 + 2H^+ + 2e^- \rightleftharpoons TcO_2 + H_2O$	0.757				1	8
4 HCl 0.05M		$TcO_4 + 4H^+ + 3e^- \rightleftharpoons TcO_2 + 2H_2O$	0.782	± 11		2	3	9
22 (NaOH)		$TcO_4^- + 2H_2O + 3e^- \rightleftharpoons TcO_2 + 4OH^-$	-0.322	± 11		2	3	10
		$HTcO_4 + H^+ + e^- \rightleftharpoons TcO_3 + H_2O$	0.707				1	11
		$TcO_4^- + 2H^+ + e^- \rightleftharpoons TcO_3 + H_2O$	0.700				1	12

Technetium 1

Standard value V	Solvent	Electrode Reference Number	Standard value V	Solvent	Electrode Reference Number	Standard value V	Solvent	Electrode Reference Number
-0.322	1	10						
0.144	1	5						
0.272	1	2						
0.400	1	1						
0.472	1	3						
0.500	1	6						
0.501	1	4						
0.700	1	12						
0.707	1	11						
0.740	1	7						
0.757	1	8						
0.782	1	9						

NOTES : Technetium

1) The value is an average of those given by several authors when they all are in the same range of uncertainty.

2) Liquid junction : KCl 1.8 M + KNO_3 1.8 M

BIBLIOGRAPHY : Technetium

1) Data reported by M. Pourbaix "Atlas d'Equilibres Electrochimiques à 25 °C" (Gauthier-Villars Editeur, Paris), pp. 294 - 299

2) G.H. Cartledge, W.T. Smith, Jr., J. Phys. Chem., 59 (1955), 1111 - 1112

3) J.W. Cobble, W.T. Smith, Jr., G.E. Boyd, J. Am. Chem. Soc., 75 (1953), 5777 - 5782

Electronically conducting phase	Intermediate species	Composition of the solution	Solvent	Temperature °C	Pressure Torr	Measuring method
a	b	c	d		e	d
Re			1	25	760	IV
Re	ReO_2		1	25	760	IV
Re	ReO_2		1	25	760	IV
Re	Re_2O_3		1	25	760	IV
Re			1	25	760	IV
Re			1	25	760	IV
Re			1	25	760	IV
Ind.			1	25	760	IV
Ind.	Re_2O_3		1	25	760	IV
Ind.			1	25	760	IV
Ind.			1	25	760	IV
Ind.	ReO_2		1	25	760	IV
Ind.	ReO_3		1	25	760	IV
Ind.			1	25	760	IV
Ind.			1	25	760	IV
Ind.	ReO_2, ReO_3		1	25	760	IV
Ind.	ReO_2		1	25	760	IV
Ind.	ReO_2		1	25	760	IV
Pt	ReO_3	1) $HReO_4$: 0.00054 up to 0.054 M; 2) $NaReO_4$: 0.00312 up to 0.0250 M; NaOH: $5.0 \cdot 10^{-6}$ up to $2.0 \cdot 10^{-4}$ M	1	25 ± 0.05		I
Ind.	Re_2O_3, ReO_2		1	25	760	IV
Ind.	Re_2O_3		1	25	760	IV
Ind.			1	25	760	IV

Comparison electrode	Liquid junction	Electrode reaction	Standard value V	Uncertainty mV	Temperature Coefficient μV/°C	Notes	Reference	Electrode Reference Number
d			f		g	h		
		$Re^{3+} + 3e^- \rightleftharpoons Re$	0.300				1	1
		$ReO_2 + 4H^+ + 4e^- \rightleftharpoons Re + 2H_2O$	0.2513				2	2
		$ReO_2 + 2H_2O + 4e^- \rightleftharpoons Re + 4OH^-$	-0.577				2	3
		$Re_2O_3 + 6H^+ + 6e^- \rightleftharpoons 2Re + 3H_2O$	0.227				1	4
		$ReO_4^- + 8H^+ + 7e^- \rightleftharpoons Re + 4H_2O$	0.368		-510	1	1,2,3	5
		$ReO_4^- + 4H_2O + 7e^- \rightleftharpoons Re + 8OH^-$	-0.584		-1460		2	6
		$Re + e^- \rightleftharpoons Re^-$	-0.400				1	7
		$Re^{3+} + 4e^- \rightleftharpoons Re^-$	0.125				1	8
		$Re_2O_3 + 6H^+ + 8e^- \rightleftharpoons 2Re^- + 3H_2O$	0.070				1	9
		$ReO_4^- + 8H^+ + 8e^- \rightleftharpoons Re^- + 4H_2O$	0.273				1	10
		$ReO_4^{2-} + 8H^+ + 7e^- \rightleftharpoons Re^- + 4H_2O$	0.412				1	11
		$ReO_2 + 4H^+ + e^- \rightleftharpoons Re^{3+} + 2H_2O$	0.157				1	12
		$ReO_3 + 6H^+ + 3e^- \rightleftharpoons Re^{3+} + 3H_2O$	0.318				1	13
		$ReO_4^- + 8H^+ + 4e^- \rightleftharpoons Re^{3+} + 4H_2O$	0.422				1	14
		$ReO_4^{2-} + 8H^+ + 3e^- \rightleftharpoons Re^{3+} + 4H_2O$	0.795				1	15
		$ReO_3 + 2H^+ + 2e^- \rightleftharpoons ReO_2 + H_2O$	0.392			1	1,4	16
		$ReO_4^- + 4H^+ + 3e^- \rightleftharpoons ReO_2 + 2H_2O$	0.510			1	1,3,4	17
		$ReO_4^- + 2H_2O + 3e^- \rightleftharpoons ReO_2 + 4OH^-$	-0.594				2	18
	1) 1 2) 3	$ReO_4^- + 2H^+ + e^- \rightleftharpoons ReO_3 + H_2O$	0.768	± 5	-12100	2	4	19
		$2ReO_2 + 2H^+ + 2e^- \rightleftharpoons Re_2O_3 + H_2O$	0.375				1	20
		$2ReO_4^- + 10H^+ + 8e^- \rightleftharpoons Re_2O_3 + 5H_2O$	0.476				1	21
		$ReO_4^- + e^- \rightleftharpoons ReO_4^{2-}$	-0.700				1	22

Rhenium 1

Standard value V	Solvent	Electrode Reference Number	Standard value V	Solvent	Electrode Reference Number	Standard value V	Solvent	Electrode Reference Number
-0.700	1	22						
-0.594	1	18						
-0.584	1	6						
-0.577	1	3						
-0.400	1	7						
0.070	1	9						
0.125	1	8						
0.157	1	12						
0.227	1	4						
0.2513	1	2						
0.273	1	10						
0.300	1	1						
0.318	1	13						
0.368	1	5						
0.375	1	20						
0.392	1	16						
0.412	1	11						
0.422	1	14						
0.476	1	21						
0.510	1	17						
0.768	1	19						
0.795	1	15						

NOTES : Rhenium

1) The value is an average of those given by several authors when they all are in the same range of uncertainty

2) Liquid junction : KCl 1.8 M + KNO_3 1.8 M or KCl saturated

BIBLIOGRAPHY : Rhenium

1) Data reported by M. Pourbaix "Atlas d'Equilibres Electrochimiques à 25 °C" (Gauthier-Villars Editeur, Paris), pp. 300 - 306

2) Data reported by A.J. De Bethune, N.A. Swendeman-Loud "Standard Aqueous Electrode Potentials and Temperature Coefficients at 25 °C" (Clifford A. Hampel, Ill.)

3) G.E. Boyd, J.W. Cobble, W.T. Smith, Jr., J. Am. Chem. Soc., 75 (1953), 5783 - 5784

4) J.P. King, J.W. Cobble, J. Am. Chem. Soc., 79 (1957), 1559 - 1563

Group VIII

Electronically conducting phase	Intermediate species	Composition of the solution	Solvent	Temperature °C	Pressure Torr	Measuring method
a	b	c	d		e	d
Fe		1) $FeSO_4$ or $FeCl_2$: 0.0160 up to 0.0864 M; 2) $FeCl_2$: 0.00132 up to 0.0983 m; 3) $FeCl_2$: up to 0.5 m; HCl : up to 1.0 m; KCl up to 0.999 m	1	25		I III IV
Fe			1	25	760	IV
Fe		$K_4[Fe(CN)_6]$	1	25		IV
Fe	$FeCO_3$		1	25	760	IV
Fe	FeO		1	25	760	IV
Fe	Fe_2O_3		1	25	760	IV
Fe	Fe_3O_4		1	25	760	IV
Fe	$Fe(OH)_2$		1	25	760	IV
Fe	Fe_3O_4		1	25	760	IV
Fe	FeS (α)		1	25	760	IV
Fe			1	25	760	IV
Pt			1	25		I
Ind.	Fe_2O_3		1	25	760	IV
Pt	Fe_3O_4		1	25		I
Ind.			1	25	760	IV
Ind.			1	25	760	IV
Pt	$Fe(OH)_3$		1	25		I
Ind.			1	25	760	IV
1) Pt 2) Au		1) $K_4[Fe(CN)_6]$: $K_3[Fe(CN)_6]$: 0.003 up to 0.1 M 2) $K_4[Fe(CN)_6]$: $2 \cdot 10^{-4}$ up to 10^{-2} M; $K_3[Fe(CN)_6]$: $2 \cdot 10^{-4}$ up to $2 \cdot 10^{-2}$ M	1	25		I
Pt		$K_3[Fe(CN)_5H_2O]$: 0.00849 up to 0.05 M				I
Hg (D.M.E.)		Britton-Robinson buffers: pH 4 - 10; $NaClO_4$: I = 0.3	1	25		II
Pt	$[Fe(CO)_4]_3$	$[Fe(CO)_4]Na_2$ or $[Fe(CO)_4]Ba$; $[Fe(CO)_4]_3$ total [Fe]: 0.0015 up to 0.15 N	1	20		I
Pt	$[Fe(CO)_4]_3$	$[Fe(CO)_4]Na_2$ or $[Fe(CO)_4]Ba$; $[Fe(CO)_4]_3$ total [Fe]: 0.0015 up to 0.15 N	1	20		I
Ind.	FeO, Fe_2O_3		1	25	760	IV
Ind.	FeO, Fe_3O_4		1	25	760	IV
Ind.	FeO, $Fe(OH)_3$		1	25	760	IV
Ind.	Fe_3O_4, Fe_2O_3		1	25	760	IV

Comparison electrode	Liquid junction	Electrode reaction	Standard value V	Uncertainty mV	Temperature Coefficient $\mu V/^{o}C$	Notes	Reference	Electrode Reference Number
d			f		g	h		
3 1		$Fe^{2+} + 2e^- \rightleftharpoons Fe$	-0.447		+52	1	1,2 3 4	1
		$Fe^{3+} + 3e^- \rightleftharpoons Fe$	-0.037				6	2
		$[Fe(CN)_6]^{4-} + 2e^- \rightleftharpoons Fe + 6CN^-$	-1.16			2	7	3
		$FeCO_3 + 2e^- \rightleftharpoons Fe + CO_3^{2-}$	-0.756		-1293		5	4
		$FeO + 2H^+ + 2e^- \rightleftharpoons Fe + H_2O$	-0.047				6	5
		$Fe_2O_3 + 6H^+ + 6e^- \rightleftharpoons 2Fe + 3H_2O$	-0.051				6	6
		$Fe_3O_4 + 8H^+ + 8e^- \rightleftharpoons 3Fe + 4H_2O$	-0.085				6	7
		$Fe(OH)_2 + 2e^- \rightleftharpoons Fe + 2OH^-$	-0.877		-1060		5	8
		$Fe(OH)_3 + 3H^+ + 3e^- \rightleftharpoons Fe + 3H_2O$	0.059				6	9
		$FeS + 2e^- \rightleftharpoons Fe + S^{2-}$	-0.95		-970		5	10
		$HFeO_2^- + 3H^+ + 2e^- \rightleftharpoons Fe + 2H_2O$	0.493				6	11
		$Fe^{3+} + e^- \rightleftharpoons Fe^{2+}$	0.771		+1188		8	12
		$Fe_2O_3 + 6H^+ + 2e^- \rightleftharpoons 2Fe^{2+} + 3H_2O$	0.728				6	13
		$Fe_3O_4 + 8H^+ + 2e^- \rightleftharpoons 3Fe^{2+} + 4H_2O$	1.230				8	14
		$Fe(OH)^{2+} + H^+ + e^- \rightleftharpoons Fe^{2+} + H_2O$	0.914				6	15
		$Fe(OH)_2^+ + 2H^+ + e^- \rightleftharpoons Fe^{2+} + 2H_2O$	1.191				6	16
		$Fe(OH)_3 + 3H^+ + e^- \rightleftharpoons Fe^{2+} + 3H_2O$	0.939				8	17
		$FeO_4^{2-} + 8H^+ + 3e^- \rightleftharpoons Fe^{3+} + 4H_2O$	2.20		-850		5	18
3 34		$[Fe(CN)_6]^{3-} + e^- \rightleftharpoons [Fe(CN)_6]^{4-}$	0.358	± 3		1	9,10	19
3		$[Fe(CN)_5H_2O]^{2-} + e^- \rightleftharpoons [Fe(CN)_5H_2O]^{3-}$	0.412				11	20
2		$[Fe(CN)_5NO]^{2-} + e^- \rightleftharpoons [Fe(CN)_5NO]^{3-}$	-0.075				12	21
3		$[Fe(CO)_4]_3 + 6e^- \rightleftharpoons 3[Fe(CO)_4]^{2-}$	-0.747				13	22
3		$[Fe(CO)_4]_3 + 3H^+ + 6e^- \rightleftharpoons 3[Fe(CO)_4H]^-$	-0.35				13	23
		$Fe_2O_3 + 2H^+ + 2e^- \rightleftharpoons 2FeO + H_2O$	-0.057				6	24
		$Fe_3O_4 + 2H^+ + 2e^- \rightleftharpoons 3FeO + H_2O$	-0.197				6	25
		$Fe(OH)_3 + H^+ + e^- \rightleftharpoons FeO + 2H_2O$	0.271				6	26
		$3Fe_2O_3 + 2H^+ + 2e^- \rightleftharpoons 2Fe_3O_4 + H_2O$	0.221				6	27

Electronically conducting phase	Intermediate species	Composition of the solution	Solvent	Temperature °C	Pressure Torr	Measuring method
a	b	c	d		e	d
Ind.	Fe_3O_4 , $Fe(OH)_3$		1	25	760	IV
Ind.			1	25	760	IV
Ind.	$Fe(OH)_2$, $Fe(OH)_3$		1	25	760	IV
Ind.			1	25	760	IV
Ind.	$Fe(OH)_3$		1	25	760	IV
Ind.	$FeO(OH)$		1	25	760	IV
Ind.	FeS (α), Fe_2S_3		1	25	760	IV
Ind.	Fe_2O_3		1	25	760	IV
Ind.			1	25	760	IV
Ind.	Fe_3O_4		1	25	760	IV
Ind.			1	25	760	IV
Ind.	$Fe(OH)_3$		1	25	760	IV
		Fe^{2+} : 0.005 M; KCl : 1 M	1	20 ± 0.1		I
		Fe^{2+} : 0.05 M; KCl: 1 M	1	20 ± 0.1		I
		Fe^{2+} : 0.005 M; KCl : 1 M	1	20± 0.1		I
		Fe^{2+} : 0.005 M; KCl: 1 M	1	20 ± 0.1		I
		Fe^{2+} : 0.005 M; KCl : 1 M	1	20 ± 0.1		I
		Fe^{2+} : 0.005 M; KCl : 1 M	1	20 ± 0.1		I
Fe		$FeCl_2$: (5.00 up to 10.04) · 10^{-3} M; cysteine : (4.94 up to 10.03) · 10^{-3} M	1	25		V
Fe		α, α'-Dipyridyl : 4 · 10^{-3} M; $FeCl_2$: (3.75 up to 74) · 10^{-3} M	1	25 ± 0.3		V
Rh		HNO_3 : 4.5 · 10^{-3} N	1	25		I
Rh		HCl : 0.248 · 10^{-3} N; [Fe(II, III) (dipyridyl) $(CN)_4$]: 10^{-3} M	1	25		I
Pt	[Fe(dipyridyl)$_2$$(CN)_2$]	H_2SO_4 : 0.01 up to 12 M	1	25 ± 0.5		I
Au		Total [Fe]: 8.69 · 10^{-4} M; KCl : 0.1 N; phosphate, acetate, or veronal buffers : pH 2.29 up to 8.36	1	20		I
	Ferrocene		1	25		
Ind.	Ferrocene	$(Butyl)_4NClO_4$: 0.1 M	30	20	760	V
	Ferrocene		7	25		
	Ferrocene		18	25		

Comparison electrode	Liquid junction	Electrode reaction	Standard value	Uncertainty	Temperature Coefficient	Notes	Reference	Electrode Reference Number
			V	mV	µV/°C			
d			f		g	h		
		$3Fe(OH)_3 + H^+ + e^- \rightleftharpoons Fe_3O_4 + 5H_2O$	1.208				6	28
		$FeO_4^{2-} + 7H^+ + 3e^- \rightleftharpoons FeOH^{2+} + 3H_2O$	1.652				6	29
		$Fe(OH)_3 + e^- \rightleftharpoons Fe(OH)_2 + OH^-$	-0.56		-960		5	30
		$FcO_4^{2-} + 6H^+ + 3e^- \rightleftharpoons Fe(OH)_2^+ + 2H_2O$	1.559				6	31
		$FeO_4^{2-} + 4H_2O + 3e^- \rightleftharpoons Fe(OH)_3 + 5OH^-$	0.72		-1620		6,14,5	32
		$FeO_4^{2-} + 3H_2O + 3e^- \rightleftharpoons FeO(OH) + 5OH^-$	0.7				14	33
		$Fe_2S_3 + 2e^- \rightleftharpoons 2FeS + S^{2-}$	-0.715				5	34
		$Fe_2O_3 + H_2O + 2e^- \rightleftharpoons 2HFeO_2^-$	-1.139				6	35
		$FeO_4^{2-} + 5H^+ + 4e^- \rightleftharpoons HFeO_2^- + 2H_2O$	1.001				6	36
		$Fe_3O_4 + 2H_2O + 2e^- \rightleftharpoons 3HFeO_2^- + H^+$	-1.819				6	37
		$Fe(OH)_2^+ + e^- \rightleftharpoons HFeO_2^- + H^+$	-0.675				6	38
		$Fe(OH)_3 + e^- \rightleftharpoons HFeO_2^- + H_2O$	-0.810				6	39
		$[Fe(alanine)]^{3+} + e^- \rightleftharpoons [Fe(alanine)]^{2+}$	0.340			3	15	40
		$[Fe(\alpha\text{-amino-n-butyric acid})]^{3+} + e^- \rightleftharpoons [Fe(\alpha\text{-amino-n-butyric acid})]^{2+}$	0.370			3	15	41
		$[Fe(\alpha\text{-amino-iso-butyric acid})]^{3+} + e^- \rightleftharpoons [Fe(\alpha\text{-amino-iso-butyric acid})]^{2+}$	0.345			3	15	42
		$[Fe(arginine)]^{3+} + e^- \rightleftharpoons [Fe(arginine)]^{2+}$	0.420			3	15	43
		$[Fe(asparagine)]^{3+} + e^- \rightleftharpoons [Fe(asparagine)]^{2+}$	0.440			3	15	44
		$[Fe(aspartic\ acid)]^{3+} + e^- \rightleftharpoons [Fe(aspartic\ acid)]^{2+}$	0.330			3	15	45
		$[Fe(cysteine)_2]^{2-} + 2e^- \rightleftharpoons Fe + 2(cysteine)^{2-}$	-0.793			4	16	46
		$[Fe(dipyridyl)]^{2+} + 2e^- \rightleftharpoons Fe + dipyridyl$	-0.571			5	17	47
3	Agar-KCl	$[Fe(dipyridyl)_3]^{3+} + e^- \rightleftharpoons [Fe(dipyridyl)_3]^{2+}$	1.120	± 5	-1000		18	48
3	Agar-KCl	$[Fe(dipyridyl)(CN)_4]^- + e^- \rightleftharpoons [Fe(dipyridyl)(CN)_4]^{2-}$	0.541	± 5	-2180	6	18	49
3		$[Fe(dipyridyl)_2(CN)_2]^+ + e^- \rightleftharpoons [Fe(dipyridyl)_2(CN)_2]$	0.776			7	19	50
4		$[Fe(EDTA)]^- + e^- \rightleftharpoons [Fe(EDTA)]^{2-}$	0.1172	± 2			20	51
		$(Ferricinium)^+ + e^- \rightleftharpoons ferrocene$	0.400	± 7			21	52
35		$(Ferricinium)^+ + e^- \rightleftharpoons ferrocene$	0.07	± 2		8	22	53
		$(Ferricinium)^+ + e^- \rightleftharpoons ferrocene$	0.410	± 12			21	54
		$(Ferricinium)^+ + e^- \rightleftharpoons ferrocene$	0.539	± 11			21	55

Electronically conducting phase	Intermediate species	Composition of the solution	Solvent	Temperature °C	Pressure Torr	Measuring method
a	b	c	d		e	d
Pt	Ferrocene	$(C_2H_5)_4NClO_4$: 0.1 N; ferricinium picrate, ferrocene	23	25 ± 0.02		I
Ind.	Ferrocene	$(Butyl)_4NClO_4$: 0.1 M	31	20	760	V
Pt	Ferrocene	Ferricinium picrate and ferrocene (equimolar mixtures) : $1.0 \cdot 10^{-3}$ M; $2.5 \cdot 10^{-3}$ M; $5.0 \cdot 10^{-3}$ M	27	25 ± 0.05		I
Pt	Acetylferrocene		1	25± 0.1		I
	Aminoferrocene hydrochloride	HCl : 0.033 M	1	25± 0.1		I
Pt	(m-Aminophenyl) ferrocene		1	25± 0.1		I
Pt	(p-Aminophenyl) ferrocene	HCl : 0.033 M	1	25± 0.1		I
Pt	Benzoxyferrocene	HCl : 0.033 M	1	25± 0.1		I
Pt	Bromoferrocene	HCl : 0.033 M	1	25± 0.1		I
Pt	(2'-Bromoaminoethyl) ferrocene	HCl : 0.033 M	1	25± 0.1		I
Pt	Butenylferrocene	HCl : 0.033 M	1	25± 0.1		I
	Carbacylferrocene	HCl : 0.033 M	1	25± 0.1		I
	(p-Carbacylphenyl) ferrocene		1			I
Pt	Carbomethoxyferrocene	HCl : 0.033 M	1	25 ± 0.1		I
Pt	Carboxyferrocene	HCl : 0.033 M	1	25± 0.1		I
Pt	(Carboxymethyl) ferrocene	HCl : 0.033 M	1	25± 0.1		I
	Chloroferrocene	HCl : 0.033 M	1	25± 0.1		I
Pt	(Chloromercuri) ferrocene	HCl : 0.033 M	1	25± 0.1		I
Pt	(2'-Dimethylamino-ethyl) ferrocene	HCl : 0.033 M	1	25± 0.1		I
	(Dimethylamino-ethyl) ferrocene	HCl : 0.033 M	1	25± 0.1		I
Pt	Ehtylferrocene	HCl : 0.033 M	1	25± 0.1		I
	1-Ethyl-1'-carbacyl-ferrocene		1			I
	1-Ethyl-2-carbacyl-ferrocene		1			I
	1-Ethyl-3-carbacyl-ferrocene		1			I

Comparison electrode	Liquid junction	Electrode reaction	Standard value V	Uncertainty mV	Temperature Coefficient μV/°C	Notes	Reference	Electrode Reference Number
d			f		g	h		
7		(Ferricinium)$^+$ + e^- ⇌ ferrocene	0.19				23	56
35		(Ferricinium)$^+$ + e^- ⇌ ferrocene	0.12	± 20		8	22	57
24		(Ferricinium)$^+$ + e^- ⇌ ferrocene	0.628	± 2			35	58
2		(Acetylferricinium)$^+$ + e^- ⇌ acetylferrocene	-0.044			9,10	24	59
2		(Aminoferricinium)$^+$ hydrochloride + e^- ⇌ aminoferrocene hydrochloride	-0.280			9,10	24	60
2		[(m-Aminophenyl) ferricinium]$^+$ + e^- ⇌ (m-aminophenyl) ferrocene	-0.056			9,10	24	61
2		[(p-Aminophenyl) ferricinium]$^+$ + e^- ⇌ (p-aminophenyl) ferrocene	-0.061			9,10	24	62
2		(Benzoxyferricinium)$^+$ + e^- ⇌ benzoxyferrocene	-0.056			9,10	24	63
2		(Bromoferricinium)$^+$ + e^- ⇌ bromoferrocene	-0.130			9,10	24	64
2		[(2'-Bromoaminoethyl) ferricinium]$^+$ + e^- ⇌ (2'-bromoaminoethyl) ferrocene	-0.022			9,10	24	65
2		(Butenylferricinium)$^+$ + e^- ⇌ butenylferrocene	0.069			9,10	24	66
2		(Carbacylferricinium)$^+$ + e^- ⇌ carbacylferrocene	-0.204			9,10	24	67
3		[(p-Carbacylphenyl) ferricinium]$^+$ + e^- ⇌ (p-carbacylphenyl) ferrocene	-0.111			9	25	68
2		(Carbomethothyferricinium)$^+$ + e^- ⇌ carbomethoxyferrocene	-0.231			9,10	24	69
2		(Carboxyferricinium)$^+$ + e^- ⇌ carboxyferrocene	-0.219			9,10	24	70
2		[(Carboxymethyl)ferricinium]$^+$ + e^- ⇌ (carboxymethyl) ferrocene	-0.008			9,10	24	71
2		(Chloroferricinium)$^+$ + e^- ⇌ chloroferrocene	-0.128			9,10	24	72
2		[(Chloromercuri) ferricinium]$^+$ + e^- ⇌ (chloromercuri) ferrocene	-0.002			9,10	24	73
2		[(2'-Dimethylaminoethyl) ferricinium]$^+$ + e^- ⇌ (2'-dimethylaminoethyl) ferrocene	-0.016			9,10	24	74
2		[(Dimethylaminoethyl) ferricinium]$^+$ + e^- ⇌ (dimethylaminoethyl) ferrocene	-0.165			9,10	24	75
2		(Ethylferricinium)$^+$ + e^- ⇌ ethylferrocene	0.081			9,10	24	76
3		(1-Ethyl-1'-carbacylferricinium)$^+$ + e^- ⇌ 1-ethyl-1'-carbacylferrocene	-0.277			9	25	77
3		(1-Ethyl-2-carbacylferricinium)$^+$ + e^- ⇌ 1-ethyl-2-carbacylferrocene	-0.164			9	25	78
3		(1-Ethyl-3-carbacylferricinium)$^+$ + e^- ⇌ 1-ethyl-3-carbacylferrocene	-0.188			9	25	79

Electronically conducting phase	Intermediate species	Composition of the solution	Solvent	Temperature °C	Pressure Torr	Measuring method
a	b	c	d		e	d
	1-Ethyl-2-hydroxy-ethylferrocene		1			I
Pt	(1'-Hydroxyethyl) ferrocene	HCl : 0.033 M	1	25± 0.1		I
Pt	(Hydroxymethyl) ferrocene		1	25± 0.1		I
Pt	Iodoferrocene	HCl : 0.033 M	1	25± 0.1		I
	Iso-propylferrocene	HCl : 0.033 M	1	25± 0.1		I
	Methoxyferrocene	HCl : 0.033 M	1	25± 0.1		I
Pt	Methylferrocene	HCl : 0.033 M	1	25± 0.1		I
	1-Methyl-1'-carbacyl ferrocene		1			I
	1-Methyl-2-carbacyl-ferrocene		1			I
	1-Methyl-3-carbacyl-ferrocene		1			I
	1-Methyl-2-ethyl-ferrocene		1			I
	1-Methyl-3-ethyl-ferrocene		1			I
Pt	(m-Nitrophenyl) ferrocene	HCl : 0.033 M	1	25± 0.1		I
Pt	(p-Nitrophenyl) ferrocene	HCl : 0.033 M	1	25± 0.1		I
Pt	Phenylferrocene	HCl : 0.033 M	1	25± 0.1		I
Pt	(m-Phenylazophenyl) ferrocene	HCl : 0.033 M	1	25± 0.1		I
	1-Phenyl-1'-carbacyl ferrocene		1			I
	1-Phenyl-2-carbacyl-ferrocene		1			I
	(Phenylmethoxyme-thyl) ferrocene	HCl : 0.033 M	1	25± 0.1		I
Pt	Propylferrocene	HCl : 0.033 M	1	25± 0.1		I
Pt		KCl : 1 N	1	25		I
Pt	Vinylferrocene	HCl : 0.033 M	1	25± 0.1		I
		Fe^{2+} : 0.005 M; KCl : 1 M	1	25± 0.1		I

Comparison electrode	Liquid junction	Electrode reaction	Standard value V	Uncertainty mV	Temperature Coefficient µV/°C	Notes	Reference	Electrode Reference Number
d			f		g	h		
3		(1-Ethyl-2-hydroxyethylferricinium)$^+$ + e$^-$ ⇌ 1-ethyl-2-hydroxyethylferrocene	-0.020			9	25	80
2		[(1'-Hydroxyethyl) ferricinium]$^+$ + e$^-$ ⇌ (1'-hydroxyethyl) ferrocene	0.003			9, 10	24	81
2		[(Hydroxymethyl) ferricinium]$^+$ + e$^-$ ⇌ (hydroxymethyl) ferrocene	-0.029			9,10	24	82
2		(Iodoferricinium)$^+$ + e$^-$ ⇌ iodoferrocene	-0.104			9,10	24	83
2		(iso-propylferricinium)$^+$ + e$^-$ ⇌ iso-propylferrocene	0.082			9,10	24	84
2		(Methoxyferricinium)$^+$ + e$^-$ ⇌ methoxyferrocene	0.114			9,10	24	85
2		(Methylferricinium)$^+$ + e$^-$ ⇌ methylferrocene	0.082			9,10	24	86
3		(1-Methyl-1'-carbacylferricinium)$^+$ + e$^-$ ⇌ 1-methyl-1'-carbacylferrocene	-0.179			9	25	87
3		(1-Methyl-2-carbacylferricinium)$^+$ + e$^-$ ⇌ 1-methyl-2-carbacylferrocene	-0.157			9	25	88
3		(1-Methyl-3-carbacylferricinium)$^+$ + e$^-$ ⇌ 1-methyl-3-carbacylferrocene	-0.188			9	25	89
3		(1-Methyl-2-ethylferricinium)$^+$ + e$^-$ ⇌ 1-methyl-2-ethylferrocene	0.098			9	25	90
3		(1-Methyl-3-ethylferricinium)$^+$ + e$^-$ ⇌ 1-methyl-3-ethylferrocene	0.097			9	25	91
2		[(m-Nitrophenyl) ferricinium]$^+$ + e$^-$ ⇌ (m-nitrophenyl) ferrocene	-0.078			9,10	24	92
2		[(p-Nitrophenyl) ferricinium]$^+$ + e$^-$ ⇌ (p-nitrophenyl) ferrocene	-0.110			9,10	24	93
2		(Phenylferricinium)$^+$ + e$^-$ ⇌ phenylferrocene	-0.004			9,10	24	94
2		[(m-Phenylazophenyl) ferricinium]$^+$ + e$^-$ ⇌ (m-phenylazophenyl) ferrocene	-0.049			9,10	24	95
3		(1-Phenyl-1'-carbacylferricinium)$^+$ + e$^-$ ⇌ 1-phenyl-1'-carbacylferrocene	-0.271			9	25	96
3		(1-Phenyl-2-carbacylferricinium)$^+$ + e$^-$ ⇌ 1-phenyl-2-carbacylferrocene	-0.262			9	25	97
2		[(Phenylmethoxyethyl) ferricinium]$^+$ + e$^-$ ⇌ (phenylmethoxyethyl) ferrocene	-0.022			9,10	24	98
2		(Propylferricinium)$^+$ + e$^-$ ⇌ propylferrocene	0.079			9,10	24	99
1		[(Ferriciniummethyl) trimethylammonium]$^{2+}$ + e$^-$ ⇌ [(ferrocenylmethyl) trimethylammonium]$^+$	0.659	±1	+1000		26	100
2		(Vinylferricinium)$^+$ + e$^-$ ⇌ vinylferrocene	0.000			9,10	24	101
		[Fe (glutamic acid)]$^{3+}$ + e$^-$ ⇌ [Fe (glutamic acid)]$^{2+}$	0.240			3	15	102

Electronically conducting phase	Intermediate species	Composition of the solution	Solvent	Temperature °C	Pressure Torr	Measuring method
a	b	c	d		e	d
		Fe^{2+}: 0.005 M; KCl : 1 M	1	20 ± 0.1		I
		Fe^{2+} : 0.005 M; KCl : 1 M	1	20 ± 0.1		I
		Fe^{2+}: 0.005 M: KCl : 1 M	1	20± 0.1		I
		Fe^{2+}: 0.005 M; KCl : 1 M	1	20± 0.1		I
Fe		$FeCl_2$: (4.77 up to 5.22) · 10^{-3} M; NaOH (6.07 up to 25.7) · 10^{-3} M; mercaptoacetic acid : (3.04 up to 10.13) · 10^{-3} M	1	25± 1		V
		Fe^{2+}: 0.005 M; KCl : 1 M	1	20± 0.1		I
Au	[Fe(nitrolotriacetate)]	KCl :0.1 N; acetate or veronal buffers :pH 3.89 up to 8.05	1	20		I
		Fe^{2+}: 0.005 M; KCl : 1 M	1	20± 0.1		I
Pt		H_2SO_4 : 1 M	1			I
Rh		HNO_3 : 4.5 · 10^{-3} N; [Fe (II, III)(tris-o-phenanthroline)]: 10^{-3} M	1	25.5		
Pt		1) $(C_2H_5)_4NClO_4$: 0.1 N 2) $NaClO_4$	23	25		I
Pt		H_2SO_4 : 1 M	1			I
Pt			1	25		I
Hg (D.M.E.)		$LiClO_4$: 0.1 M	1			II
Pt		H_2SO_4 : 1 M	1			I
Pt		H_2SO_4 : 1 M	1			I
Pt		H_2SO_4 : 1 M	1			I
		Fe^{2+}: 0.005 M; KCl : 1 M	1	20± 0.1		I
		Fe^{2+}: 0.005 M; KCl : 1 M	1	20± 0.1		I
Au or Rh		[Fe (II, III)(pyridine-2-aldoxime)$_3$]: 1 · 10^{-4} up to 5 · 10^{-4} M; NaOH, NaCl, NaH_2PO_4 : pH 6.0 up to 7.5	1	20±0.05		I
Pt			1	25		I
		Fe^{2+}: 0.005 M; KCl : 1 M	1	20± 0.1		I

Comparison electrode	Liquid junction	Electrode reaction	Standard value V	Uncertainty mV	Temperature Coefficient μV/°C	Notes*	Reference	Electrode Reference Number
d			f		g	h		
		$[Fe(glycine)]^{3+} + e^- \rightleftharpoons [Fe(glycine)]^{2+}$	0.380			3	15	103
		$[Fe(glycylglycine)]^{3+} + e^- \rightleftharpoons [Fe(glycylglycine)]^{2+}$	0.360			3	15	104
		$[Fe(hydroxyproline)]^{3+} + e^- \rightleftharpoons [Fe(hydroxyproline)]^{2+}$	0.445			3	15	105
		$[Fe(leucine)]^{3+} + e^- \rightleftharpoons [Fe(leucine)]^{2+}$	0.362			3	15	106
		$[Fe(mercaptoacetate)_2]^{2-} + 2e^- \rightleftharpoons Fe + 2(mercaptoacetate)^{2-}$	-0.768			4	27	107
		$[Fe(methionine)]^{3+} + e^- \rightleftharpoons [Fe(methionine)]^{2+}$	0.400			3	15	108
1		$[Fe(nitrolotriacetate)] + e^- \rightleftharpoons [Fe(nitrolotriacetate)]^-$	0.3295	±5			36	109
		$[Fe(ornithine)]^{3+} + e^- \rightleftharpoons [Fe(ornithine)]^{2+}$	0.415			3	15	110
3		$[Fe(1,10\text{-}phenanthroline)]^{3+} + e^- \rightleftharpoons [Fe(1,10\text{-}phenanthroline)]^{2+}$	1.06			9,10	28	111
3		$[Fe(1,10\text{-}phenanthroline)_3]^{3+} + e^- \rightleftharpoons [Fe(1,10\text{-}phenanthroline)_3]^{2+}$	1.147	±5	-900		18	112
7		$[Fe(1,10\text{-}phenanthroline)_3]^{3+} + e^- \rightleftharpoons [Fe(1,10\text{-}phenanthroline)_3]^{2+}$	0.805	±11	-890		23 29	113
3		$[Fe(5\text{-}chloro\text{-}1,10\text{-}phenanthroline)]^{3+} + e^- \rightleftharpoons [Fe(5\text{-}chloro\text{-}1,10\text{-}phenanthroline)]^{2+}$	1.12			9,10	28	114
		$[Fe(4,7\text{-}dihydroxyphenanthroline)_3]^{3-} + e^- \rightleftharpoons [Fe(4,7\text{-}dihydroxyphenanthroline)_3]^{2-}$	-0.20				30	115
3		$[Fe(4,7\text{-}dimethyl\text{-}1,10\text{-}phenanthroline)_3]^{3+} + e^- \rightleftharpoons [Fe(4,7\text{-}dimethyl\text{-}1,10\text{-}phenanthroline)_3]^{2+}$	0.936				31	116
3		$[Fe(5\text{-}methyl\text{-}1,10\text{-}phenanthroline)]^{3+} + e^- \rightleftharpoons [Fe(5\text{-}methyl\text{-}1,10\text{-}phenanthroline)]^{2+}$	1.02			9,10	28	117
3		$[Fe(5\text{-}nitro\text{-}1,10\text{-}phenanthroline)]^{3+} + e^- \rightleftharpoons [Fe(5\text{-}nitro\text{-}1,10\text{-}phenanthroline)]^{2+}$	1.25			9,10	28	118
3		$[Fe(5\text{-}phenyl\text{-}1,10\text{-}phenanthroline)]^{3+} + e^- \rightleftharpoons [Fe(5\text{-}phenyl\text{-}1,10\text{-}phenanthroline)]^{2+}$	1.08			9,10	28	119
		$[Fe(\beta\text{-}phenylalanine)]^{3+} + e^- \rightleftharpoons [Fe(\beta\text{-}phenylalanine)]^{2+}$	0.415			3	15	120
		$[Fe(proline)]^{3+} + e^- \rightleftharpoons [Fe(proline)]^{2+}$	0.395			3	15	121
3		$[Fe(pyridine\text{-}2\text{-}aldoxime)_3]^{2+} + e^- \rightleftharpoons [Fe(pyridine\text{-}2\text{-}aldoxime)_3]^{+}$	0.348	±1	-1500	11, 12	32	122
1		$[Fe(pyridine\text{-}2,6\text{-}dialdoxime)_2]^- + e^- \rightleftharpoons [Fe(pyridine\text{-}2,6\text{-}dialdoxime)_2]^{2-}$	0.204	±2			33	123
		$[Fe(sarcosine)]^{3+} + e^- \rightleftharpoons [Fe(sarcosine)]^{2+}$	0.380			3	15	124

Electronically conducting phase	Intermediate species	Composition of the solution	Solvent	Temperature °C	Pressure Torr	Measuring method
a	b	c	d		e	d
		Fe^{2+}: 0.005 M; KCl : 1 M	1	20± 0.1		I
Pt		H_2SO_4 : 0.1 up to 2.0 M	1	25		I
		Fe^{2+}: 0.005 M; KCl : 1 M	1	20± 0.1		I
		Fe^{2+}: 0.005 M; KCl: 1 M	1	20± 0.1		I
		Fe^{2+}: 0.005 M; KCl: 1 M	1	20± 0.1		I

Comparison electrode	Liquid junction	Electrode reaction	Standard value V	Uncertainty mV	Temperature Coefficient μV/°C	Notes	Reference	Electrode Reference Number
d			f		g	h		
		$[Fe(serine)]^{3+} + e^- \rightleftharpoons [Fe(serine)]^{2+}$	0.405			3	15	125
		$[Fe(2,2',2''\text{-ter-pyridyl})_2]^{3+} + e^- \rightleftharpoons$ $[Fe(2,2',2''\text{-ter-pyridyl})_2]^{2+}$	-0.927	± 2		12, 13	34	126
		$[Fe(threonine)]^{3+} + e^- \rightleftharpoons [Fe(threonine)]^{2+}$	0.430			3	15	127
		$[Fe(tryptophan)]^{3+} + e^- \rightleftharpoons [Fe(tryptophan)]^{2+}$	0.415			3	15	128
		$[Fe(valine)]^{3+} + e^- \rightleftharpoons [Fe(valine)]^{2+}$	0.380			3	15	129

Iron 1

Standard value V	Solvent	Electrode Reference Number
-1.819	1	37
-1.16	1	3
-1.139	1	35
-0.95	1	10
-0.927	1	126
-0.877	1	8
-0.810	1	39
-0.793	1	46
-0.768	1	107
-0.756	1	4
-0.747	1	22
-0.715	1	34
-0.675	1	38
-0.571	1	47
-0.56	1	30
-0.447	1	1
-0.35	1	23
-0.280	1	60
-0.277	1	77
-0.271	1	96
-0.262	1	97
-0.231	1	69
-0.219	1	70
-0.204	1	67
-0.20	1	115
-0.197	1	25
-0.188	1	79
-0.188	1	89
-0.179	1	87
-0.165	1	75
-0.164	1	78
-0.157	1	88
-0.130	1	64
-0.128	1	72
-0.111	1	68
-0.110	1	93
-0.104	1	83
-0.085	1	7
-0.078	1	92
-0.075	1	21
-0.061	1	62
-0.057	1	24
-0.056	1	63
-0.056	1	61
-0.051	1	6
-0.049	1	95
-0.047	1	5
-0.044	1	59
-0.037	1	2
-0.029	1	82
-0.022	1	65
-0.022	1	98
-0.020	1	80
-0.016	1	74
-0.008	1	71
-0.004	1	94
0.000	1	101
0.002	1	73
0.003	1	81
0.059	1	9
0.069	1	66
0.079	1	99
0.081	1	76
0.082	1	86
0.082	1	84
0.097	1	91
0.098	1	90
0.114	1	85
0.1172	1	51
0.204	1	123
0.221	1	27
0.240	1	102
0.271	1	26
0.3295	1	109
0.330	1	45
0.340	1	40
0.345	1	42
0.348	1	122
0.358	1	19
0.360	1	104
0.362	1	106
0.370	1	41
0.380	1	103
0.380	1	124
0.380	1	129
0.395	1	121
0.400	1	52
0.400	1	108
0.405	1	125
0.412	1	20
0.415	1	110
0.415	1	120
0.415	1	128
0.420	1	43
0.430	1	127
0.440	1	44
0.445	1	105
0.493	1	11
0.541	1	49
0.659	1	100
0.7	1	33
0.72	1	32
0.728	1	13
0.771	1	12
0.776	1	50

Standard value V	Solvent	Electrode Reference Number
0.914	1	15
0.936	1	116
0.939	1	17
1.001	1	36
1.02	1	117
1.06	1	11
1.08	1	119
1.12	1	114
1.120	1	48
1.147	1	112
1.191	1	16
1.208	1	28
1.230	1	14
1.25	1	118
1.559	1	31
1.652	1	29
2.20	1	18
0.07	30	53
0.410	7	54
0.539	18	55
0.19	23	56
0.805	23	113
0.12	31	57
0.628	27	58

Standard value V	Solvent	Electrode Reference Number

Standard value V	Solvent	Electrode Reference Number

NOTES: Iron

1) Potentiometric, voltammetric and slow amperostatic data
2) Heat capacity measurements
3) Potentiometric titrations at I = 1 M (KCl)
4) Solubility data
5) Spectrophotometric data
6) Formal electric tension in $0.248 \cdot 10^{-3}$ N HCl
7) Formal electric tension in 0.5 M H_2SO_4
8) Calculated from dissociation constants
9) Potentiometric titrations
10) Formal electric tension
11) Temperature ranges from 14.5 to 33.2 °C
12) Formal electric tension in 1.0 M H_2SO_4
13) Agar-saturated KCl liquid junction

BIBLIOGRAPHY: Iron

1) W.A. Patrick, W.E. Thompson, J. Am. Chem. Soc., 75 (1953), 1184 - 1187
2) K.F. Bonhoeffer, W. Jena, Z. Electrochem., 55(1951), 151- 154
3) T. Hurlen, Acta Chem. Scand., 14 (1960), 1533 - 1554
4) J.W. Larson, P. Cerutti, H . K. Garber, L.G. Hepler, J. Phys. Chem., 72 (1968), 2902 - 2907
5) Data reported by A.J. De Bethune, N.A. Swendeman-Loud, Standard Aqueous Electrode Potentials and Temperature Coefficients (Clifford A. Hampel, Ill.)
6) Data reported by M. Pourbaix, Atlas d'Equilibres Electrochimiques à 25 °C, (Gauthier-Villars Editeur, Paris, 1963), pp. 307 - 321
7) C.C. Stephenson, J.C. Morrow, J. Am. Chem. Soc., 78 (1956), 275 - 277
8) B.N. Mattoo, Zh. Priklad. Khim., 33(1960), 2015 - 2020
9) G.I.H. Hanania, D.H. Irvine, W.A. Eaton, P. George, J. Phys. Chem., 71 (1967), 2022 - 2030

BIBLIOGRAPHY: Iron

10) R.C. Murray, Jr., P.A. Rock, Electrochim. Acta, 13 (1968), 969 - 975

11) W.U. Malik, Indian J. Chem., 4 (1966), 106 - 108

12) J. Mašek, J. Dempir, Collect. Czech. Chem. Commun., 34 (1969), 727 - 741

13) W. Hieber, W. Hübel, Z. Elektrochem., 57 (1953), 331 - 338

14) R.H. Wood, J. Am. Chem. Soc., 80 (1958), 2038 - 2041

15) D.D. Perin, J. Chem. Soc., 1959, 290 - 296

16) N. Tanacka, I.M. Kolthoff, W. Strieks, J. Am. Chem. Soc., 77 (1955), 1996

17) P. Krumholz, J. Am. Chem. Soc., 71 (1949), 3654 - 3656

18) P. George, G.I.H. Hanania, D.H. Irvine, J. Chem. Soc., 1959, 2548 - 2554

19) A.A. Schilt, Anal. Chem., 35 (1963), 1599 - 1602

20) G. Schwarzenbach, J. Heller, Helv. Chim. Acta, 34 (1951), 576 - 591

21) Data reported by J.J. Lagowski, The Chemistry of Non-aqueous Solvents, Vol. 1 (Academic Press, New York and London), p. 159

22) G.D. Guérin, A. Caillet, C.R. Acad. Sci. Paris, Ser. C, 273 (1971), 235 - 238

23) I.M. Kolthoff, F.G. Thomas, J. Phys. Chem., 69 (1965), 3049 - 3058

24) S.P. Gubin, E.G. Perevalova, Dokl. Akad. Nauk SSSR, 143 (1962), 1351 - 1354

25) A.N. Nesmeyanov, E.G. Perevalova, L.P. Yureva, S.P. Gubin, Izv. Akad. Nauk SSSR, Ser. Khim., 1965, 909 - 911

26) T.I. L'vova, A.A. Pendin, K.D. Shirko, B.P. Nikol'skii, J. General Chem. USSR, 36 (1966), 2048 - 2052

27) D.L. Leussing, I.M. Kolthoff, J. Am. Chem. Soc., 75 (1953), 3904 - 3911

28) W.W. Brandt, D.K. Gullstrom, J. Am. Chem. Soc., 74 (1952), 3532 - 3535

29) B. Kratochvil, J. Krock, J. Phys. Chem., 70 (1966), 944 - 966

30) G.I.H. Hanania, M.W. Makinen, P. George, W.A. Eaton, (unpublished results)

31) F. Farba, R.T. Iwamoto, Anal. Chem., 38 (1966), 143

32) G.I.H. Hanania, D.H. Irvine, F.R. Shurayh, J. Phys. Chem., 72 (1968), 1355 - 1361

33) G.I.H. Hanania, D.H. Irvine, M.S. Michaelides, F.S. Shurayh, Proc. 9th Int. Cong. Coord. Chem., St. Moritz Bad, 1965, 1966, 222

34) F.P. Dwyer, E.C. Gyarfas, J. Am. Chem. Soc., 76 (1954), 6320 - 6321

35) I.L.M. Mukherjee, J.J. Kelly, W. Baranetzky, J. Sica, J. Phys. Chem., 72 (1968), 3410

36) G. Schwarzenbach, J. Heller, Helv. Chim. Acta, 34 (1951), 1889 - 1900

Electronically conducting phase	Intermediate species	Composition of the solution	Solvent	Temperature °C	Pressure Torr	Measuring method
a	b	c	d		e	d
Ind.		$CoSO_4$	1	25	760	V,IV
Co		$CoCl_2$	7	25		I
Co	$CoCO_3$		1	25	760	IV
Co (Pt)	$Co_2[Fe(CN)_6]$	$K_4[Fe(CN)_6]$: 0.0125 up to 0.1 M	1	25		I
Co	CoO		1	25	760	IV
Co	$Co(OH)_2$		1	25	760	IV
Co	$Co(OH)_2$		1	25	760	IV
Ind.	$Co(S_2O_3)$ (in sol.)	$CoCl_2$: (10.36 up tó 22.83) · 10^{-3} M; BaS_2O_3 : (14.09 up to 16.47) · 10^{-3} M	1	25		V
Co			1	25	760	IV
Ind.		H_2SO_4 : 1 N	1	25		I,IV
Ind.	CoO_2		1	25	760	IV
Ind.	Co_2O_3		1	25	760	IV
Ind.	Co_3O_4		1	25	760	IV
Ind.	CoO_2		1	25	760	IV
Pt		$[Co(CO)_4]Na$, $[Co(CO)_4]_2$: 0.009 up to 0.013 M; pH : 8.5 up to 10	1	20		I
Ind.		NH_4NO_3 : 1 N	1	25		IV
Ind.			1	25	760	IV
Ind.	CoO, Co_3O_4		1	25	760	IV
Ind.	Co_2O_3, CoO_2		1	25	760	IV
Co	Co_2O_3, CoO_2	NaOH : 0.1 and 1 N; Na_2CO_3 : 1 N; $Na_2B_4O_7$: 0.1 M; pH : 9.2 ÷ 14.0	1	18		III
Ind.	Co_3O_4, Co_2O_3		1	25	760	IV
Pt	$Co(OH)_2$	$Co(OH)_2$: satd.; KOH : 0.01 up to 10 N	1			III
Ind.	$Co(OH)_2$, Co_3O_4		1	25	760	IV
Ind.	$Co(OH)_2$, $Co(OH)_3$		1	25	760	IV
Ind.			1	25	760	IV
Ind.	Co_2O_3		1	25	760	IV
Ind.	Co_3O_4		1	25	760	IV
Glass		$CoCl_2$: (4.32 up to 4.88) · 10^{-3} M; alanine : (39.49 up to 44.61) · 10^{-3} M; NaOH : (0.74 up to 6.48) · 10^{-3} M	1	25± 0.02		I

Comparison electrode	Liquid junction	Electrode reaction	Standard value	Uncertainty	Temperature Coefficient	Notes	Reference	Electrode Reference Number
			V	mV	μV/°C			
d			f		g	h		
		$Co^{2+} + 2e^- \rightleftharpoons Co$	-0.28		+60	1,2	1,2	1
3		$Co^{2+} + 2e^- \rightleftharpoons Co$	-0.233				3	2
		$CoCO_3 + 2e^- \rightleftharpoons Co + CO_3^{2-}$	-0.64				2	3
		$Co_2[Fe(CN)_6] + 4e^- \rightleftharpoons 2Co + [Fe(CN)_6]^{4-}$	0.2691				4	4
		$CoO + 2H^+ + 2e^- \rightleftharpoons Co + H_2O$	0.166				5	5
		$Co(OH)_2 + 2H^+ + 2e^- \rightleftharpoons Co + 2H_2O$	0.095				5	6
		$Co(OH)_2 + 2e^- \rightleftharpoons Co + 2OH^-$	-0.73		-1064		2	7
		$Co(S_2O_3) + 2e^- \rightleftharpoons Co + S_2O_3^{2-}$	-0.337			3	6	8
		$HCoO_2^- + 3H^+ + 2e^- \rightleftharpoons Co + 2H_2O$	0.659				5	9
		$Co^{3+} + e^- \rightleftharpoons Co^{2+}$	1.83			2,4	7,8,2	10
		$CoO_2 + 4H^+ + 2e^- \rightleftharpoons Co^{2+} + 2H_2O$	1.612				5	11
		$Co_2O_3 + 6H^+ + 2e^- \rightleftharpoons 2Co^{2+} + 3H_2O$	1.746				5	12
		$Co_3O_4 + 8H^+ + 2e^- \rightleftharpoons 3Co^{2+} + 4H_2O$	2.112				5	13
		$CoO_2 + 4H^+ + e^- \rightleftharpoons Co^{3+} + H_2O$	1.416				5	14
3	KCl satd.	$[Co(CO)_4]_2 + 2e^- \rightleftharpoons 2[Co(CO)_4]^-$	-0.40				9	15
		$[Co(NH_3)_5(H_2O)]^{3+} + e^- \rightleftharpoons [Co(NH_3)_5(H_2O)]^{2+}$	0.37			5,6 7	10	16
		$[Co(NH_3)_6]^{3+} + e^- \rightleftharpoons [Co(NH_3)_6]^{2+}$	0.108				2	17
		$Co_3O_4 + 2H^+ + 2e^- \rightleftharpoons 3CoO + H_2O$	0.777				5	18
		$2CoO_2 + 2H^+ + 2e^- \rightleftharpoons Co_2O_3 + H_2O$	1.477				5	19
		$2CoO_2 + H_2O + 2e^- \rightleftharpoons Co_2O_3 + 2OH^-$	0.58			8	11	20
		$3Co_2O_3 + 2H^+ + 2e^- \rightleftharpoons 2Co_3O_4 + H_2O$	1.018				5	21
22		$CoO_2^- + e^- + 2H_2O \rightleftharpoons Co(OH)_2 + 2OH^-$	0.22				12	22
		$Co_3O_4 + 2H_2O + 2H^+ + 2e^- \rightleftharpoons 3Co(OH)_2$	0.993				5	23
		$Co(OH)_3 + e^- \rightleftharpoons Co(OH)_2 + OH^-$	0.17		-800		2	24
		$Co^{3+} + 2H_2O + e^- \rightleftharpoons HCoO_2^- + 3H^+$	-0.065				5	25
		$Co_2O_3 + H_2O + 2e^- \rightleftharpoons 2HCoO_2^-$	-0.128				5	26
		$Co_3O_4 + 2H_2O + 2e^- \rightleftharpoons 3HCoO_2^- + H^+$	-0.700				5	27
3		$[Co(alanine)]^{2+} + 2e^- \rightleftharpoons Co + alanine$	-0.419				13	28

Electronically conducting phase	Intermediate species	Composition of the solution	Solvent	Temperature °C	Pressure Torr	Measuring method
a	b	c	d		e	d
Glass		$CoCl_2$: (4.32 up to 4.88) · 10^{-3} M; alanine : (39.49 up to 44.61) · 10^{-3} M; NaOH : (0.74 up to 6.48) · 10^{-3} M	1	25 ± 0.02		I
Glass		$CoCl_2$: (4.27 up to 4.76) · 10^{-3} M; glycine : (68.33 up to 76.06) · 10^{-3} M; NaOH : (5.21 up to 6.85) · 10^{-3} M	1	25± 0.02		I
Glass		$CoCl_2$: (4.27 up to 4.76) · 10^{-3} M; glycine : (68.33 up to 76.06) · 10^{-3} M; NaOH : (5.21 up to 6.85) · 10^{-3} M	1	25 ± 0.02		I
Glass		$CoCl_2$: (4.19 up to 4.40) · 10^{-3} M; glycylglycine : (14.62 up to 15.36) · 10^{-3} M; NaOH : (0.99 up to 3.37) · 10^{-3} M	1	25± 0.02		I
Glass		$CoCl_2$: (4.19 up to 4.40) · 10^{-3} M; glycylglycine : (14.62 up to 15.36) · 10^{-3} M; NaOH : (0.99 up to 3.37) · 10^{-3} M	1	25 ± 0.02		I
Pt, H_2		$CoCl_2$: (1.978 up to 4.824) · 10^{-3} M; glycylglycylglycine : (24.95 up to 34.16) · 10^{-3} M; NaOH : (1.934 up to 4.854) · 10^{-3} M	1	25± 0.03		I
Pt, H_2		$CoCl_2$: (1.978 up to 4.824) · 10^{-3} M; glycyglycylglycine : (24.95 up to 34.16) · 10^{-3} M; NaOH : (1.934 up to 4.854) · 10^{-3} M	1	25 ± 0.03		I
Ind.	[Co (malonate)]	$CoCl_2$: 0.01233 M; NaH malonate : 0.004931 M;	1	25		V
Ind.		Co oxalate : up to 2 · 10^{-4} m; K_2 oxalate : up to 6 · 10^{-3} m	1	25± 0.3		V
Glass	[Co (8-hydroxyquinoline-5-sulfonate)]	(8-hydroxyquinoline-5-sulfonate)$^{2-}$: 2.954 · 10^{-3} M; NaOH : 1.4 · 10^{-3} M; Co $(NO_3)_2$	1	25		I
Glass		(8-hydroxyquinoline-5-sulfonate)$^{2-}$: 2.954 · 10^{-3} M; NaOH : 1.4 · 10^{-3} M; Co $(NO)_3)_2$	1	25		I
Hg (D.M.E.)		KNO_3 : 0.2 M ; [Co (propylenediaminetetraacetate)]$^{2-}$: [Co (propylenediaminetetraacetate)]$^-$; acetate buffer : pH 4.6 ÷ 5.4	1	25		II
Hg (D.M.E.)		KNO_3 : 0.2 M; [Co (trimethylenediaminotetraacetate)]$^{2-}$; [Co (trimethylenediaminotetraacetate)]$^-$; acetate buffer: pH 4.6 ÷ 5.4	1	25		II
Pt		[Co (diethylenediamine)$_2$]$^{2+}$: 1 · 10^{-3} M; [Co (diethylenediamine)$_2$]$^{3+}$: 1 · 10^{-3} M; diethylenediamine : 0.1 M; KCl : 1 N	1	25		III
Pt		2,2'-Dipyridyl: 0.01 M; Co $(NO_3)_2$: 0.01 M; monochloroacetic acid + NH_3 buffer at pH = 2	1	25		I
Hg (D.M.E.		KNO_3 : 0.2 M; [Co (EDTA)]$^{2-}$; [Co (EDTA)]$^-$; acetate buffer : pH 4.6 ÷ 5.4	1	25		II

Comparison electrode	Liquid junction	Electrode reaction	Standard value V	Uncertainty mV	Temperature Coefficient µV/°C	Notes	Reference	Electrode Reference Number
d			f		g	h		
3		$[Co(alanine)_2]^{2+} + 2e^- \rightleftharpoons Co + 2(alanine)$	-0.527				13	29
3		$[Co(glycine)]^{2+} + 2e^- \rightleftharpoons Co + glycine$	-0.431				13	30
3		$[Co(glycine)_2]^{2+} + 2e^- \rightleftharpoons Co + 2(glycine)$	-0.550				13	31
3		$[Co(glycylglycine)]^{2+} + 2e^- \rightleftharpoons Co + glycylglycine$	-0.380				13	32
3		$[Co(glycyglycine)_2]^{2+} + 2e^- \rightleftharpoons Co + 2(glycylglycine)$	-0.450				13	33
4		$[Co(glycylglycylglycine)]^{2+} + 2e^- \rightleftharpoons$ $Co + glycylglycylglycine$	-0.365				14	34
4		$[Co(glycylglycylglycine)_2]^{2+} + 2e^- \rightleftharpoons$ $Co + 2(glycylglycylglycine)$	-0.442				14	35
		$[Co(malonate)] + 2e^- \rightleftharpoons Co + (malonate)^{2-}$	-0.380			9	15	36
		$[Co(oxalate)_2]^{2-} + 2e^- \rightleftharpoons Co + 2(oxalate)^{2-}$	-0.474			3	16	37
4		$[Co(8\text{-}hydroxyquinoline\text{-}5\text{-}sulfonate)] + 2e^- \rightleftharpoons$ $Co + (8\text{-}hydroxyquinoline\text{-}5\text{-}sulfonate)^{2-}$	-0.537			4	17	38
4		$[Co(8\text{-}hydroxyquinoline\text{-}5\text{-}sulfonate)_2]^{2-} + 2e^- \rightleftharpoons$ $Co + 2(8\text{-}hydroxyquinoline\text{-}5\text{-}sulfonate)^{2-}$	-0.746			4	17	39
3		$[Co(propylenediaminetetraacetate)]^- + e^- \rightleftharpoons$ $[Co(propylenediaminetetraacetate)]^{2-}$	0.36				18	40
3		$[Co(trimethylenediaminetetraacetate)]^- + e^- \rightleftharpoons$ $[Co(trimethylenediaminetetraacetate)]^{2-}$	0.29				18	41
3		$[Co(diethylenediamine)_2]^{3+} + e^- \rightleftharpoons$ $[Co(diethylenediamine)_2]^{2+}$	-0.232	± 2		10	19	42
3		$[Co(dipyridyl)_3]^{3+} + e^- \rightleftharpoons [Co(dipyridyl)_3]^{2+}$	0.31			5	20	43
3		$[Co(EDTA)]^- + e^- \rightleftharpoons [Co(EDTA)]^{2-}$	0.37				18	44

Electronically conducting phase	Intermediate species	Composition of the solution	Solvent	Temperature °C	Pressure Torr	Measuring method
a	b	c	d		e	d
Au		[Co (ethylenediamine)$_3$]Cl$_3$: 0.001001 up to 0.000176 m;[Co (ehtylenediamine)$_3$]Cl$_2$: 0.000280 up to 0.00100 m; NaCl : 0.00172 up to 0.9048: ehtylenediamine : 0.0995 m	1	25 ± 0.1		I
Pt	Trans-[Co (glycine)$_3$]	KCl: 1.0 M	1	25		I
Pt		KCl : 1.0 M	1	25		I
Pt		KCl : 1.0 M	1	25		I

Comparison electrode	Liquid junction	Electrode reaction	Standard value V	Uncertainty mV	Temperature Coefficient µV/°C	Notes	Reference	Electrode Reference Number
d			f		g	h		
3		[Co (ehtylenediamine)$_3$]$^{3+}$ + 3e$^-$ ⇌ [Co (ethylenediamine)$_3$]$^{2+}$	0.180	± 2	-1070 ± 90		21	45
3		Trans- [Co (glycine)$_3$] + e$^-$ ⇌ trans-[Co (glycine)$_3$]$^-$	0.20			5	22	46
3		[Co (oxalate)$_3$]$^{3-}$ + e$^-$ ⇌ [Co (oxalate)$_3$]$^{4-}$	0.57	± 20		5	22	47
3		[Co (oxalate)$_2$ (H$_2$O)$_2$]$^-$ + e$^-$ ⇌ [Co (oxalate)$_2$ (H$_2$O)$_2$]$^{2-}$	0.78	± 40		5	22	48

Cobalt 1

Standard value V	Solvent	Electrode Reference Number
-0.746	1	39
-0.73	1	7
-0.700	1	27
-0.64	1	3
-0.550	1	31
-0.537	1	38
-0.527	1	29
-0.474	1	37
-0.450	1	33
-0.442	1	35
-0.431	1	30
-0.419	1	18
-0.40	1	15
-0.380	1	32
-0.380	1	36
-0.365	1	34
-0.337	1	8
-0.28	1	1
-0.232	1	42
-0.128	1	26
-0.065	1	25
0.095	1	6
0.108	1	17
0.166	1	5
0.17	1	24
0.180	1	45
0.20	1	46
0.22	1	22
0.2691	1	4
0.29	1	41
0.31	1	43
0.36	1	40
0.37	1	44
0.37	1	16
0.57	1	47

Standard value V	Solvent	Electrode Reference Number
0.58	1	20
0.659	1	9
0.777	1	18
0.78	1	48
0.993	1	23
1.018	1	21
1.416	1	14
1.477	1	19
1.612	1	11
1.746	1	12
1.83	1	10
2.112	1	13
-0.233	1	2

Standard value V	Solvent	Electrode Reference Number

1) Calorimetric data
2) Thermodynamic data
3) Solubility data
4) Potentiometric data
5) Formal electric tension
6) Spectrophotometric data
7) Kinetic data
8) Cathode-ray oscillopolarographic data
9) Colorimetric data
10) Voltammetric data

BIBLIOGRAPHY: Cobalt

1) R.N. Goldberg, R.G. Riddell, M.R. Wingard, H.P. Hopkins, C.A. Wuffs, L.G. Heple, J. Phys. Chem., 70 (1966), 706 - 710

2) Data reported by A.J. De Bethune, N.A. Swendeman-Loud, Standard Aqueous Electrode Potentials and Temperature Coefficients at 25 °C, (Clifford A. Hampel, Ill.)

3) Taniewska-Osinska, Acta Chim. Soc. Sci. Łodz, 7 (1961), 23 - 29

4) W.U. Malik, A. Das, Indinan J. Chem., 4 (1966), 203 - 205

5) Data reported by M. Pourbaix; Atlas d'Equilibres Electrochimiques à 25 °C (Gauthier-Villars Editeur, Paris, 1963), pp. 322 - 329

6) T.O. Denney, C.B. Monk, Trans. Faraday Soc., 47 (1951), 992 - 998

7) F. Verbeek, Z. Eechebaut, Bull. Soc. Chim. Belges, 67 (1958), 204 - 224

8) Z.M. Chebrikova, Zh. V. Belaya, M.A. Loshkarev, Tr. Dnepropetr. Khim. Tekhnol. Inst., 1963, 55 - 62

9) W. Hieber, W. Hübel, Z. Elektrochem., 57 (1953), 331 - 338

10) R.G. Yalman, Inorg. Chem., 1 (1962), 16 - 19

11) S.E.S. El-Wakkad, A. Hickling, Trans. Faraday Soc., 46 (1950), 820 - 824

12) R.V. Boldin, Elektrokhim., 3 (1967), 1259 - 1262

13) C.B. Monk, Trans. Faraday Soc., 47 (1951), 297 - 302

14) J.I. Evans, C.B. Monk, Trans. Faraday Soc., 51 (1955), 1244

15) D.I. Stock, C.W. Davies, J. Chem. Soc., 1949, 1371 - 1374

16) J.E. Barney, W.J. Argersinger, Jr., C.A. Reynolds, J. Am. Chem. Soc., 73 (1951), 3785 - 3788

17) R. Näsänen, E. Uisitalo, Acta Chem. Scand., 8 (1954), 112 - 118

18) N. Tanaka, H. Ogino, Bull. Chem. Soc. Japan, 38 (1965), 1054

19) H. Bartlet, M. Prügel, Electrochim. Acta, 16 (1971), 1815 - 1821

20) F. Vydra, R. Pribil, Talanta, 8 (1961), 824 - 828

21) J.J. Kim, P.A. Rock, Inorg. Chem., 8 (1969), 563 - 566

22) L. Hin-Fat, W.C.E. Higginson, J.Chem. Soc. A, 1967, 298 - 301

Nickel 1

Electronically conducting phase	Intermediate species	Composition of the solution	Solvent	Temperature °C	Pressure Torr	Measuring method
a	b	c	d		e	d
Ind.	Ni_3O_4		1	25	760	IV
Ni			1	25	760	IV
Ni		$NiSO_4$	1	25		I
Ni		LiCl: 0.01 m	7	25		I
Ni	$NiCO_3$		1	25	760	IV
Ni			1	25	760	IV
Ni	NiO		1	25	760	IV
Ni	NiO_2		1	25	760	IV
Ni	$Ni(OH)_2$		1	25	760	IV
Ni	$Ni(OH)_2$		1	25	760	IV
		$NiCl_2$: (0.2155 up to 1.7259) · 10^{-3} M; $NaPO_3$: (0.6146 up to 0.6267) · 10^{-3} M	1	25		III
Ni	NiS (α)		1	25	760	IV
Ind.	NiS (β)		1	25	760	IV
Ni	NiS (γ)		1	25	760	IV
	$Ni(S_2O_3)$	$NiCl_2$: (10.48 up to 23.59) · 10^{-3} M; BaS_2O_3 : (14.12 up to 16.67) · 10^{-3} M	1	25 ± 0.03		V
Ind.	NiO_2		1	25	760	IV
Ind.	Ni_2O_3		1	25	760	IV
Ind.	Ni_3O_4		1	25	760	IV
Ind.	$Ni(OH)_3$		1	25	760	IV
Hg (D.M.E.)		$K_2[Ni(CN)_3]$: 0.333 · 10^{-2} M up to 2.697 · 10^{-2} M; $K_2[Ni(CN)_4]$: 0.0533 · 10^{-2} M up to 3 · 10^{-2} M	1	25		II
Ind.	NiO, Ni_2O_3		1	25	760	IV
Ind.	NiO, Ni_3O_4		1	25	760	IV
Ind.	Ni_2O_3, NiO_2		1	25	760	IV
Ind.	Ni_3O_4, Ni_2O_3		1	25	760	IV
Ind.	$Ni(OH)_2$, NiO_2		1	25	760	IV
Ind.	Ni_2O_3, $Ni(OH)_2$		1	25	760	IV
Ind.	Ni_3O_4, $Ni(OH)_2$		1	25	760	IV
Ind.	$Ni(OH)_2$, $Ni(OH)_3$		1	25	760	IV
Ind.	$Ni(OH)_3$, $Ni(OH)_4$		1	25	760	IV
Glass		Ni^{2+} : 10^{-3} M; acetylacetone : 0.02 M	1	25± 0.02		I

Comparison electrode	Liquid junction	Electrode reaction	Standard value V	Uncertainty mV	Temperature Coefficient μV/°C	Notes	Reference	Electrode Reference Number
d			f		g	h		
		$Ni_3O_4 + 2H_2O + 2e^- \rightleftharpoons 3HNiO_2^- + H^+$	-0.718				1	1
		$HNIO_2^- + 3H^+ + 2e^- \rightleftharpoons Ni + 2H_2O$	0.648				1	2
19		$Ni^{2+} + 2e^- \rightleftharpoons Ni$	-0.257	± 8	+60		2,3,4	3
1		$Ni^{2+} + 2e^- \rightleftharpoons Ni$	0.088				5	4
		$NiCO_3 + 2e^- \rightleftharpoons Ni + CO_3^{2-}$	-0.45		-1271		2	5
		$[Ni(NH_3)_6]^{2+} + 2e^- \rightleftharpoons Ni + 6NH_3$	-0.476				2	6
		$NiO + 2H^+ + 2e^- \rightleftharpoons Ni + H_2O$	0.116				1	7
		$NiO_2 + 4H^+ + 2e^- \rightleftharpoons Ni^{2+} + 2H_2O$	1.678				2	8
		$Ni(OH)_2 + 2H^+ + 2e^- \rightleftharpoons Ni + 2H_2O$	0.110				1	9
		$Ni(OH)_2 + 2e^- \rightleftharpoons Ni + 2OH^-$	-0.72		-1040		2	10
		$[Ni(P_3O_9)]^- + 2e^- \rightleftharpoons Ni + P_3O_9^{3-}$	-0.345			1	6	11
		$NiS + 2e^- \rightleftharpoons Ni + S^{2-}$	-0.814				2,7	12
		$NiS + 2e^- \rightleftharpoons Ni + S^{2-}$	-0.960				7	13
		$NiS + 2e^- \rightleftharpoons Ni + S^{2-}$	-1.07				2,7	14
		$Ni(S_2O_3) + 2e^- \rightleftharpoons Ni + S_2O_3^{2-}$	-0.372			2	8	15
		$NiO_2 + 4H^+ + 2e^- \rightleftharpoons Ni^{2+} + 2H_2O$	1.593				1	16
		$Ni_2O_3 + 6H^+ + 2e^- \rightleftharpoons 2Ni^{2+} + 3H_2O$	1.753				1	17
		$Ni_3O_4 + 8H^+ + 2e^- \rightleftharpoons 3Ni^{2+} + 4H_2O$	1.977				1	18
		$Ni(OH)_3 + 3H^+ + 3e^- \rightleftharpoons Ni^{2+} + 3H_2O$	2.08				9	19
3		$[Ni(CN)_4]^{2-} + e^- \rightleftharpoons [Ni(CN)_3]^{2-} + CN^-$	-0.401			3	10	20
		$Ni_2O_3 + 2H^+ + 2e^- \rightleftharpoons 2NiO + H_2O$	1.020				1	21
		$Ni_3O_4 + 2H^+ + 2e^- \rightleftharpoons 3NiO + H_2O$	0.876				1	22
		$2NiO_2 + 2H^+ + 2e^- \rightleftharpoons Ni_2O_3 + H_2O$	1.434				1	23
		$3Ni_2O_3 + 2H^+ + 2e^- \rightleftharpoons 2Ni_3O_4 + H_2O$	1.305				1	24
		$NiO_2 + 2H_2O + 2e^- \rightleftharpoons Ni(OH)_2 + 2OH^-$	-0.490				2	25
		$Ni_2O_3 + H_2O + 2H^+ + 2e^- \rightleftharpoons 2Ni(OH)_2$	1.032				1	26
		$Ni_3O_4 + 2H_2O + 2H^+ + 2e^- \rightleftharpoons 3Ni(OH)_2$	0.897				1	27
		$Ni(OH)_3 + e^- \rightleftharpoons Ni(OH)_2 + OH^-$	0.48				9	28
		$Ni(OH)_4 + e^- \rightleftharpoons Ni(OH)_3 + OH^-$	0.6				9	29
		$[Ni(acetylacetone)]^{2+} + 2e^- \rightleftharpoons Ni + acetylacetone$	-0.425				11	30

Electronically conducting phase	Intermediate species	Composition of the solution	Solvent	Temperature °C	Pressure Torr	Measuring method
a	b	c	d		e	d
Glass		Ni^{2+} : 10^{-3} M; acetylacetone : 0.02 M	1	25 ± 0.02		I
Glass		$NiCl_2$: (3.40 up to 23.51) · 10^{-3} M; glycine : (56.13 up to 62.34) · 10^{-3} M; NaOH : (0.35 up to 6.99) · 10^{-3} M	1	25± 0.02	760	I, V
Glass		$NiCl_2$: (3.40 up to 23.51) · 10^{-3} M; glycine : (56.13 up to 62.34) · 10^{-3} M; NaOH : (0.35 up to 6.99) · 10^{-3} M	1	25± 0.02		I, V
Glass		$NiCl_2$: 4.25 up to 30.07) · 10^{-3} M; glycylglycine : (24.60 up to 29.68) · 10^{-3} M; NaOH : (2.49 up to 11.20) · 10^{-3} M	1	25± 0.02		I
Glass		$NiCl_2$: (4.25 up to 30.07) · 10^{-3} M; glycylglycine : (24.60 up to 29.68) · 10^{-3} M; NaOH : (2.49 up to 11.20) · 10^{-3} M	1	25± 0.02		I
Pt, H_2		$NiCl_2$: (9.21 up to 10.97) · 10^{-3} M; glycylglycylglycine: (34.16 up to 40.16) · 10^{-3} M; NaOH : (1.113 up to 4.215) . 10^{-3} M	1	25± 0.03		I
Pt, H_2		$NiCl_2$: (9.21 up to 10.97) . 10^{-3} M; glycylglycylglycine: (34.16 up to 40.16) · 10^{-3} M; NaOH : (1.113 up to 4.215) · 10^{-3} M	1	25± 0.03		I
Glass	[Ni (8-hydroxyquinoline-5-sulfonate)]	(8-hydroxyquinoline-5-sulfonate)$^{2-}$: 2.954 · 10^{-3} M; NaOH : 1.4 · 10^{-3} M; Ni $(NO_3)_2$	1	25		I
Glass		(8-hydroxyquinoline-5-sulfonate)$^{2-}$: 2.954 . 10^{-3} M; NaOH :1.4 · 10^{-3} M; Ni $(NO_3)_2$	1	25		I
	Ni (malonate)	$NiCl_2$: 12.33 · 10^{-3} M; NaH malonate : 4.931 · 10^{-3} M	1	25		V
Ni			1	25	760	V

Comparison electrode	Liquid junction	Electrode reaction	Standard value	Uncertainty	Temperature Coefficient	Notes	Reference	Electrode Reference Number
			V	mV	μV/°C			
d			f		g	h		
		$[Ni(acetylacetone)_2]^{2+} + 2e^- \rightleftharpoons Ni + 2(acetylacetone)$	-0.559				11	31
3		$[Ni(glycine)]^{2+} + e^- \rightleftharpoons Ni + glycine$	- 0.433			4, 2	12	32
3		$[Ni(glycine)_2]^{2+} + 2e^- \rightleftharpoons Ni + 2(glycine)$	-0.578			4, 2	12	33
3		$[Ni(glycylglycine)]^{2+} + 2e^- \rightleftharpoons Ni + glycylglycine$	-0.383			4	12	34
3		$[Ni(glycylglycine)_2]^{2+} + 2e^- \rightleftharpoons Ni + 2(glycylglycine)$	-0.483			4	12	35
4		$[Ni(glycylglycylglycine)]^{2+} + 2e^- \rightleftharpoons Ni + glycylglycylglycine$	-0.368				13	36
4		$[Ni(glycylglycylglycine)_2]^{2+} + 2e^- \rightleftharpoons Ni + 2(glycyglycylglycine)$	-0.459				13	37
4		$[Ni(8\text{-hydroxyquinoline-5-sulfonate})] + 2e^- \rightleftharpoons Ni + (8\text{-hydroxyquinoline-5-sulfonate})^{2-}$	-0.538				14	38
4		$[Ni(8\text{-hydroxyquinoline-5-sulfonate})_2]^{2-} + 2e^- \rightleftharpoons Ni + 2(8\text{-hydroxyquinoline-5-sulfonate})^{2-}$	-0.796				14	39
		$Ni(malonate) + 2e^- \rightleftharpoons Ni + (malonate)^{2-}$	-0.368			5	15	40
		$[Ni(oxalate)_2]^{2-} + 2e^- \rightleftharpoons Ni + 2(oxalate)^{2-}$	-0.442			2	16	41

Nickel 1

Standard value V	Solvent	Electrode Reference Number
-1.07	1	14
-0.960	1	13
-0.814	1	12
-0.796	1	39
-0.72	1	10
-0.718	1	1
-0.578	1	33
-0.559	1	31
-0.538	1	38
-0.490	1	25
-0.483	1	35
-0.476	1	6
-0.459	1	37
-0.45	1	5
-0.442	1	41
-0.433	1	32
-0.425	1	30
-0.401	1	20
-0.383	1	34
-0.372	1	15
-0.368	1	36
-0.368	1	40
-0.345	1	11
-0.257	1	3
0.110	1	9
0.116	1	7
0.48	1	28
0.6	1	29
0.648	1	2
0.876	1	22
0.897	1	27
1.020	1	21
1.032	1	26
1.305	1	24
1.434	1	23
1.593	1	16
1.678	1	8
1.753	1	17
1.977	1	18
2.08	1	19
0.088	7	4

Standard value V	Solvent	Electrode Reference Number

NOTES: Nickel

1) Conductometric data
2) Solubility data
3) Normal electric tension
4) Potentiometric data
5) Colorimetric data

BIBLIOGRAPHY: Nickel

1) Data reported by M. Pourbaix, Atlas d'Equilibres Electrochimiques à 25 °C (Gauthier-Villars Editeur, Paris, 1963), pp. 330 - 349
2) Data reported by A.J. De Bethune, N.A. Swendeman-Loud, Standard Aqueous Electrode Potentials and Temperature Coefficients at 25 °C (Clifford A. Hampel, Ill.)
3) W. Palczewska, Roczniki Chem., 29 (1955), 594 - 602
4) A.T. Vagramyan, L.A. Uvarov, Izv. Akad. Nauk SSSR, Otd. Khim. Nauk, 1962, 1520 - 1524
5) B. Jakuszewski, Z. Kozlowskii, Roczniki Chem., 38 (1964), 93
6) H.W. Jones, C.B. Monk, C.W. Davies, J. Chem. Soc., 1964, 2693 - 2695
7) A. Ringbom, Solubility of Sulfides, Report to Analytical Section, I.U.P.A.C., July 1953
8) T.O. Denney, C.B. Monk, Trans. Faraday Soc., 47 (1951), 992 - 998
9) A. Prokopiokas, Liet. TSR Mosklu Akad. Darbai Ser. B, 2 (1962), 31
10) V. Caglioti, G. Sartori, P. Silvestroni, Ric. Scient., 17 (1947), 624 - 626
11) L.E. Maley, D.P. Mellor, Austral. J. Sci. Res., 2A (1949), 92
12) C.B. Monk, Trans. Faraday Soc., 47 (1951), 297 - 302
13) J.I. Evans, C.B. Monk, Trans. Faraday Soc., 51 (1955), 1244 - 1250
14) R. Näsänen, E. Uisitalo, Acta Chem. Scand., 8 (1954), 112
15) D.I. Stock, C.W. Davies, J. Chem. Soc., 1949, 1371 - 1374
16) J.E. Barney, W.J. Argersinger, Jr., C.A. Reynolds, J. Am. Chem. Soc., 73 (1951), 3785

Electronically conducting phase	Intermediate species	Composition of the solution	Solvent	Temperature °C	Pressure Torr	Measuring method
a	b	c	d		e	d
Ru			1	25	760	IV
Ru			1	25	760	IV
Ru	Ru $(OH)_4$		1	25	760	IV
Ru	Ru $(OH)_4$		1	25	760	IV
Ru	Ru_2O_3		1	25	760	IV
Ru	RuO_4 (in sol.)		1	25	760	IV
Ru			1	25	760	IV
Ind.	H_2RuO_5 (in sol.)		1	25	760	IV
Au		HCl : (1.25 up to 9.56) $\cdot 10^{-2}$ M; $RuCl_2$: (0.245 up to 3.07) $\cdot 10^{-3}$ M; $RuCl_3$: (0.245 up to 3.07) $\cdot 10^{-3}$ M	1	25		I
Ind.	RuO_2		1	25	760	IV
Ind.	Ru_2O_3		1	25	760	IV
Ind.			1	25	760	IV
Ind.			1	25	760	IV
Ind.	Ru $(OH)_4$		1	25	760	IV
Pt		Voltammetric measurements : $K_4[Ru(CN)_6]$: 0.593 $\cdot 10^{-3}$ M; KCl : 0.2 M; potentiometric measurements : $K_4[Ru(CN)_6]$: 0.0485 M; H_2SO_4 : 0.5 M	1	27 ± 0.5		I, III
Ag		$NaClO_4$: 0.036 M; Ru^{2+}, Ru^{3+}: 10^{-4} up to 10^{-3} M	1	25 ± 0.2		I
Ag		$NaClO_4$: 0.036 M; Ru^{2+}, Ru^{3+}: 10^{-4} up to 10^{-3} M	1	25 ± 0.2		I
Ind.	RuO_2, H_2RuO_5(in sol.)		1	25	760	IV
Ind.	Ru $(OH)_4$		1	25	760	IV
Ind.	RuO_2, RuO_4 (s)		1	25	760	IV
Ind.	Ru $(OH)_4$, RuO_4 (in sol.)		1	25	760	IV
Ind.	RuO_2		1	25	760	IV
Ind.	Ru $(OH)_4$		1	25	760	IV
Ind.	RuO_2		1	25	760	IV
Ind.	Ru $(OH)_4$		1	25	760	IV
Ind.	Ru_2O_3, RuO_2		1	25	760	IV
Ind.	Ru_2O_3		1	25	760	IV
Ind.	H_2RuO_5 (in sol.)		1	25	760	IV
Ind.			1	25	760	IV

Comparison electrode	Liquid junction	Electrode reaction	Standard value	Uncertainty	Temperature Coefficient	Notes	Reference	Electrode Reference Number
			V	mV	μV/°C			
d			f		g	h		
		$Ru^{2+} + 2e^- \rightleftharpoons Ru$	0.455				1	1
		$RuCl_5^{2-} + 3e^- \rightleftharpoons Ru + 5Cl^-$	0.601				2	2
		$Ru(OH)_4 + 4H^+ + 4e^- \rightleftharpoons Ru + 4H_2O$	0.68				3	3
		$Ru(OH)_4 + 4e^- \rightleftharpoons Ru + 4OH^-$	-0.15				3	4
		$Ru_2O_3 + 6H^+ + 6e^- \rightleftharpoons 2Ru + 3H_2O$	0.738				1	5
		$RuO_4 + 8H^+ + 8e^- \rightleftharpoons Ru + 4H_2O$	1.04				3	6
		$RuO_4^{2-} + 8H^+ + 6e^- \rightleftharpoons Ru + 4H_2O$	1.193				1	7
		$H_2RuO_5 + 8H^+ + 6e^- \rightleftharpoons Ru^{2+} + 5H_2O$	1.307				1	8
34		$Ru^{3+} + e^- \rightleftharpoons Ru^{2+}$	0.2487	± 800			4	9
		$RuO_2 + 4H^+ + 2e^- \rightleftharpoons Ru^{2+} + 2H_2O$	1.120				1	10
		$Ru_2O_3 + 6H^+ + 2e^- \rightleftharpoons 2Ru^{2+} + 3H_2O$	1.304				1	11
		$RuO_4^- + 8H^+ + 5e^- \rightleftharpoons Ru^{2+} + 4H_2O$	1.368				1	12
		$RuO_4^{2-} + 8H^+ + 4e^- \rightleftharpoons Ru^{2+} + 4H_2O$	1.563				1	13
		$Ru(OH)_4 + 4H^+ + e^- \rightleftharpoons Ru^{3+} + 4H_2O$	1.6				3	14
3		$[Ru(CN)_6]^{3-} + e^- \rightleftharpoons [Ru(CN)_6]^{4-}$	0.86	± 50		1	5	15
3		$[Ru(NH_3)_6]^{3+} + e^- \rightleftharpoons [Ru(NH_3)_6]^{2+}$	0.214				6	16
3		$[Ru(NH)_3)_5(H_2O)]^{3+} + e^- \rightleftharpoons [Ru(NH_3)_5]^{2+} + H_2O$	0.20				6	17
		$H_2RuO_5 + 4H^+ + 4e^- \rightleftharpoons RuO_2 + 3H_2O$	1.400				1	18
		$HRuO_5^- + 4H_2O + 4e^- \rightleftharpoons Ru(OH)_4 + 5OH^-$	0.54				3	19
		$RuO_4 + 4H^+ + 4e^- \rightleftharpoons RuO_2 + 2H_2O$	1.387				1	20
		$RuO_4 + 4H_2O + 4e^- \rightleftharpoons Ru(OH)_4 + 4OH^-$	0.57				3	21
		$RuO_4^- + 4H^+ + 3e^- \rightleftharpoons RuO_2 + 2H_2O$	1.533				1	22
		$RuO_4^- + 4H_2O + 3e^- \rightleftharpoons Ru(OH)_4 + 4OH^-$	0.43				3	23
		$RuO_4^{2-} + 4H^+ + 2e^- \rightleftharpoons RuO_2 + 2H_2O$	2.005				1	24
		$RuO_4^{2-} + 4H_2O + 2e^- \rightleftharpoons Ru(OH)_4 + 4OH^-$	0.35				3	25
		$2RuO_2 + 2H^+ + 2e^- \rightleftharpoons Ru_2O_3 + H_2O$	0.937				1	26
		$2RuO_4^{2-} + 10H^+ + 6e^- \rightleftharpoons Ru_2O_3 + 5H_2O$	1.649				1	27
		$H_2RuO_5 + e^- \rightleftharpoons RuO_4^- + H_2O$	1.001				1	28
		$HRuO_5^- + H^+ + e^- \rightleftharpoons RuO_4^- + H_2O$	1.660				1	29

Electronically conducting phase	Intermediate species	Composition of the solution	Solvent	Temperature °C	Pressure Torr	Measuring method
a	b	c	d		e	d
Ind.			1	25	760	IV
1) Glass 2) Pt	RuO_4 (in sol.)	1) RuO_4^-: (0.7 up to 3.91) · 10^{-5} M; RuO_4: (2.02 up to 15.1) · 10^{-5} M; pH : 5.01 up to 5.72 2) $NaClO_4$, $KRuO_4$, RuO_4 , pH : 10.6 up to 12.1	1	25		I, III
Pt		1) NaOH + Na_3PO_4 + K_2HPO_4 or NaOH + Na_2CO_3 ; pH: 11.43 up to 12.60 2) $NaClO_4$, K_2RuO_4 , RuO_4 , pH : 10.6 up to 12.1	1	25		I, III
1) Rh 2,3) Pt		1) [Ru (dipyridyl)$_3$] Cl_2 : 10^{-3} M 2) H_2SO_4 : 0.1 up to 16.0 M 3) H_2SO_4 : 0.01 up to 12.0 M	1	25		I
Pt	[Ru (dipyridyl)$_2$ $(CN)_2$]	H_2SO_4 : 0.01 up to 12 M; or HNO_3 : 1 up to 16 M	1	25 ± 0.05		I
Pt		H_2SO_4 : 0.1 up to 2.0 M	1	25		I

Comparison electrode	Liquid junction	Electrode reaction	Standard value	Uncertainty	Temperature Coefficient	Notes	Reference	Electrode Reference Number
			V	mV	μV/°C			
d			f		g	h		
		$HRuO_5^- + e^- \rightleftharpoons RuO_4^- + OH^-$	0.85				3	30
3		$RuO_4 + e^- \rightleftharpoons RuO_4^-$	1.00	± 20		1,2	7,8	31
3		$RuO_4^- + e^- \rightleftharpoons RuO_4^{2-}$	0.59	± 20		2	7,8	32
3		$[Ru(dipyridyl)_3]^{3+} + e^- \rightleftharpoons [Ru(dipyridyl)_3]^{2+}$	1.29			2,4	5, 9, 10	33
3		$[Ru(dipyridyl)_2(CN)_2]^+ + e^- \rightleftharpoons [Ru(dipyridyl)_2(CN)_2]$	1.10			4	10	34
		$[Ru(terpyridyl)_2]^{3+} + e^- \rightleftharpoons [Ru(terpyridyl)_2]^{2+}$	1.202	± 2		4	9	35

Ruthenium 1

Standard value V	Solvent	Electrode Reference Number
-0.15	1	4
0.20	1	17
0.214	1	16
0.2487	1	9
0.35	1	25
0.43	1	43
0.455	1	1
0.54	1	19
0.57	1	21
0.59	1	32
0.601	1	2
0.68	1	3
0.738	1	5
0.85	1	30
0.86	1	15
0.937	1	26
1.00	1	31
1.001	1	28
1.04	1	6
1.10	1	34
1.120	1	10
1.193	1	7
1.202	1	35
1.29	1	33
1.304	1	11
1.307	1	8
1.368	1	12
1.387	1	20
1.400	1	18
1.533	1	22
1.563	1	13
1.6	1	14
1.649	1	27
1.660	1	29
2.005	1	24

NOTES: Ruthenium

1) Potentiometric and voltammetric data

2) The value is an average of those given by two authors; they are in the same range of uncertainty

3) Voltammetric data

4) Formal electric tension values

BIBLIOGRAPHY: Ruthenium

1) Data reported by M. Pourbaix, Atlas d'Equilibres Electrochimiques à 25 °C (Gauthier-Villars Editeur, Paris, 1963), pp. 343 - 349

2) Data reported by A.J. De Bethune, N.A. Swendeman-Loud, Standard Aqueous Electrode Potentials and Temperature Coefficients at 25 °C (Clifford A. Hampel, Ill.)

3) R.N. Goldberg, L.G. Hepler, Chem. Rev., 60 (1960), 229 - 252

4) R.R. Buckley, E.E. Mercer, J. Phys. Chem., 70 (1966), 3103 - 3106

5) P. George, G.I.H. Hanania, D.H. Irvine, J. Chem. Soc., 1959, 2548 - 2554

6) J.F. Andicott, H. Taube, Inorg. Chem., 4 (1965), 437 - 445

7) R.E. Connick, C.R. Hurley, J. Phys. Chem., 61 (1957), 1018 - 1020

8) M.D. Silvermann, H.A. Levy, J. Am. Chem. Soc., 76 (1954), 3319 - 3321

9) W.W. Brandt, G.F. Smith, Anal. Chem., 21 (1949), 1313 - 1319

10) A.A. Schilt, Anal. Chem., 35 (1963), 1599 - 1602

Electronically conducting phase	Intermediate species	Composition of the solution	Solvent	Temperature °C	Pressure Torr	Measuring method
a	b	c	d		e	d
Rh			1	25	760	IV
Rh			1	25	760	IV
Rh		$Rh(ClO_4)_3$: 1.08 up to 17.02 · 10^{-4} M; $HClO_4$:pH5	1	25 ± 0.1		I
Rh			1	25	760	IV
Rh	RhO		1	25	760	IV
Rh	Rh_2O		1	25	760	IV
Rh	Rh_2O_3		1	25	760	IV
Rh			1	25	760	IV
Rh			1	25	760	IV
Rh	$Rh(OH)_3$		1	25	760	IV
Ind.			1	25	760	IV
Ind.	Rh_2O_3		1	25	760	IV
Ind.			1	25	760	IV
Ind.			1	25	760	IV
Ind.	Rh_2O_3		1	25	760	IV
Ind.			1	25	760	IV
Ind.			1	25	760	IV
Ind.	RhO_2		1	25	760	IV
Ind.			1	25	760	IV
Ind.			1	25	760	IV
Ind.	Rh_2O		1	25	760	IV
Ind.	Rh_2O		1	25	760	IV
Ind.	Rh_2O, RhO		1	25	760	IV
Ind.	Rh_2O, Rh_2O_3		1	25	760	IV
Ind.	RhO, Rh_2O_3		1	25	760	IV
Ind.			1	25	760	IV
Ind.	RhO_2		1	25	760	IV
Ind.	Rh_2O_3 , RhO_2		1	25	760	IV
Ind.	Rh_2O_3		1	25	760	IV
Ind.	$Rh(OH)_3$		1	25	760	IV

Comparison electrode	Liquid junction	Electrode reaction	Standard value V	Uncertainty mV	Temperature Coefficient μV/°C	Notes	Reference	Electrode Reference Number
d			f		g	h		
		$Rh^{+} + e^{-} \rightleftharpoons Rh$	0.600				1	1
		$Rh^{2+} + 2e^{-} \rightleftharpoons Rh$	0.600				1	2
3		$Rh^{3+} + 3e^{-} \rightleftharpoons Rh$	0.758	±2		1	2	3
		$[RhCl_6]^{3-} + 3e^{-} \rightleftharpoons Rh + 6Cl^{-}$	0.431		-145		3	4
		$RhO + 2H^{+} + 2e^{-} \rightleftharpoons Rh + H_2O$	0.81				4	5
		$Rh_2O + 2H^{+} + 2e^{-} \rightleftharpoons 2Rh + H_2O$	0.796				1	6
		$Rh_2O_3 + 3H_2O + 6e^{-} \rightleftharpoons 2Rh + 6OH^{-}$	0.04		-1230		3	7
		$RhO_4^{2-} + 8H^{+} + 6e^{-} \rightleftharpoons Rh + 4H_2O$	1.1				4	8
		$RhO_4^{2-} + 4H_2O + 6e^{-} \rightleftharpoons Rh + 8OH^{-}$	0.0				4	9
		$Rh(OH)_3 + 3e^{-} \rightleftharpoons Rh + 3OH^{-}$	0.0				4	10
		$Rh^{2+} + e^{-} \rightleftharpoons Rh^{+}$	0.600				1	11
		$Rh_2O_3 + 6H^{+} + 4e^{-} \rightleftharpoons 2Rh^{+} + 3H_2O$	0.975				1	12
		$RhO_4^{2-} + 8H^{+} + 5e^{-} \rightleftharpoons Rh^{+} + 4H_2O$	1.717				1	13
		$Rh^{3+} + e^{-} \rightleftharpoons Rh^{2+}$	1.198				1	14
		$Rh_2O_3 + 6H^{+} + 2e^{-} \rightleftharpoons 2Rh^{2+} + 3H_2O$	1.349				1	15
		$RhO_4^{2-} + 8H^{+} + 4e^{-} \rightleftharpoons Rh^{2+} + 4H_2O$	1.995				1	16
		$RhO^{2+} + 2H^{+} + e^{-} \rightleftharpoons Rh^{3+} + H_2O$	1.4				4	17
		$RhO_2 + 4H^{+} + e^{-} \rightleftharpoons Rh^{3+} + 2H_2O$	1.881				1	18
		$RhO_4^{2-} + 8H^{+} + 3e^{-} \rightleftharpoons Rh^{3+} + 4H_2O$	2.261				1	19
		$[RhCl_6]^{2-} + e^{-} \rightleftharpoons [RhCl_6]^{3-}$	1.2				4	20
		$2Rh^{2+} + H_2O + 2e^{-} \rightleftharpoons Rh_2O + 2H^{+}$	0.396				1	21
		$2Rh^{3+} + H_2O + 4e^{-} \rightleftharpoons Rh_2O + 2H^{+}$	0.801				1	22
		$2RhO + 2H^{+} + 2e^{-} \rightleftharpoons Rh_2O + H_2O$	0.882				1	23
		$Rh_2O_3 + 4H^{+} + 4e^{-} \rightleftharpoons Rh_2O + 2H_2O$	0.877				1	24
		$Rh_2O_3 + 2H^{+} + 2e^{-} \rightleftharpoons 2RhO + H_2O$	0.871				1	25
		$RhO_4^{2-} + 6H^{+} + 2e^{-} \rightleftharpoons RhO^{2+} + 3H_2O$	1.5				4	26
		$RhO_4^{2-} + 4H^{+} + 2e^{-} \rightleftharpoons RhO_2 + 2H_2O$	2.452				1	27
		$2RhO_2 + 2H^{+} + 2e^{-} \rightleftharpoons Rh_2O_3 + H_2O$	1.730				1	28
		$2RhO_4^{2-} + 10H^{+} + 6e^{-} \rightleftharpoons Rh_2O_3 + 5H_2O$	2.211				1	29
		$RhO_4^{2-} + 4H_2O + 3e^{-} \rightleftharpoons Rh(OH)_3 + 5OH^{-}$	-0.1				4	30

Rhodium 1

Standard value V	Solvent	Electrode Reference Number	Standard value V	Solvent	Electrode Reference Number	Standard value V	Solvent	Electrode Reference Number
-0.1	1	30						
0.0	1	10						
0.0	1	9						
0.04	1	7						
0.396	1	21						
0.431	1	4						
0.600	1	1						
0.600	1	2						
0.600	1	11						
0.758	1	3						
0.796	1	6						
0.801	1	22						
0.81	1	5						
0.871	1	25						
0.877	1	24						
0.882	1	23						
0.975	1	12						
1.1	1	8						
1.198	1	14						
1.2	1	20						
1.349	1	15						
1.4	1	17						
1.5	1	26						
1.717	1	13						
1.730	1	28						
1.881	1	18						
1.995	1	16						
2.211	1	29						
2.261	1	19						
2.452	1	27						

NOTES: Rhodium

1) $NaClO_4$ 1 N

BIBLIOGRAPHY: Rhodium

1) Data reported by M. Pourbaix, Atlas d'Equilibres Electrochimiques à 25 ^oC (Gauthier-Villars Editeur, Paris 1963), pp. 350 - 357

2) J. Amosse, M. Ruband, M.J. Barbier, Compt. Rend., 273 C (1971), 1708 - 1710

3) Data reported by A.J. De Bethune, N.A. Swendeman-Loud, Standard Aqueous Electrode Potentials and Temperature Coefficients at 25 ^oC (Clifford A. Hampel, Ill.)

4) R.N. Goldberg, L.G. Hepler, Chem. Rev., 68 (1968), 229 - 252

Electronically conducting phase	Intermediate species	Composition of the solution	Solvent	Temperature °C	Pressure Torr	Measuring method
a	b	c	d		e	d
Pd		1) K halides 2) $Pd(ClO_4)_2$, $HClO_4$	1	25		I
Pd			1	25	760	IV
Pd			1	25	760	IV
Pd			1	25		V
Pd	$[PdCl_2]$		1	25		V
Pd			1	25		V
Pd		$[PdCl_4]$: 0.02 up to 0.1 M; Cl^- : 0.4 up to 2.0 M	1	25		III
Pd			1	25		V
Pd			1	25		V
Pd			1	25	760	IV
Pd			1	25	760	IV
Pd			1	25	760	IV
Pd			1	25	760	IV
Pd			1	25	760	IV
Pd			1	25	760	IV
Pd	PdO		1	25	760	IV
Pd	$Pd(OH)_2$		1	25	760	IV
Pd	$Pd(OH)_2$		1	25	760	IV
Pd	$Pd(OH)_4$		1	25	760	IV
Pd			1	25	760	IV
Ind.	PdO_2		1	25	760	IV
Ind.			1	25	760	IV
Ind.			1	25	760	IV
Pd	Pd_2H		1	25	760	IV
Ind.			1	25	760	IV
Ind.	PdO, PdO_2		1	25	760	IV
Ind.	PdO_2, PdO_3		1	25	760	IV
Ind.	$Pd(OH)_2$, PdO_2		1	25	760	IV
Ind.	$Pd(OH)_2$, $Pd(OH)_4$		1	25	760	IV
Pt		$[Pd(\text{ehtylenediamine})_2]^{2+}$ and $[Pd(\text{ethylenediamine})_2Br_2]$: 10^{-6} up to 10^{-5} M	1	25		I
Pt		$[Pd(\text{ethylenediamine})_2]^{2+}$ and $[Pd(\text{ethylenediamine})_2Cl_2]^{2+}$: 10^{-6} up to 10^{-5} M	1	25		I

Comparison electrode	Liquid junction	Electrode reaction	Standard value V	Uncertainty mV	Temperature Coefficient μV/°C	Notes	Reference	Electrode Reference Number
d			f		g	h		
3		$Pd^{2+} + 2e^- \rightleftharpoons Pd$	0.951			1	1,2	1
		$[PdBr_4]^{2-} + 2e^- \rightleftharpoons Pd + 4Br^-$	0.56		+235	1	3,4,5	2
		$[PdBr_6]^{2-} + 2e^- \rightleftharpoons Pd + 6Br^-$	0.74				4	3
		$[PdCl]^+ + e^- \rightleftharpoons Pd + Cl^-$	0.771			2	6	4
		$[PdCl_2] + 2e^- \rightleftharpoons Pd + 2Cl^-$	0.638			2	6	5
		$[PdCl_3]^- + 2e^- \rightleftharpoons Pd + 3Cl^-$	0.567			2	6	6
3		$[PdCl_4]^{2-} + 2e^- \rightleftharpoons Pd + 4Cl^-$	0.591	± 9		3	7	7
		$[PdCl_5]^{3-} + 2e^- \rightleftharpoons Pd + 5Cl^-$	0.553			2	6	8
		$[PdCl_6]^{4-} + 2e^- \rightleftharpoons Pd + 6Cl^-$	0.615			2	6	9
		$[PdCl_6]^{2-} + 2e^- \rightleftharpoons Pd + 6Cl^-$	0.92				4	10
		$[Pd(CN)_4]^{2-} + 2e^- \rightleftharpoons Pd + 4CN^-$	-0.4				4	11
		$[PdI_4]^{2-} + 2e^- \rightleftharpoons Pd + 4I^-$	0.18				4	12
		$[PdI_6]^{2-} + 4e^- \rightleftharpoons Pd + 6I^-$	0.33				4	13
		$[Pd(NH_3)_4]^{2+} + 2e^- \rightleftharpoons Pd + 4NH_3$	0.0				4	14
		$[Pd(NO_2)_4]^{2-} + 2e^- \rightleftharpoons Pd + 4NO_2^-$	0.34				4	15
		$PdO + 2H^+ + 2e^- \rightleftharpoons Pd + H_2O$	0.917				8	16
		$Pd(OH)_2 + 2H^+ + 2e^- \rightleftharpoons Pd + 2H_2O$	0.897				8	17
		$Pd(OH)_2 + 2e^- \rightleftharpoons Pd + 2OH^-$	0.07		-1064	1	3,4	18
		$Pd(OH)_4 + 4e^- \rightleftharpoons Pd + 4OH^-$	0.4				4	19
		$[Pd(SCN)_4]^{2-} + 2e^- \rightleftharpoons Pd + 4SCN^-$	0.14				4	20
		$PdO_2 + 4H^+ + 2e^- \rightleftharpoons Pd^{2+} + 2H_2O$	1.194				8	21
		$[PdBr_6]^{2-} + 2e^- \rightleftharpoons [PdBr_4]^{2-} + 2Br^-$	0.99				4	22
		$[PdCl_6]^{2-} + 2e^- \rightleftharpoons [PdCl_4]^{2-} + 2e^-$	1.288		-450		3	23
		$2Pd + H^+ + e^- \rightleftharpoons Pd_2H$	0.048				8	24
		$[PdI_6]^{2-} + 2e^- \rightleftharpoons [PdI_4]^{2-} + 2I^-$	0.48				4	25
		$PdO_2 + 2H^+ + 2e^- \rightleftharpoons PdO + H_2O$	1.263				8	26
		$PdO_3 + 2H^+ + 2e^- \rightleftharpoons PdO_2 + H_2O$	2.030				8	27
		$PdO_2 + 2H^+ + 2e^- \rightleftharpoons Pd(OH)_2$	1.283				8	28
		$Pd(OH)_4 + 2e^- \rightleftharpoons Pd(OH)_2 + 2OH^-$	0.7				4	29
3		$[Pd(\text{ehtylenediamine})_2Br_2]^{2+} + 2e^- \rightleftharpoons [Pd(\text{ehtylenediamine})_2]^{2+} + 2Br^-$	0.692	± 0.5		4	9	30
3		$[Pd(\text{ethylenediamine})_2Cl_2]^{2+} + 2e^- \rightleftharpoons [Pd(\text{ethylenediamine})_2]^{2+} + 2Cl^-$	1.13	± 0.5		4	9	31

Electronically conducting phase	Intermediate species	Composition of the solution	Solvent	Temperature °C	Pressure Torr	Measuring method
a	b	c	d		e	d
Pt		$[Pd(ethylenediamine)_2]^{2+}$ and $[Pd(ethylenediamine)_2I_2]^{2+}$: 10^{-6} up to 10^{-5} M	1	25		I

Comparison electrode	Liquid junction	Electrode reaction	Standard value	Uncertainty	Temperature Coefficient	Notes	Reference	Electrode Reference Number
			V	mV	μV/°C			
d			f		g	h		
3		$[Pd\,(ethylenediamine)_2\,I_2]^{2+} + 2e^- \rightleftharpoons [Pd\,(ethylenediamine)_2]^{2+} + 2I^-$	0.625	± 0.5		4	9	32

Palladium 1

Standard value V	Solvent	Electrode Reference Number
-0.4	1	11
0.0	1	14
0.048	1	24
0.07	1	18
0.14	1	20
0.18	1	12
0.33	1	13
0.34	1	15
0.4	1	19
0.48	1	25
0.553	1	8
0.56	1	2
0.567	1	6
0.591	1	7
0.615	1	9
0.625	1	32
0.638	1	5
0.692	1	30
0.7	1	29
0.74	1	3
0.771	1	4
0.897	1	17
0.917	1	16
0.92	1	10
0.951	1	1
0.99	1	22
1.13	1	31
1.194	1	21
1.263	1	26
1.283	1	28
1.288	1	23
2.030	1	27

Standard value V	Solvent	Electrode Reference Number

Standard value V	Solvent	Electrode Reference Number

NOTES: Palladium

1) The value is an average of those given by several authors; they all are in the same range of uncertainty
2) Spectrophotometric data
3) Voltammetric data
4) Formal electric tension

BIBLIOGRAPHY: Palladium

1) A.A. Grinberg, N.N. Kiseleva, M.I. Gelfman, Dokl. Akad. Nauk SSSR, 153 (1963), 1327 - 1329
2) R.M. Izatt, D. Eatough, J.J. Christensen, J. Chem. Soc. A, 1967, 1301 - 1304
3) Data reported by A.J. De Bethune, N.A. Swendeman-Loud, Standard Aqueous Electrode Potentials and Temperature Coefficients at 25 °C (Clifford A. Hampel, Ill.)
4) R.N. Goldberg, L.G. Hepler, Chem. Rev., 68 (1968), 229 - 252
5) W.M. Latimer, Oxidation Potentials (Prentice Hall, New York, 1938), pp. 202 - 204
6) H.A. Droll, Thesis, Pennsylvania State University, 1956, University Microfilms 16705
7) V.I. Kravtsov, M.I. Zelenskii, Elektrokhim., 2 (1966), 1138 - 1143
8) Data reported by M. Pourbaix, Atlas d'Equilibres Electrochimiques à 25 °C (Gauthier Villars Editeur, Paris, 1963), pp. 358 - 363
9) A.V. Babaeva, E.Ya. Khananova, Zh. Neorgan. Khim., 10 (1965), 2579 - 2581

Electronically conducting phase	Intermediate species	Composition of the solution	Solvent	Temperature °C	Pressure Torr	Measuring method
a	b	c	d		e	d
Os	H_2OsO_5 (in sol.)		1	25	760	IV
Os			1	25	760	IV
Os	OsO_2		1	25	760	IV
Os	$Os(OH)_4$		1	25	760	IV
Os	OsO_4		1	25	760	IV
Os			1	25	760	IV
		HBr	1	25		I
		HCl	1	25		I
Ind.	OsO_2, H_2OsO_5 (in sol)		1	25	760	IV
Ind.	$Os(OH)_4$		1	25	760	IV
Ind.	OsO_2, OsO_4		1	25	760	IV
Ind.	OsO_2, OsO_4 (gas)		1	25	760	IV
Ind.	OsO_2		1	25	760	IV
Ind.	H_2OsO_5 (in sol.)		1	25	760	IV
Ind.			1	25	760	IV
Ind.	OsO_4		1	25	760	IV
Ind.	OsO_4 (gas)		1	25	760	IV
Ind.			1	25	760	IV
1) Pt 2) Au		1) [Os (dipyridyl)$_3$]$(ClO_4)_2$ · $2H_2O$: 5 · 10^{-4}M; [Os (dipyridyl)$_3$]$(ClO_4)_3$ · H_2O : 5· 10^{-4}M; KNO_3 or KCl or HCl 2) [Os (dipyridyl)$_3$]$(ClO_4)_2$: 2.5 · 10^{-4}M; [Os (dipyridyl)$_3$]$(ClO_4)_2$: 2.5 · 10^{-4} M; KCl	1	25		I
Pt	[Os (dipyridyl)$_2$ $(CN)_2$]	H_2SO_4 : 0.01 up to 12 M; [Os (dipyridyl)$_2$ $(CN)_2$]		25 ± 0.5		I
Au		[Os (dipyridyl)$_2$ (2,4- pentanedione)]$(ClO_4)_2$: 1.25 · 10^{-4} M; [Os (dipyridyl)$_2$ (2,4-pentanedione)] ClO_4 : 1.25 · 10^{-4} M; KCl	1	25± 0.005		I
Au		[Os (dipyridyl) (pyridyl)$_4$]$(ClO_4)_2$: 2.5 · 10^{-4} M; [Os (dipyridyl) (pyridyl)$_4$]$(ClO_4)_3$: 2.5 · 10^{-4} M; KCl	1	25± 0.005		I
Au		[Os (dipyridyl)$_2$ (pyridyl)$_2$]$(ClO_4)_2$: 2.5 · 10^{-4} M; [Os (dipyridyl)$_2$ (pyridyl)$_2$]$(ClO_4)_3$: 2.5 · 10^{-4} M; KCl	1	25± 0.005		I
Au		[Os (dipyridyl) (pyridyl)$_3$ Br]$(ClO_4)_2$: 1.25 · 10^{-4} M; [Os (dipyridyl)(pyridyl)$_3$ Br]Cl : 1.25 · 10^{-4} M; KCl	1	25± 0.005		I

Comparison electrode	Liquid junction	Electrode reaction	Standard value V	Uncertainty mV	Temperature Coefficient μV/°C	Notes	Reference	Electrode Reference Number
d			f		g	h		
		$H_2OsO_5 + 8H^+ + 8e^- \rightleftharpoons Os + 5H_2O$	0.850				1	1
		$HOsO_5^- + 4H_2O + 8e^- \rightleftharpoons Os + 9OH^-$	0.015				2	2
		$OsO_2 + 4H^+ + 4e^- \rightleftharpoons Os + 2H_2O$	0.687				1	3
		$Os(OH)_4 + 4e^- \rightleftharpoons Os + 4OH^-$	-0.12				3	4
		$OsO_4 + 8H^+ + 8e^- \rightleftharpoons Os + 4H_2O$	0.85		-433		2	5
		$OsO_4^{2-} + 8H^+ + 6e^- \rightleftharpoons Os + 4H_2O$	0.994				1	6
		$[OsBr_6]^{2-} + e^- \rightleftharpoons [OsBr_6]^{3-}$	0.452				4	7
		$[OsCl_6]^{2-} + e^- \rightleftharpoons [OsCl_6]^{3-}$	0.452				4	8
		$H_2OsO_5 + 4H^+ + 4e^- \rightleftharpoons OsO_2 + 3H_2O$	1.013				1	9
		$HOsO_5^- + 4H_2O + 4e^- \rightleftharpoons Os(OH)_4 + 5OH^-$	0.10				3	10
		$OsO_4 + 4H^+ + 4e^- \rightleftharpoons OsO_2 + 2H_2O$	1.005				1	11
		$OsO_4 + 4H^+ + 4e^- \rightleftharpoons OsO_2 + 2H_2O$	1.035				1	12
		$OsO_4^{2-} + 4H^+ + 2e^- \rightleftharpoons OsO_2 + 2H_2O$	1.607				1	13
		$H_2OsO_5 + 2e^- \rightleftharpoons OsO_4^{2-} + H_2O$	0.418				1	14
		$HOsO_5^- + H^+ + 2e^- \rightleftharpoons OsO_4^{2-} + H_2O$	0.714				1	15
		$OsO_4 + 2e^- \rightleftharpoons OsO_4^{2-}$	0.402				1	16
		$OsO_4 + 2e^- \rightleftharpoons OsO_4^{2-}$	0.463				1	17
		$OsO_5^{2-} + 2H^+ + 2e^- \rightleftharpoons OsO_4^{2-} + H_2O$	1.142				1	18
3	KCl satd.	$[Os(dipyridyl)_3]^{3+} + e^- \rightleftharpoons [Os(dipyridyl)_3]^{2+}$	0.8821			1	5,6	19
3		$[Os(dipyridyl)_2(CN_2]^+ + e^- \rightleftharpoons$ $[Os(dipyridyl)_2(CN)_2]$	0.792				8	20
3	KCl satd.	$[Os(dipyridyl)_2(2,4\text{-pentanedione})]^{2+} + e^- \rightleftharpoons$ $[Os(dipyridyl)_2(2,4\text{-pentanedione})]^+$	0.1530	± 0.1			6	21
3	KCl satd.	$[Os(dipyridyl)(pyridyl)_4]^{3+} + e^- \rightleftharpoons$ $[Os(dipyridyl)(pyridyl)_4]^{2+}$	0.8052	± 0.1			5	22
3	KCl satd.	$[Os(dipyridyl)_2(pyridyl)_2]^{3+} + e^- \rightleftharpoons$ $[Os(dipyridyl)_2(pyridyl)_2]^{2+}$	0.8345	± 0.1			5	23
3	KCl satd.	$[Os(dipyridyl)(pyridyl)_3Br]^{2+} + e^- \rightleftharpoons$ $[Os(dipyridyl)(pyridyl)_3Br]^+$	0.4437	± 0.1			5	24

Electronically conducting phase	Intermediate species	Composition of the solution	Solvent	Temperature °C	Pressure Torr	Measuring method
a	b	c	d		e	d
Au		[Os (dipyridyl)$_2$(pyridyl) Br](ClO$_4$)$_2$: 1.25 · 10^{-4} M; [Os (dipyridyl)$_2$(pyridyl) Br]Cl : 1.25 · 10^{-4} M; KCl	1	25± 0.005		I
Au		[Os (dipyridyl)(pyridyl)$_3$ Cl](ClO$_4$)$_2$: 2.5 · 10^{-4} M; [Os (dipyridyl)(pyridyl)$_3$ Cl]Cl : 2.5 · 10^{-4} M; KCl	1	25± 0.005		I
Au		[Os (dipyridyl)$_2$(pyridyl) Cl](ClO$_4$)$_2$: 2.5 · 10^{-4} M; [Os (dipyridyl)$_2$(pyridyl) Cl]Cl: 2.5 · 10^{-4} M; KCl	1	25± 0.005		I
Au		[Os (dipyridyl)(pyridyl)$_3$ Br](ClO$_4$)$_2$: 1.25 · 10^{-4} M; [Os (dipyridyl)(pyridyl)$_3$ Br]Cl : 1.25 · 10^{-4}M; KCl	1	25± 0.005		I
Au		[Os (dipyridyl)$_2$(pyridyl) I](ClO$_4$)$_2$: 1.25 · 10^{-4} M; [Os (dipyridyl)$_2$(pyridyl) I]Cl : 1.25 · 10^{-4} M; KCl	1	25± 0.005		I
Au		[Os (tripyridyl)$_2$](ClO$_4$)$_2$; [Os (tripyridyl)$_2$](ClO$_4$)$_3$ H_2SO_4 : 0.1 up to 2.0 M	1	25		I
Au		[Os (tripyridyl)(dipyridyl) Br] (ClO$_4$)$_2$: 1.25 · 10^{-4}M; [Os (tripyridyl)(dipyridyl)Br]Cl : 1.25 · 10^{-4} M;KCl	1	25± 0.005		I
Au		[Os (tripyridyl)(dipyridyl) Cl](ClO$_4$)$_2$: 2.5 · 10^{-4}M; [Os (tripyridyl)(dipyridyl) Cl]Cl : 2.5 · 10^{-4} M; KCl	1	25± 0.005		I
Au		[Os (tripyridyl)(dipyridyl) I]ClO$_4$)$_2$: 1.25 · 10^{-4} M; [Os (tripyridyl)(dipyridyl) I]Cl : 1.25 · 10^{-4} M; KCl	1	25± 0.005		I
Au		[Os (tripyridyl)(dipyridyl)(pyridyl)](ClO$_4$)$_2$: 2.5 · 10^{-4} M; [Os (tripyridyl)(dipyridyl)(pyridyl)](ClO$_4$)$_3$: 2.5 · 10^{-4} M; KCl	1	25± 0.005		I
Au		[Os (tripyridyl)(dipyridyl)(4-ethylpyridyl)](ClO$_4$)$_2$: 2.5 · 10^{-4} M; [Os (tripyridyl)(dipyridyl)(4-ethylpy-ridyl)](ClO$_4$)$_3$: 2.5 · 10^{-4} M; KCl	1	25± 0.005		I
Au		[Os (tripyridyl)(dipyridyl)(3-methylpyridyl)](ClO$_4$)$_2$: 2.5 · 10^{-4} M; [Os (tripyridyl)(dipyridyl)(3-methylpy-ridyl)](ClO$_4$)$_2$: 2.5 · 10^{-4} M; KCl	1	25± 0.005		I
Au		[Os (tripyridyl)(dipyridyl)(4-methylpyridyl)](ClO$_4$)$_2$: 2.5 · 10^{-4} M; [Os (tripyridyl)(dipyridyl)(4-methyl-pyridyl)](ClO$_4$)$_3$: 2.5 · 10^{-4} M; KCl	1	25± 0.005		I
Au		[Os (tripyridyl)(dipyridyl)(4-propylpyridyl)](ClO$_4$)$_2$: 2.5 · 10^{-4} M; [Os (tripyridyl)(dipyridyl)(4-propyl-pyridyl)](ClO$_4$)$_3$: 2.5 · 10^{-4} M; KCl	1	25± 0.005		I
Au		[Os (tripyridyl)(pyridyl)$_3$](ClO$_4$)$_2$: 2.5 · 10^{-4} M; [Os (tripyridyl) (pyridyl)$_3$](ClO$_4$)$_3$: 2.5 · 10^{-4} M; KCl	1	25± 0.005		I

Comparison electrode	Liquid junction	Electrode reaction	Standard value V	Uncertainty mV	Temperature Coefficient μV/°C	Notes	Reference	Electrode Reference Number
d			f		g	h		
3	KCl satd.	$[Os(dipyridyl)_2(pyridyl)Br]^{2+} + e^- \rightleftharpoons [Os(dipyridyl)_2(pyridyl)Br]^+$	0.4871	± 0.1			5	25
3	KCl satd.	$[Os(dipyridyl)_3Cl]^{2+} + e^- \rightleftharpoons [Os(dipyridyl)(pyridyl)_3Cl]^+$	0.4255	± 0.1			5	26
3	KCl satd.	$[Os(dipyridyl)_2(pyridyl)Cl]^{2+} + e^- \rightleftharpoons [Os(dipyridyl)_2(pyridyl)Cl]^+$	0.4836	± 0.1			5	27
3	KCl satd.	$[Os(dipyridyl)(pyridyl)_3I]^{2+} + e^- \rightleftharpoons [Os(dipyridyl)(pyridyl)_3I]^+$	0.4508	± 0.1			5	28
3	KCl satd.	$[Os(dipyridyl)_2(pyridyl)I]^{2+} + e^- \rightleftharpoons [Os(dipyridyl)_2(pyridyl)I]^+$	0.4888	± 0.1			5	29
3		$[Os(tripyridyl)_2]^{3+} + e^- \rightleftharpoons [Os(tripyridyl)_2]^{2+}$	0.9866	± 0.5			7	30
3	KCl satd.	$[Os(tripyridyl)(dipyridyl)Br]^{2+} + e^- \rightleftharpoons [Os(tripyridyl)(dipyridyl)Br]^+$	0.5675	± 0.1			5	31
3	KCl satd.	$[Os(tripyridyl)(dipyridyl)Cl]^{2+} + e^- \rightleftharpoons [Os(tripyridyl)(dipyridyl)Cl]^+$	0.5628	± 0.1			5	32
3	KCl satd.	$[Os(tripyridyl)(dipyridyl)I]^{2+} + e^- \rightleftharpoons [Os(tripyridyl)(dipyridyl)I]^+$	0.5666	± 0.1			5	33
3	KCl satd.	$[Os(tripyridyl)(dipyridyl)(pyridyl)]^{3+} + e^- \rightleftharpoons [Os(tripyridyl)(dipyridyl)(pyridyl)]^{2+}$	0.8713	± 0.1			5	34
3	KCl satd.	$[Os(tripyridyl)(dipyridyl)(4\text{-}ethylpyridyl)]^{3+} + e^- \rightleftharpoons [Os(tripyridyl)(dipyridyl)(4\text{-}ethylpyridyl)]^{2+}$	0.8587	± 0.1			5	35
3	KCl satd.	$[Os(tripyridyl)(dipyridyl)(3\text{-}methylpyridyl)]^{3+} + e^- \rightleftharpoons [Os(tripyridyl)(dipyridyl)(3\text{-}methylpyridyl)]^{2+}$	0.8672	± 0.1			5	36
3	KCl satd.	$[Os(tripyridyl)(dipyridyl)(4\text{-}methylpyridyl)]^{3+} + e^- \rightleftharpoons [Os(tripyridyl)(dipyridyl)(4\text{-}methylpyridyl)]^{2+}$	0.8506	± 0.1			5	37
3	KCl satd.	$[Os(tripyridyl)(dipyridyl)(4\text{-}propylpyridyl)]^{3+} + e^- \rightleftharpoons [Os(tripyridyl)(dipyridyl)(4\text{-}propylpyridyl)]^{2+}$	0.8609	± 0.1			5	38
3	KCl satd.	$[Os(tripyridyl)(pyridyl)_3]^{3+} + e^- \rightleftharpoons [Os(tripyridyl)(pyridyl)_3]^{2+}$	0.7999	± 0.1			5	39

Osmium 1

Standard value V	Solvent	Electrode Reference Number
-0.12	1	4
0.015	1	2
0.10	1	10
0.1530	1	21
0.402	1	16
0.418	1	14
0.4255	1	26
0.4437	1	24
0.4508	1	28
0.452	1	7
0.452	1	8
0.463	1	17
0.4836	1	27
0.4871	1	25
0.4888	1	29
0.5628	1	32
0.5666	1	33
0.5675	1	31
0.687	1	3
0.714	1	15
0.792	1	20
0.7999	1	39
0.8052	1	22
0.8345	1	23
0.850	1	1
0.85	1	5
0.8506	1	37
0.8587	1	35
0.8609	1	38
0.8672	1	36
0.8713	1	34
0.8821	1	19
0.9866	1	30
0.994	1	6
1.005	1	11
1.013	1	9
1.035	1	12
1.142	1	18
1.607	1	13

Standard value V	Solvent	Electrode Reference Number

NOTES: Osmium

1) The value is an average of those given by several authors; they all are in the same range of uncertainty

BIBLIOGRAPHY: Osmium

1) Data reported by M. Pourbaix, "Atlas d'Equilibres Electrochimiques à 25 °C" (Gauthier-Villars Editeur, Paris, 1963), pp. 364 - 372

2) Data reported by A.J. De Bethune, N.A. Swendeman-Loud, "Standard Aqueous Electrode Potentials and Temperature Coefficients at 25 °C" (Clifford A. Hampel, Ill.)

3) R.N. Goldberg, L.G. Hepler, Chem. Rev., 68 (1968), 229 - 252

4) F.P. Dwyer, H.A. Mckenzie, R.S. Nyholm, J. Proc. Roy. Soc. N.S. Wales, 80 (1947), 183 - 186; 242 - 246

5) G.T. Barnes, F.P. Dwyer, E.C. Gyarfas, Trans. Faraday Soc., 48 (1952), 269 - 273

6) D.A. Buckingham, F.P. Dwyer, A.M. Sargensen, Inorg. Chem., 5 (1966), 1243 - 1249

7) F.P. Dwyer, E.C. Gyarfas, J. Am. Chem. Soc., 76 (1954), 6320 - 6321

8) A.A. Schilt, Anal. Chem., 35 (1963), 1599 - 1602

Electronically conducting phase	Intermediate species	Composition of the solution	Solvent	Temperature °C	Pressure Torr	Measuring method
a	b	c	d		e	d
Ir			1	25	760	IV
Ir			1	25	760	IV
Ir			1	25	760	IV
Ir	IrO		1	25	760	IV
Ir	IrO_2		1	25	760	IV
Ir	Ir_2O_3		1	25	760	IV
Ir	Ir_2O_3		1	25	760	IV
Ind.	IrO_2		1	25	760	IV
Ind.			1	25	760	IV
Ind.			1	25	760	IV
		NaBr : 1 N	1	25		I
Rh		$K_3[IrCl_6]$: $2 \cdot 10^{-3}$ M; HCl : 0.31 N	1	25		I
Rh		$(NH_4)_2[Ir(H_2O)Cl_5]$: (4.86 up to 10.8) $\cdot 10^{-3}$ M; HNO_3 : 0.2 N	1	25		I
Rh	Trans-$[Ir(H_2O)_2Cl_4]$	$HClO_4$: 0.4 N; NaH_2PO_4 : 0.2 N	1	25		I
Rh	$[Ir(H_2O)_3Cl_3]$	$HClO_4$: 0.4 N, or HNO_3 : 0.4 N	1	25		I
			1	25		I
Ind.	IrO_2		1	25	760	IV
Ind.	Ir_2O_3, IrO_2		1	25	760	IV
Ind.	Ir_2O_3		1	25	760	IV

Comparison electrode	Liquid junction	Electrode reaction	Standard value V	Uncertainty mV	Temperature Coefficient μV/°C	Notes	Reference	Electrode Reference Number
d			f		g	h		
		$Ir^{3+} + 3e^- \rightleftharpoons Ir$	1.156				1	1
		$[IrCl_6]^{3-} + 3e^- \rightleftharpoons Ir + 6Cl^-$	0.77		-30		2	2
		$[IrCl_6]^{2-} + 4e^- \rightleftharpoons Ir + 6Cl^-$	0.86				3	3
		$IrO + 2H^+ + 2e^- \rightleftharpoons Ir + H_2O$	0.87				3	4
		$IrO_2 + 4H^+ + 4e^- \rightleftharpoons Ir + 2H_2O$	0.926				1	5
		$Ir_2O_3 + 6H^+ + 6e^- \rightleftharpoons 2Ir + 3H_2O$	0.926				1	6
		$Ir_2O_3 + 3H_2O + 6e^- \rightleftharpoons 2Ir + 6OH^-$	0.098				2	7
		$IrO_2 + 4H^+ + e^- \rightleftharpoons Ir^{3+} + 2H_2O$	0.233				1	8
		$IrO_4^{2-} + 8H^+ + 3e^- \rightleftharpoons Ir^{3+} + 4H_2O$	1.448				1	9
		$[IrBr_6]^{3-} + e^- \rightleftharpoons [IrBr_6]^{4-}$	0.99				2	10
		$[IrBr_6]^{2-} + e^- \rightleftharpoons [IrBr_6]^{3-}$	0.97			1,2	4,5	11
3		$[IrCl_6]^{2-} + e^- \rightleftharpoons [IrCl_6]^{3-}$	0.8665		-1310		6	12
3		$[Ir(H_2O)Cl_5]^- + e^- \rightleftharpoons [Ir(H_2O)Cl_5]^{2-}$	1.0				7	13
3		Trans-$[Ir(H_2O)_2Cl_4] + e^- \rightleftharpoons$ trans-$[Ir(H_2O)_2Cl_4]^-$	1.22	±20			8	14
3		$[Ir(H_2O)_3Cl_3]^+ + e^- \rightleftharpoons [Ir(H_2O)_3Cl_3]$	1.30	±10			8	15
		$[IrI_6]^{2-} + e^- \rightleftharpoons [IrI_6]^{3-}$	0.485			1	5	16
		$IrO_4^{2-} + 4H^+ + 2e^- \rightleftharpoons IrO_2 + 2H_2O$	2.057				1	17
		$2IrO_2 + 2H^+ + 2e^- \rightleftharpoons Ir_2O_3 + H_2O$	0.926				1	18
		$2IrO_4^{2-} + 10H^+ + 6e^- \rightleftharpoons Ir_2O_3 + 5H_2O$	1.680				1	19

Iridium 1

Standard value V	Solvent	Electrode Reference Number
0.098	1	7
0.233	1	8
0.485	1	16
0.77	1	2
0.86	1	3
0.8665	1	12
0.87	1	4
0.926	1	5
0.926	1	6
0.926	1	18
0.97	1	11
0.99	1	10
1.0	1	13
1.156	1	1
1.22	1	14
1.30	1	15
1.448	1	9
1.680	1	19
2.057	1	17

NOTES : Iridium

1) Formal electric tension

2) The value is an average of those given by several authors; they all are in the same range of uncertainty

BIBLIOGRAPHY: Iridium

1) Data reported by M. Pourbaix, Atlas d'Equilibres Electrochimiques à 25 °C (Gauthier-Villars Editeurs, Paris, 1963), pp. 373 - 377

2) Data reported by A.J. De Bethune, N.A. Swendeman-Loud, Standard Aqueous Electrode Potentials and Temperature Coefficients at 25 °C (Clifford A. Hampel, Ill.)

3) R.N. Goldberg, L.G. Hepler, Chem. Rev., 68 (1968), 229 - 252

4) F.P. Dwyer, H.A. McKenzie, R.S. Nyholm, J. Proc. Roy. Soc. N.S. Wales, 81 (1947), 216 - 220

5) B.V. Ptitsyn, Zh. Neorgan. Khim., 15 (1945), 277 - 282

6) P. George, G.I. Hanania, D.H. Irvine, J. Chem. Soc., 1957, 3048 - 3052

7) J.C. Chang, C.S. Garner, Inorg. Chem., 4 (1965), 209 - 215

8) A.A. El-Awady, E.J. Bounsall, C.S. Garner, Inorg. Chem., 6 (1967), 79 - 86

Electronically conducting phase	Intermediate species	Composition of the solution	Solvent	Temperature °C	Pressure Torr	Measuring method
a	b	c	d		e	d
Pt			1	25	760	IV
Pt			1	25	760	IV
Pt		H_2SO_4 : 1 M	1	25		I
Pt			1	25	760	IV
Pt		NaI : 0.5 M	1			III
Pt			1	25	760	IV
Pt			1	25	760	IV
Pt			1	25	760	IV
Pt	Cis-$[Pt(NH_3)_2Cl_2]$		1	25	760	IV
Pt	Trans-$[Pt(NH_3)_2Cl_2]$		1	25	760	IV
Pt			1	25	760	IV
Pt	Cis-$[Pt(NH_3)_2I_2]$		1	25	760	IV
Pt	Trans-$[Pt(NH_3)_2I_2]$		1	25	760	IV
Pt	PtO		1	25	760	IV
Pt	$Pt(OH)_2$		1	25	760	IV
Pt	$Pt(OH)_2$	K_2SO_4 : 0.2 M	1	25		III
Pt	PtS, H_2S (gas)		1	25	760	IV
Pt	PtS, H_2S (in sol.)		1	25	760	IV
Ind.	PtO_2		1	25	760	IV
Pt			1	25	760	IV
Pt	$[\{PtCl(NCl_2)(NH_3)_4\}Cl_2]$, $[PtCl(NCl_2)(NH_3)_4]$	$[\{PtCl(NCl_2)(NH_3)\}Cl_2]$: $2 \cdot 10^{-3}$ M; H_2SO_4 : 1 M	1	25 ± 0.1		I
Pt		$[\{PtCl_3(NCl_2)(NH_3)_2\}Cl]^-$: $5 \cdot 10^{-4}$ M; H_2SO_4 : 1 M	1	25		I
Pt	$[Pt(NO_2)(NH_3)_2Cl]$, $[Pt\{Cl(NO_2)\}(NH_3)_2]$	$[Pt(NO_2)(NH_3)_2Cl]$: 10^{-4} M; $[Pt\{Cl(NO_2)\}_2(NH_3)_2]$: 10^{-4}	1	25		I
Ind.			1	25	760	IV
Pt		NaI : 0.5 M	1			III
Pt		$[Pt(NH_3)_4]^{2-}$: 10^{-4} M; $[Pt(NH_3)_4Br_2]^{2+}$: 10^{-4} M	1	25		I
Ind.			1	25	760	IV
Pt		$[Pt(NH_3)_4]$: 10^{-4} M; $[Pt(NH_3)_4I_2]$: 10^{-4} M	1	25		I
Ind.			1	25	760	IV

Comparison electrode	Liquid junction	Electrode reaction	Standard value V	Uncertainty mV	Temperature Coefficient µV/°C	Notes	Reference	Electrode Reference Number
d			f		g	h		
		$Pt^{2+} + 2e^- \rightleftharpoons Pt$	1.188				1	1
		$[PtBr_4]^{2-} + 2e^- \rightleftharpoons Pt + 4Br^-$	0.581		+150		2	2
3		$[PtCl_4]^{2-} + 2e^- \rightleftharpoons Pt + 4Cl^-$	0.755		-230		2,3	3
		$[Pt(CN)_4]^{2-} + 2e^- \rightleftharpoons Pt + 4CN^-$	0.09				4	4
3		$[PtI_4]^{2-} + 2e^- \rightleftharpoons Pt + 4I^-$	0.40			1	5	5
		$[Pt(NH_3)_4]^{2-} + 2e^- \rightleftharpoons Pt + 4NH_3$	0.25				4	6
		$[Pt(NH_3)Cl_3]^- + 2e^- \rightleftharpoons Pt + NH_3 + 3Cl^-$	0.61				4	7
		$[Pt(NH_3)Cl_5]^- + 4e^- \rightleftharpoons Pt + NH_3 + 5Cl^-$	0.66				4	8
		Cis-$[Pt(NH_3)_2Cl_2] + 2e^- \rightleftharpoons Pt + 2NH_3 + 2Cl^-$	0.49				4	9
		Trans-$[Pt(NH_3)_2Cl_2] + 2e^- \rightleftharpoons Pt + 2NH_3 + 2Cl^-$	0.53				4	10
		$[Pt(NH_3)_3Cl_3]^+ + 2e^- \rightleftharpoons Pt + 3NH_3 + 3Cl^-$	0.54				4	11
		Cis-$[Pt(NH_3)_2I_2] + 2e^- \rightleftharpoons Pt + 2NH_3 + 2I^-$	0.29				4	12
		Trans-$[Pt(NH_3)_2I_2] + 2e^- \rightleftharpoons Pt + 2NH_3 + 2I^-$	0.34				4	13
		$PtO + 2H^+ + 2e^- \rightleftharpoons Pt + H_2O$	0.980				1	14
		$Pt(OH)_2 + 2H^+ + 2e^- \rightleftharpoons Pt + 2H_2O$	0.98		-310		2	15
1		$Pt(OH)_2 + 2e^- \rightleftharpoons Pt + 2OH^-$	0.14		-1144	1,2	6,2	16
		$PtS + 2H^+ + 2e^- \rightleftharpoons Pt + H_2S$	-0.297		+168		2	17
		$PtS + 2H^+ + 2e^- \rightleftharpoons Pt + H_2S$	-0.327		-265		2	18
		$PtO_2 + 4H^+ + 2e^- \rightleftharpoons Pt^{2+} + 2H_2O$	0.837				1	19
		$[PtCl_6]^{2-} + 2e^- \rightleftharpoons [PtCl_4]^{2-} + 2Cl^-$	0.68		-330		2	20
3		$[\{PtCl(NCl_2)(NH_3)_4\}Cl] + 2e^- \rightleftharpoons$ $[PtCl(NCl_2)(NH_3)_4] + 2Cl^-$	1.28	± 30		3	7	21
3		$[\{PtCl_3(NCl_2)(NH_3)_2\}Cl]^- + 2e^- \rightleftharpoons$ $[\{PtCl(NCl_2)(NH_3)_2\}Cl]^- + 2Cl^-$	0.95	± 40		3	7	22
3		$[Pt\{Cl(NO_2)\}_2(NH_3)_2] + 2e^- \rightleftharpoons$ $[PtCl(NO_2)(NH_3)_2] + NO^{2-} + Cl^-$	0.722				8	23
		$[Pt(CN)_4Cl_2]^{2-} + 2e^- \rightleftharpoons [Pt(CN)_4]^{2-} + 2Cl^-$	0.89				4	24
3		$[PtI_6]^{2-} + 2e^- \rightleftharpoons [PtI_4]^{2-} + 2I^-$	0.39			1	5	25
1		$[Pt(NH_3)_4Br_2]^{2+} + 2e^- \rightleftharpoons [Pt(NH_3)_4]^{2-} + 2Br^-$	0.583			3	9	26
		$[Pt(NH_3)_4Cl_2]^{2+} + 2e^- \rightleftharpoons [Pt(NH_3)_4]^{2+} + 2Cl^-$	0.60				4	27
1		$[Pt(NH_3)_4I_2]^{2+} + 2e^- \rightleftharpoons [Pt(NH_3)_4]^{2+} + 2I^-$	0.431			3	9	28
		$[Pt(NH_3)_4(SCN)_2]^{2+} + 2e^- \rightleftharpoons [Pt(NH_3)_4]^{2+} + 2SCN^-$	0.49				4	29

Electronically conducting phase	Intermediate species	Composition of the solution	Solvent	Temperature °C	Pressure Torr	Measuring method
a	b	c	d		e	d
Pt	Cis-$[Pt(NH_3)_2Br_2]$, cis-$[Pt(NH_3)_2Br_4]$	Cis-$[Pt(NH_3)_2Br_2]$: 10^{-4} M; cis-$[Pt(NH_3)_2Br_4]$: 10^{-4} M	1	25		I
Pt	Trans-$[Pt(NH_3)_2Br_2]$; trans-$[Pt(NH_3)_2Br_4]$	Trans-$[Pt(NH_3)_2Br_2]$: $3.3 \cdot 10^{-4}$; trans-$[Pt(NH_3)_2Br_4]$: $3.3 \cdot 10^{-4}$	1	25		I
Ind.			1	25	760	IV
Ind.	Cis-$[Pt(NH_3)_2Cl_2]$; cis-$[Pt(NH_3)_2Cl_4]$		1	25	760	IV
Ind.	Trans-$[Pt(NH_3)_2Cl_2]$; trans-$[Pt(NH_3)_2Cl_4]$		1	25	760	IV
Ind.			1	25	760	IV
Pt	Cis-$[Pt(NH_3)_2I_2]$; cis-$[Pt(NH_3)_2I_4]$	Cis-$[Pt(NH_3)_2I_2]$: 10^{-4} M; cis-$[Pt(NH_3)_2I_4]$: 10^{-4} M	1	25		I
Pt	Trans-$[Pt(NH_3)_2I_2]$; trans-$[Pt(NH_3)_2I_4]$	Trans-$[Pt(NH_3)_2I_2]$: 10^{-4} M; trans-$[Pt(NH_3)_2I_4]$: 10^{-4} M	1	25		I
Pt	$[Pt\{(NH_3)(NO_2)\}_2]$, $[Pt\{(NH_3)(NO_2)\}_2Br_2]$	$[Pt\{(NH_3)(NO_2)\}_2]$: 10^{-} M; $[Pt\{(NH_3)(NO_2)\}_2Br_2]$: 10^{-} M	1	25		I
Pt	$[Pt(NH_3)_2(NO_2)_2]$, $[Pt(NH_3)_2(NO_2)_2Br_2]$	$[Pt(NH_3)_2(NO_2)_2]$: 10^{-4} M; $[Pt(NH_3)_2(NO_2)_2Br_2]$: 10^{-4} M	1	25		I
Pt	$[Pt(NH_3)(NO_2)(NH_3)Br]$, $[Pt(NH_3)(NO_2)(NH_3)Br(NO_2)Br]$	$[Pt(NH_3)(NO_2)(NH_3)Br]$: 10^{-4} M; $[Pt(NH_3)(NO_2)(NH_3)Br(NO_2)Br]$: 10^{-4} M	1	25		I
Pt	$[Pt\{(NH_3)(NO_2)\}_2]$, $[Pt\{(NH_3)(NO_2)\}_2Cl_2]$	$[Pt\{(NH_3)(NO_2)\}_2]$: 10^{-4} M; $[Pt\{(NH_3)(NO_2)\}_2Cl_2]$: 10^{-4} M	1	25		I
Pt	$[Pt(NH_3)_2(NO_2)_2]$, $[Pt(NH_3)_2(NO_2)_2Cl_2]$	$[Pt(NH_3)_2(NO_2)_2]$: 10^{-4} M; $[Pt(NH_3)_2(NO_2)_2Cl_2]$: 10^{-4} M	1	25		I
Pt	$[Pt(NH_3)_2(NO_2)Cl]$, $[Pt(NH_3)_2(NO_2)_2Cl_2]$	$[Pt(NH_3)_2(NO_2)Cl]$: 10^{-4} M; $[Pt(NH_3)_2(NO_2)_2Cl_2]$: 10^{-4} M	1	25		I
Ind.	PtO, PtO_2		1	25	760	IV
Ind.	PtO_2, PtO_3		1	25	760	IV
Pt	$Pt(OH)_2$		1	25	760	IV
Pt			1	25	760	IV
Pt	Cis-[Pt(methylamine)$_2$ Cl_2]		1	25	760	IV
Pt	Trans-[Pt(methylamine)$_2$ Cl_2]		1	25	760	IV
Pt	Cis-[Pt(methylamine)$_2$ I_2]		1	25	760	IV

Comparison electrode	Liquid junction	Electrode reaction	Standard value	Uncertainty	Temperature Coefficient	Notes	Reference	Electrode Reference Number
			V	mV	μV/°C			
d			f		g	h		
1		Cis-$[Pt(NH_3)_2Br_4] + 2e^- \rightleftharpoons$ cis-$[Pt(NH_3)_2Br_2] + 2Br^-$	0.597			3	9	30
1		Trans-$[Pt(NH_3)_2Br_4] + 2e^- \rightleftharpoons$ trans-$[Pt(NH_3)_2Br_2] + 2Br^-$	0.600			3	9	31
		$[Pt(NH_3)Cl_5]^- + 2e^- \rightleftharpoons [Pt(NH_3)Cl_3]^- + 2Cl^-$	0.70				4	32
		Cis-$[Pt(NH_3)_2Cl_4] + 2e^- \rightleftharpoons$ cis-$[Pt(NH_3)_2Cl_2] + 2Cl^-$	0.69				4	33
		Trans-$[Pt(NH_3)_2Cl_4] + 2e^- \rightleftharpoons$ trans-$[Pt(NH_3)_2Cl_2] + 2Cl^-$	0.69				4	34
		$[Pt(NH_3)_3Cl_5]^+ + 2e^- \rightleftharpoons [Pt(NH_3)_3Cl_3]^+ + 2Cl^-$	0.64				4	35
1		Cis-$[Pt(NH_3)_2I_4] + 2e^- \rightleftharpoons$ cis-$[Pt(NH_3)_2I_2] + 2I^-$	0.390			3	4	36
1		Trans-$[Pt(NH_3)_2I_4] + 2e^- \rightleftharpoons$ trans-$[Pt(NH_3)_2I_2] + 2I^-$	0.383			3	9	37
3		$[Pt\{(NH_3)(NO_2)\}_2Br_2] + 2e^- \rightleftharpoons [Pt\{(NH_3)(NO_2)\}_2] + 2Br^-$	0.782				8	38
3		$[Pt(NH_3)_2(NO_2)_2Br_2] + 2e^- \rightleftharpoons [Pt(NH_3)_2(NO_2)_2] + 2Br^-$	0.786				8	39
3		$[Pt(NH_3)(NO_2)(NH_3)Br(NO_2)Br] + 2e^- \rightleftharpoons [Pt(NH_3)(NO_2)(NH_3)Br] + Br^- + NO_2^-$	0.807				8	40
3		$[Pt\{(NH_3)(NO_2)\}_2Cl_2] + 2e^- \rightleftharpoons [Pt\{(NH_3)(NO_2)\}_2] + 2Cl^-$	0.759				8	41
3		$[Pt(NH_3)_2(NO_2)_2Cl_2] + 2e^- \rightleftharpoons [Pt(NH_3)_2(NO_2)_2] + 2Cl^-$	0.763				8	42
3		$[Pt(NH_3)_2(NO_2)_2Cl_2] + 2e^- \rightleftharpoons [Pt(NH_3)_2(NO_2)Cl] + NO_2^- + Cl^-$	0.778				8	43
		$PtO_2 + 2H^+ + 2e^- \rightleftharpoons PtO + H_2O$	1.045				1	44
		$PtO_3 + 2H^+ + 2e^- \rightleftharpoons PtO_2 + H_2O$	2.000				1	45
		$PtO_2 + 2H^+ + 2e^- \rightleftharpoons Pt(OH)_2$	1.1				2	46
		$[Pt(\text{methylamine})_4]^{2+} + 2e^- \rightleftharpoons$ Pt + 4 methylamine	0.20				4	47
		Cis-$[Pt(\text{methylamine})_2Cl_2] + 2e^- \rightleftharpoons$ Pt + 2 methylamine + $2Cl^-$	0.52				4	48
		Trans-$[Pt(\text{methylamine})_2Cl_2] + 2e^- \rightleftharpoons$ Pt + 2 methylamine + $2Cl^-$	0.51				4	49
		Cis-$[Pt(\text{methylamine})_2I_2] + 2e^- \rightleftharpoons$ Pt + 2 methylamine + $2I^-$	0.29				4	50

Electronically conducting phase	Intermediate species	Composition of the solution	Solvent	Temperature °C	Pressure Torr	Measuring method
a	b	c	d		e	d
Pt	Trans-[Pt (methylamine)$_2$ I_2]		1	25	760	IV
Pt	[{PtCl$_2$ (NCl$_2$)(NH$_3$) (pyridine)$_2$ }Cl]; [PtCl(NCl$_2$)(NH$_3$) (pyridine)$_2$]	[{PtCl(NCl$_2$)(NH$_3$)(pyridine)$_2$ }Cl$_2$]: 10^{-3} M; H_2SO_4 : 1 M	1	25 ± 0.1		I
Pt		[Pt (butylamine)$_4$]$^{2+}$: 10^{-4} M; [Pt (butylamine)$_4$ Cl$_2$]$^{2+}$: 10^{-4} M; H_2SO_4 : 1 M	1	25		I
Pt		[Pt (ethylamine)$_4$]$^{2+}$: 10^{-3} M; [Pt (ethylamine)$_4$ Cl$_2$]$^{2+}$: 10^{-3} M; HCl : 1 M	1	25		I
Pt		[Pt (ethylenediamine)$_2$]$^{2+}$: 10^{-4} M; [Pt (ethylenediamine)$_2$ Cl$_2$]$^{2+}$: 10^{-4} M; HCl : 1 M	1	25		I
Pt		[Pt (methylamine)$_4$]$^{2+}$: 10^{-3} M; [Pt (methylamine)$_4$ Cl$_2$]$^{2+}$: 10^{-3} M; HCl : 1 M	1	25		I
Pt		[Pt (propylamine)$_4$]$^{2+}$: 10^{-4} M; [Pt (propylamine)$_4$ Cl$_2$]$^{2+}$: 10^{-4} M; H_2SO_4 : 1 M	1	25		I

Comparison electrode	Liquid junction	Electrode reaction	Standard value V	Uncertainty mV	Temperature Coefficient µV/°C	Notes	Reference	Electrode Reference Number
d			f		g	h		
		Trans-[Pt (methylamine)$_2$ I$_2$] + 2e$^-$ ⇌ Pt + 2 methylamine + 2I$^-$	0.34				4	51
3		[{PtCl$_2$ (NCl$_2$)(NH$_3$)(pyridine)$_2$ }Cl$_2$] + 2e$^-$ ⇌ [PtCl (NCl$_2$)(NH$_3$)(pyridine)$_2$] + 2Cl$^-$	1.02	± 40		3	7	52
3		[Pt (butylamine)$_4$ Cl$_2$]$^{2+}$ + 2e$^-$ ⇌ [Pt (butylamine)$_4$]$^{2+}$ + 2Cl$^-$	0.8380			3	3	53
3		[Pt (ethylamine)$_4$ Cl$_2$]$^{2+}$ + 2e$^-$ ⇌ [Pt (ethylamine)$_4$]$^{2+}$ + 2Cl$^-$	0.6919			3	3	54
3		[Pt (ethylenediamine)$_2$ Cl$_2$]$^{2+}$ + 2e$^-$ ⇌ [Pt (ethylenediamine)$_2$]$^{2+}$ + 2Cl$^-$	0.5660			3	3	55
3		[Pt (methylamine)$_4$ Cl$_2$]$^{2+}$ + 2e$^-$ ⇌ [Pt(methylamine)$_4$]$^{2+}$ + 2Cl$^-$	0.6220			3	3	56
3		[Pt (propylamine)$_4$ Cl$_2$]$^{2+}$ + 2e$^-$ ⇌ [Pt (propylamine)$_4$]$^{2+}$ + 2Cl$^-$	0.8327			3	3	57

Platinum 1

Standard value V	Solvent	Electrode Reference Number	Standard value V	Solvent	Electrode Reference Number	Standard value V	Solvent	Electrode Reference Number
-0.327	1	18	0.70	1	32			
-0.297	1	17	0.722	1	23			
0.09	1	4	0.755	1	3			
0.14	1	16	0.759	1	41			
0.20	1	47	0.763	1	42			
0.25	1	6	0.778	1	43			
0.29	1	12	0.782	1	38			
0.29	1	50	0.786	1	39			
0.34	1	13	0.807	1	40			
0.34	1	51	0.8327	1	57			
0.383	1	37	0.837	1	19			
0.39	1	25	0.8380	1	53			
0.390	1	36	0.89	1	24			
0.40	1	5	0.95	1	22			
0.431	1	28	0.98	1	15			
0.49	1	29	0.980	1	14			
0.49	1	9	1.02	1	52			
0.51	1	49	1.045	1	44			
0.52	1	48	1.1	1	46			
0.53	1	10	1.188	1	1			
0.54	1	11	1.28	1	21			
0.5660	1	55	2.000	1	45			
0.583	1	26						
0.581	1	2						
0.597	1	30						
0.60	1	27						
0.600	1	31						
0.61	1	7						
0.6220	1	56						
0.64	1	35						
0.66	1	8						
0.68	1	20						
0.69	1	33						
0.69	1	34						
0.6919	1	54						

NOTES: Platinum

1) Voltammetric data

2) Chronopotentiometric data

3) Formal electric tension

BIBLIOGRAPHY: Platinum

1) Data reported by M. Pourbaix, "Atlad d'Equilibres Electrochimiques à 25 °C" (Gauthier-Villars Editeur, Paris, 1963), pp. 378 - 383

2) Data reported by A.J. De Bethune, N.A. Swendeman-Loud, "Standard Aqueous Electrode Potentials and Temperature Coefficients at 25 °C" (Clifford A. Hampel, Ill.)

3) A.A. Grinberg, S.C. Dkhara, M.I. Gelfman, Zh. Neorgan. Khim., 13 (1968), 2199 - 2204

4) R.V. Goldberg, L.G. Hepler, Chem. Rev., 68 (1968), 229 - 252

5) A.T. Hubbard, F.C. Anson, Anal. Chem., 38 (1966), 58 - 61

6) D.T. Sawyer, L.V. Interrante, J. Electroanal. Chem., 2 (1961), 310 - 327

7) Yu. N. Kukushkin, Z.V. Drokina, Zh. Neorgan. Khim., 14 (1969), 2452 - 2454

8) I.I. Chernyaev, G.S. Muraveiskaya, L.S. Korablina, Zh. Neorgan. Khim., 11 (1966), 1339 - 1344

9) A.A. Grinberg, B.Z. Orlova, Zh. Priklad. Khim., 22 (1949), 441 - 447

Lanthanides

Electronically conducting phase	Intermediate species	Composition of the solution	Solvent	Temperature °C	Pressure Torr	Measuring method
a	b	c	d		e	d
La			1	25	760	IV
La	$La[Co(CN)_6]$		1	25		III
			1	25		I
La	$La[Fe(CN)_6]$	2) $La[Fe(CN)_6]$: 0.1 up to 0.9 m; glycine : 0.1, 0.25 and 0.5 m; 3) $La[Fe(CN)_6]$: $1.025 \cdot 10^{-4}$ m	1	25		III
La	La_2O_3		1	25	760	IV
La	$La(OH)_3$		1	25	760	IV
La	$La(OH)_3$		1	25	760	IV
La	$La(P_3O_9)$		1	25		III
La			1	25		III
La	La_2S_3		1	25	760	IV
		1) $La_2(SO_4)_3$: $(0.4$ up to $1.8) \cdot 10^{-3}$ N 2) $La_2(SO_4)_3$: up to 0.1021 N	1	25 ± 0.02		III, I
Glass		NaH adipate : $5 \cdot 10^{-3}$ M; $LaCl_3$: (5 and 15) $\cdot 10^{-3}$ M	1	25		I
Glass		NaH fumarate : $5 \cdot 10^{-3}$ M; $LaCl_3$: (5 and 15) $\cdot 10^{-3}$ M	1	25		I
Glass		NaH glutarate : $5 \cdot 10^{-3}$ M; $LaCl_3$: (5 and 15) $\cdot 10^{-3}$ M	1	25		I
Glass		NaH malonate : $15 \cdot 10^{-3}$ M; $LaCl_3$: (5 and 15) $\cdot 10^{-3}$ M	1	25		I
Glass		NaH phthalate : $5 \cdot 10^{-3}$ M; $LaCl_3$: (5 and 15) $\cdot 10^{-3}$ M	1	25		I
La			1	25		V
La			1	25		I
Glass		NaH succinate: $5 \cdot 10^{-3}$ M; $LaCl_3$: (5 and 15) $\cdot 10^{-3}$ M	1	25		I
Ce			1	25	760	IV
Ind.			1	25	760	IV
Pt		$(NH_4)_2Ce(NO_3)_6$; Log (Ce^{4+}/Ce^{3+}): -1.2 up to + 0.8	23	25 ± 0.2		I
Ce		$Ce(ClO_4)_3$: 10^{-4} M; NaBr : 0.493 up to 1.697 M	1	25		V
Ce		$Ce(ClO_4)_3$: $(0.68$ up to $680) \cdot 10^{-3}$M; $NaClO_4$: up to 4.19 M; $HClO_4$: $(0.57$ up to $570) \cdot 10^{-3}$ M	1	25		V
			1	25		V,I

Comparison electrode	Liquid junction	Electrode reaction	Standard value V	Uncertainty mV	Temperature Coefficient μV/°C	Notes	Reference	Electrode Reference Number
d			f		g	h		
		$La^{3+} + 3e^- \rightleftharpoons La$	-2.522				1	1
1		$La[Co(CN)_6] + 3e^- \rightleftharpoons La + [Co(CN)_6]^{3-}$	-2.597			1,2	2,3	2
		$La(F)^{2+} + 3e^- \rightleftharpoons La + F^-$	-2.592				4	3
1		$La[Fe(CN)_6] + 3e^- \rightleftharpoons La + [Fe(CN)_6]^{3-}$	-2.594			1,2	5,6,7	4
		$La_2O_3 + 6H^+ + 6e^- \rightleftharpoons 2La + 3H_2O$	-1.856				1	5
		$La(OH)_3 + 3H^+ + 3e^- \rightleftharpoons La + 3H_2O$	-2.069				1	6
		$La(OH)_3 + 3e^- \rightleftharpoons La + 3OH^-$	-2.90		-950		8	7
1		$La(P_3O_9) + 3e^- \rightleftharpoons La + P_3O_9^{3-}$	-2.636			2	9	8
1		$La(P_4O_{12})^- + 3e^- \rightleftharpoons La + P_4O_{12}^{4-}$	-2.655			2	9	9
		$La_2S_3 + 6e^- \rightleftharpoons 2La + 3S^{2-}$	-2.775				10	10
		$La(SO_4)^+ + 3e^- \rightleftharpoons La + SO_4^{2-}$	-2.596			2,3	11,12,13	11
		$La(adipate)^+ + 3e^- \rightleftharpoons La + adipate^{2-}$	-2.604				14	12
		$La(fumarate)^+ + 3e^- \rightleftharpoons La + fumarate^{2-}$	-2.582				14	13
		$La(glutarate)^+ + 3e^- \rightleftharpoons La + glutarate^{2-}$	-2.598				14	14
		$La(malonate)^+ + 3e^- \rightleftharpoons La + malonate^{2-}$	-2.620				14	15
		$La(phthalate)^+ + 3e^- \rightleftharpoons La + phthalate^{2-}$	-2.633				14	16
		$La(picrate)_2^+ + 3e^- \rightleftharpoons La + 2(picrate)^-$	-2.586			4	15	17
		$La(picrate)^{2+} + 3e^- \rightleftharpoons La + picrate^-$	-2.543				15	18
		$La(succinate)^+ + 3e^- \rightleftharpoons La + succinate^{2+}$	-2.601				14	19
		$Ce^{3+} + 3e^- \rightleftharpoons Ce$	-2.483				1	20
		$Ce^{4+} + e^- \rightleftharpoons Ce^{3+}$	1.61				8	21
7,3		$Ce^{4+} + e^- \rightleftharpoons Ce^{3+}$	1.300	± 2		5	4	22
1		$Ce(Br)^{2+} + 3e^- \rightleftharpoons Ce + Br^-$	-2.489			6	16	23
1		$Ce(ClO_4)^{2+} + 3e^- \rightleftharpoons Ce + ClO_4^-$	-2.521			4	17	24
		$Ce(F)^{2+} + 3e^- \rightleftharpoons Ce + F^-$	-2.563			1,3,6	19	25

Electronically conducting phase	Intermediate species	Composition of the solution	Solvent	Temperature °C	Pressure Torr	Measuring method
a	b	c	d		e	d
Ce	Ce_2O_3		1	25	760	IV
Ce	$Ce(OH)_3$		1	25	760	IV
Ind.			1	25	760	IV
Ind.			1	25	760	IV
Ce	$Ce(PO_4)$	$CeCl_3$: 0.125 M; NaH_2PO_4: 0.75 M; HCl : pH = 2	1	25		V
Ce		$CeCl_3$: 0.125 M; NaH_2PO_4: 0.75 M; HCl : pH = 2	1	25		V
Ce	Ce_2S_3		1	25	760	IV
Ce		$CeCl_3$: 0.125 M; $NaHSO_3$: 0.75 M; HCl: pH = 2	1	25		V
Ce		$Ce_2(SO_4)_3$: up to 0.5302 N	1	25± 0.02		III
Ind.	CeO_2, Ce_2O_3		1	25	760	IV
Ind.	Ce_2O_3		1	25	760	IV
Ce			1	25		V
Ce			1	25		V
Pr			1	25	760	IV
Ind.			1	25	760	IV
Ind.	PrO_2		1	25	760	IV
Pr	Pr_2O_3		1	25	760	IV
Pr	$Pr(OH)_3$		1	25	760	IV
Pr	$Pr(OH)_3$		1	25	760	IV
Pr		$Pr_2(SO_4)_3$: up to 1.1477 N	1	25 ± 0.02		III
Ind.	PrO_2, Pr_2O_3		1	25	760	IV
Ind.	PrO_2, $Pr(OH)_3$		1	25	760	IV
Nd			1	25	760	IV
Nd	Nd_2O_3		1	25	760	IV
Nd	$Nd(OH)_3$		1	25	760	IV
Nd	$Nd(OH)_3$		1	25	760	IV
Nd		$Nd_2(SO_4)_3$: up to 0.4807	1	25		III
Nd			1	25		V
Nd			1	25		V

Comparison electrode	Liquid junction	Electrode reaction	Standard value	Uncertainty	Temperature Coefficient	Notes	Reference	Electrode Reference Number
			V	mV	µV/°C			
d			f		g	h		
		$Ce_2O_3 + 6H^+ + 6e^- \rightleftharpoons 2Ce + 3H_2O$	2.046				1	26
		$Ce(OH)_3 + 3e^- \rightleftharpoons Ce + 3OH^-$	-2.87				8	27
		$Ce(OH)^{3+} + H^+ + e^- \rightleftharpoons Ce^{3+} + H_2O$	1.715				1	28
		$Ce(OH)_2^{2+} + 2H^+ + e^- \rightleftharpoons Ce^{3+} + 2H_2O$	1.731				1	29
1		$Ce(PO_4) + 3e^- \rightleftharpoons Ce + PO_4^{3-}$	-2.853			6	18	30
1		$Ce(P_2O_7)^- + 3e^- \rightleftharpoons Ce + P_2O_7^{4-}$	-2.825			6	18	31
		$Ce_2S_3 + 6e^- \rightleftharpoons 2Ce + 3S^{2-}$	-2.688				10	32
1		$Ce(SO_3)^+ + 3e^- \rightleftharpoons Ce + SO_3^{2-}$	-2.641			6	18	33
1		$Ce(SO_4)^+ + 3e^- \rightleftharpoons Ce + SO_4^{2-}$	-2.555			2	13	34
		$2CeO_2 + 2H^+ + 2e^- \rightleftharpoons Ce_2O_3 + H_2O$	1.559				1	35
		$2Ce(OH)_2^{2+} + 2e^- \rightleftharpoons Ce_2O_3 + H_2O + 2H^+$	0.422				1	36
1		Ce (2,4-dinitrophenate)$^{2+}$ + $3e^-$ $\rightleftharpoons$ Ce + 2,4-dinitrophenate$^-$	-2.504			4	15	37
		Ce (picrate)$^{2+}$ + $3e^-$ $\rightleftharpoons$ Ce + picrate$^-$	-2.504			4	15	38
		$Pr^{3+} + 3e^- \rightleftharpoons Pr$	-2.462				1	39
		$Pr^{4+} + e^- \rightleftharpoons Pr^{3+}$	2.860				1	40
		$PrO_2 + 4H^+ + e^- \rightleftharpoons Pr^{3+} + 2H_2O$	2.761				1	41
		$Pr_2O_3 + 6H^+ + 6e^- \rightleftharpoons 2Pr + 3H_2O$	-1.829				1	42
		$Pr(OH)_3 + 3H^+ + 3e^- \rightleftharpoons Pr + 3H_2O$	-2.018				1	43
		$Pr(OH)_3 + 3e^- \rightleftharpoons Pr + 3OH^-$	-2.85		-920		8	44
1		$Pr(SO_4)^+ + 3e^- \rightleftharpoons Pr + SO_4^{2-}$	-2.534			2	13	45
		$2PrO_2 + 2H^+ + 2e^- \rightleftharpoons Pr_2O_3 + H_2O$	0.863				1	46
		$PrO_2 + H_2O + H^+ + e^- \rightleftharpoons Pr(OH)_3$	1.431				1	47
		$Nd^{3+} + 3e^- \rightleftharpoons Nd$	-2.431				1	48
		$Nd_2O_3 + 6H^+ + 6e^- \rightleftharpoons 2Nd + 3H_2O$	-1.811				1	49
		$Nd(OH)_3 + 3H^+ + 3e^- \rightleftharpoons Nd + 3H_2O$	-2.013				1	50
		$Nd(OH)_3 + 3e^- \rightleftharpoons Nd + 3OH^-$	-2.84				8	51
		$Nd(SO_4)^+ + 3e^- \rightleftharpoons Nd + SO_4^{2-}$	-2.504			2	13	52
1		Nd(oxalate)$^+$ + $3e^-$ $\rightleftharpoons$ Nd + oxalate^{2-}	-2.575			7	15	53
1		Nd (oxalate)$_2^-$ + $3e^-$ $\rightleftharpoons$ Nd + 2 (oxalate)$^{2-}$	-2.661			7	15	54

Electronically conducting phase	Intermediate species	Composition of the solution	Solvent	Temperature °C	Pressure Torr	Measuring method
a	b	c	d		e	d
Pm			1	25	760	IV
Pm	Pm_2O_3		1	25	760	IV
Pm	$Pm(OH)_3$		1	25	760	IV
Sm			1	25	760	IV
Sm			1	25	760	IV
Ind.			1	25	760	IV
Sm	Sm_2O_3		1	25	760	IV
Ind.	Sm_2O_3		1	25	760	IV
Sm	$Sm(OH)_3$		1	25	760	IV
Sm		$Sm_2(SO_4)_3$: up to 0.2575 N	1	25 ± 0.02		III
Eu			1	25	760	IV
Eu			1	25	760	IV
Pt		$Eu(ClO_4)_3$: $1.60 \cdot 10^{-4}$ M; HCOOH : up to 0.562 M; $NaClO_4$ up to I = 1 m	1	25± 0.1		II, III
Eu	Eu_2O_3		1	25	760	IV
Ind.	Eu_2O_3		1	25	760	IV
Eu	$Eu(OH)_3$		1	25	760	IV
Gd			1	25	760	IV
			1	25		I
Gd	Gd_2O_3		1	25	760	IV
Gd	$Gd(OH)_3$		1	25	760	IV
Gd		$Gd_2(SO_4)_3$ up to 0.3272 N	1	25 ± 0.02		III
Tb			1	25	760	IV
Tb	Tb_2O_3		1	25	760	IV
Tb	$Tb(OH)_3$		1	25	760	IV
Dy			1	25	760	IV
Dy	Dy_2O_3		1	25	760	IV
Dy	$Dy(OH)_3$		1	25	760	IV

Comparison electrode	Liquid junction	Electrode reaction	Standard value V	Uncertainty mV	Temperature Coefficient μV/°C	Notes	Reference	Electrode Reference Number
d			f		g	h		
		$Pm^{3+} + 3e^- \rightleftharpoons Pm$	-2.423				1	55
		$Pm_2O_3 + 6H^+ + 6e^- \rightleftharpoons 2Pm + 3H_2O$	-2.008				1	56
		$Pm(OH)_3 + 3e^- \rightleftharpoons Pm + 3OH^-$	-2.84				8	57
		$Sm^{2+} + 2e^- \rightleftharpoons Sm$	-3.121				1	58
		$Sm^{3+} + 3e^- \rightleftharpoons Sm$	-2.414		+136		8	59
		$Sm^{3+} + e^- \rightleftharpoons Sm^{2+}$	-1.000				1	60
		$Sm_2O_3 + 6H^+ + 6e^- \rightleftharpoons 2Sm + 3H_2O$	-2.004				1	61
		$Sm_2O_3 + 6H^+ + 2e^- \rightleftharpoons 2Sm^{2+} + 3H_2O$	0.230				1	62
		$Sm(OH)_3 + 3e^- \rightleftharpoons Sm + 3OH^-$	-2.83				8	63
1		$Sm(SO_4)^+ + 3e^- \rightleftharpoons Sm + SO_4^{2-}$	-2.487			2	13	64
		$Eu^{2+} + 2e^- \rightleftharpoons Eu$	-3.395				1	65
		$Eu^{3+} + 3e^- \rightleftharpoons Eu$	-2.407		+137		8	66
4		$Eu^{3+} + e^- \rightleftharpoons Eu^{2+}$	-0.36			8, 9	20	67
		$Eu_2O_3 + 6H^+ + 6e^- \rightleftharpoons 2Eu + 3H_2O$	-2.002				1	68
		$Eu_2O_3 + 6H^+ + 2e^- \rightleftharpoons 2Eu^{2+} + 3H_2O$	-0.783				1	69
		$Eu(OH)_3 + 3e^- \rightleftharpoons Eu + 3OH^-$	-2.83				8	70
		$Gd^{3+} + 3e^- \rightleftharpoons Gd$	-2.397				1	71
		$Gd(F)^{2+} + 3e^- \rightleftharpoons Gd + F^-$	-2.482				19	72
		$Gd_2O_3 + 6H^+ + 6e^- \rightleftharpoons 2Gd + 3H_2O$	-1.994				1	73
		$Gd(OH)_3 + 3e^- \rightleftharpoons Gd + 3OH^-$	-2.82				8	74
1		$Gd(SO_4)^+ + 3e^- \rightleftharpoons Gd + SO_4^{2-}$	-2.470			2	13	75
		$Tb^{3+} + 3e^- \rightleftharpoons Tb$	-2.391				1	76
		$Tb_2O_3 + 6H^+ + 6e^- \rightleftharpoons 2Tb + 3H_2O$	-1.999				1	77
		$Tb(OH)_3 + 3e^- \rightleftharpoons Tb + 3OH^-$	-2.79				8	78
		$Dy^{3+} + 3e^- \rightleftharpoons Dy$	-2.353				1	79
		$Dy_2O_3 + 6H^+ + 6e^- \rightleftharpoons 2Dy + 3H_2O$	-1.956				1	80
		$Dy(OH)_3 + 3e^- \rightleftharpoons Dy + 3OH^-$	-2.78				8	81

Electronically conducting phase	Intermediate species	Composition of the solution	Solvent	Temperature °C	Pressure Torr	Measuring method
a	b	c	d		e	d
Ho			1	25	760	IV
Ho	Ho_2O_3		1	25	760	IV
Ho	$Ho(OH)_3$		1	25	760	IV
Ho		$H_2(SO_4)_3$: up to 0.6509 N	1	25 ± 0.02		III
Er			1	25	760	IV
Er	Er_2O_3		1	25	760	IV
Er	$Er(OH)_3$		1	25	760	IV
Er		$Er(SO_4)_3$: up to 14.536 N	1	25 ± 0.02		III
Tm			1	25	760	IV
Tm	Tm_2O_3		1	25	760	IV
Tm	$Tm(OH)_3$		1	25	760	IV
Yb			1	25	760	IV
Yb			1	25	760	IV
Ind.			1	25	760	IV
Yb	Yb_2O_3		1	25	760	IV
Ind.	Yb_2O_3		1	25	760	IV
Yb	$Yb(OH)_3$		1	25	760	IV
Yb		$Yb_2(SO_4)_3$: up to 3.3345 N	1	25 ± 0.02		III
Yb		Yb_2 (oxalate)$_3$: satd.	1	25		V
Yb		Yb_2(oxalate)$_3$: satd.	1	25		V
Lu			1	25	760	IV
Lu	Lu_2O_3		1	25	760	IV
Lu	$Lu(OH)_3$		1	25	760	IV

Comparison electrode	Liquid junction	Electrode reaction	Standard value V	Uncertainty mV	Temperature Coefficient μV/°C	Notes	Reference	Electrode Reference Number
d			f		g	h		
		$Ho^{3+} + 3e^- \rightleftharpoons Ho$	-2.319				1	82
		$Ho_2O_3 + 6H^+ + 6e^- \rightleftharpoons 2Ho + 3H_2O$	-1.937				1	83
		$Ho(OH)_3 + 3e^- \rightleftharpoons Ho + 3OH^-$	-2.77				8	84
1		$Ho(SO_4)^+ + 3e^- \rightleftharpoons Ho + SO_4^{2-}$	-2.391			2	13	85
		$Er^{3+} + 3e^- \rightleftharpoons Er$	-2.296				1	86
		$Er_2O_3 + 6H^+ + 6e^- \rightleftharpoons 2Er + 3H_2O$	-1.918				1	87
		$Er(OH)_3 + 3e^- \rightleftharpoons Er + 3OH^-$	-2.75				8	88
1		$Er(SO_4)^+ + 3e^- \rightleftharpoons Er + SO_4^{2-}$	-2.368			2	13	89
		$Tm^{3+} + 3e^- \rightleftharpoons Tm$	-2.278				1	90
		$Tm_2O_3 + 6H^+ + 6e^- \rightleftharpoons 2Tm + 3H_2O$	-1.913				1	91
		$Tm(OH)_3 + 3e^- \rightleftharpoons Tm + 3OH^-$	-2.74				8	92
		$Yb^{2+} + 2e^- \rightleftharpoons Yb$	-2.797				1	93
		$Yb^{3+} + 3e^- \rightleftharpoons Yb$	-2.267		+188		8	94
		$Yb^{3+} + e^- \rightleftharpoons Yb^{2+}$	-1.205				1	95
		$Yb_2O_3 + 6H^+ + 6e^- \rightleftharpoons 2Yb + 3H_2O$	-1.902				1	96
		$Yb_2O_3 + 6H^+ + 2e^- \rightleftharpoons 2Yb^{2+} + 3H_2O$	-0.114				1	97
		$Yb(OH)_3 + 3e^- \rightleftharpoons Yb + 3OH^-$	-2.73				8	98
1		$Yb(SO_4)^+ + 3e^- \rightleftharpoons Yb + SO_4^{2-}$	-2.339			2	13	99
1		$Yb(oxalate)^+ + 3e^- \rightleftharpoons Yb + oxalate^{2+}$	-2.412			7	21,22	100
1		$Yb(oxalate)_2^- + 3e^- \rightleftharpoons Yb + 2(oxalate)^{2-}$	-2.504			7	21,22	101
		$Lu^{3+} + 3e^- \rightleftharpoons Lu$	-2.255				1	102
		$Lu_2O_3 + 6H^+ + 6e^- \rightleftharpoons 2Lu + 3H_2O$	-1.892				1	103
		$Lu(OH)_3 + 3e^- \rightleftharpoons Lu + 3OH^-$	-2.72					104

Lanthanides 1

Standard value V	Solvent	Electrode Reference Number
	La	
-2.90	1	7
-2.775	1	10
-2.655	1	9
-2.636	1	8
-2.633	1	16
-2.620	1	15
-2.604	1	12
-2.601	1	19
-2.598	1	14
-2.597	1	2
-2.596	1	11
-2.594	1	4
-2.592	1	3
-2.586	1	17
-2.582	1	13
-2.543	1	18
-2.522	1	1
-2.069	1	6
-1.856	1	5
	Ce	
-2.87	1	27
-2.853	1	30
-2.825	1	31
-2.688	1	32
-2.641	1	33
-2.563	1	25
-2.555	1	34
-2.521	1	24
-2.504	1	37
-2.504	1	38
-2.489	1	23
-2.483	1	20
0.422	1	36
1.559	1	35

Standard value V	Solvent	Electrode Reference Number
1.61	1	21
1.715	1	28
1.731	1	29
2.046	1	26
1.300	23	22
	Pr	
-2.85	1	44
-2.534	1	45
-2.462	1	39
-2.018	1	43
-1.829	1	42
0.863	1	46
1.431	1	47
2.761	1	41
2.860	1	40
	Nd	
-2.84	1	51
-2.661	1	54
-2.575	1	53
-2.504	1	52
-2.431	1	48
-2.013	1	50
-1.811	1	49
	Pm	
-2.84	1	57
-2.423	1	55
-2.008	1	56
	Sm	
-3.121	1	58
-2.83	1	63
-2.487	1	64
-2.414	1	59
-2.004	1	61
-1.000	1	60
0.230	1	62

Standard value V	Solvent	Electrode Reference Number
	Eu	
-3.395	1	65
-2.83	1	70
-2.407	1	67
-2.002	1	68
-0.36	1	67
0.783	1	69
	Gd	
-2.82	1	74
-2.482	1	72
-2.470	1	75
-2.397	1	71
-1.994	1	73
	Tb	
-2.79	1	78
-2.391	1	76
-1.999	1	77
	Dy	
-2.78	1	81
-2.353	1	79
-1.956	1	80
	Ho	
-2.77	1	84
-2.391	1	85
-2.319	1	82
-1.937	1	83
	Er	
-2.75	1	88
-2.368	1	89
-2.296	1	86
-1.918	1	87
	Tm	
-2.74	1	92
-2.278	1	90
-1.913	1	91

Lanthanides 2

Standard value V	Solvent	Electrode Reference Number
	Yb	
-2.797	1	93
-2.73	1	98
-2.504	1	101
-2.412	1	100
-2.339	1	99
-2.267	1	94
-1.902	1	96
-1.205	1	95
-0.114	1	97
	Lu	
-2.72	1	104
-2.255	1	102
-1.892	1	103

Standard value V	Solvent	Electrode Reference Number

Standard value V	Solvent	Electrode Reference Number

NOTES: Lanthanides

1) The value is an average of those given by several authors; they all are in the same range of uncertainty

2) Conductometric data

3) Potentiometric data

4) Spectrophotometric data

5) The U^{O} value is versus Pt, H_2 in H_2O

6) Ion exchange data

7) Solubility data

8) Polarographic data

9) Chronopotentiometric data

BIBLIOGRAPHY: Lanthanides

1) Data reported by M. Pourbaix, "Atlas d'Equilibres Electrochimiques à 25 °C" (Gauthier-Villars Editeur, Paris, 1963), pp. 183 - 197

2) J.C. James, Thesis, London, 1947

3) J.C. James, C.B. Monk, Trans. Faraday Soc., 46 (1950), 1041

4) J.P. Rao, A.R.V. Murthy, J. Phys. Chem., 68 (1964), 1573

5) C.W. Davies, J.C. James, Proc. Royal Soc., 195 A (1948), 116

6) H.S. Dunsmore, J.C. James, J. Chem. Soc., 1951 , 2925

7) D. Berg, A. Patterson, Jr., J. Am. Chem. Soc., 75 (1953), 1484

8) Data reported by A.J. De Bethune, N.A. Swendeman-Loud, "Standard Aqueous Electrode Potentials and Temperature Coefficients at 25 °C" (Clifford A. Hampel, Ill.)

9) C.B. Monk, J. Chem. Soc., 1949, 1317

10) F.D. Rossini, Nat. Bureau Standards, Circular 500, 1952

11) J.L. Jenkins, C.B. Monk, J. Am. Chem. Soc., 72 (1950), 2695

12) H.W. Jones, C.B. Monk, Trans. Faraday Soc., 48 (1952), 929

13) F.H. Spedding, S. Jaffe, J. Am. Chem. Soc., 76 (1954), 882 - 884

14) J.N. Peacock; J.C. James, J. Chem. Soc., 1951, 2233

15) M. Babtelsky, S. Kertes, Bull. Soc. Chim. France, 1955, 328

16) E.W. Mayer, S.D. Schwartz, J. Am. Chem. Soc., 73 (1951), 222 - 224

17) L.J. Heidt, J. Berestecky, J. Am. Chem. Soc., 77 (1955), 2049 - 2054

18) S.W. Mayer, S.D. Schwartz, J. Am. Chem. Soc., 72 (1960), 5106 - 5110

19) J.W. Kury, Thesis, University of California, Berkekey 1953, UCRL, 2271

20) D.J. Macero, L.B. Anderson, P. Malachesky, J. Electroanal. Chem., 10 (1965), 76 - 81

21) C.E. Crouthamel, D.S. Martin, J. Am. Chem. Soc., 73 (1951), 569 - 573

22) C.E. Crouthamel, D.S. Martin, J. Am. Chem. Soc., 72 (1950), 1382

Actinides

Actinides 1

Electronically conducting phase	Intermediate species	Composition of the solution	Solvent	Temperature °C	Pressure Torr	Measuring method
a	b	c	d		e	d
Ac			1	25	760	IV
Th			1	25	760	IV
Th			1	25	760	IV
Th			1	25		V
Th	ThO_2		1	25	760	IV
Th	Th $(OH)_4$		1	25	760	IV
Th	Th $(OH)_4$		1	25	760	IV
Pa			1	25	760	IV
Pa			1	25	760	IV
Ind.			1	25	760	IV
Pa			1	25	760	IV
Ind.	PaO_2		1	25	760	IV
U			1	25	760	IV
U	UO		1	25	760	IV
U	UO_2		1	25	760	IV
U	UO_2		1	25	760	IV
U	U_2O_3		1	25	760	IV
U	U $(OH)_3$		1	25	760	IV
U	U $(OH)_4$		1	25	760	IV
Ind.			1	25	760	IV
Ind.	UO_2		1	25	760	IV
Ind.			1	25	760	IV
Ind.	U $(OH)_4$		1	25	760	IV
Ind.			1	25	760	IV
Pt		$U(ClO_4)_4$: 0.00388 up to 0.0970 m; $UO_2(ClO_4)$: 0.00376 up to 0.0940 m; $HClO_4$: 0.0456 up to 1.140 m	1	25 ± 0.1		I
U	UH_3		1	25	760	IV
Ind.	UH_3		1	25	760	IV
Ind.	UH_3 , UO		1	25	760	IV
Ind.	UH_3 , UO_2		1	25	760	IV

Comparison electrode	Liquid junction	Electrode reaction	Standard value V	Uncertainty mV	Temperature Coefficient µV/°C	Notes	Reference	Electrode Reference Number
d			f		g	h		
		$Ac^{3+} + 3e^- \rightleftharpoons Ac$	-2.6				1	1
		$Th^{4+} + 4e^- \rightleftharpoons Th$	-1.899				2	2
		$Th(Cl)^{3+} + 4e^- \rightleftharpoons Th + Cl^-$	-1.920			1	3	3
		$Th(F)^{3+} + 4e^- \rightleftharpoons Th + F^-$	-2.028			2	4	4
		$ThO_2 + 4H^+ + 4e^- \rightleftharpoons Th + 2H_2O$	-1.789				2	5
		$Th(OH)_4 + 4H^+ + 4e^- \rightleftharpoons Th + 4H_2O$	-1.650				2	6
		$Th(OH)_4 + 4e^- \rightleftharpoons Th + 4OH^-$	-2.48		-990		1	7
		$Pa^{3+} + 3e^- \rightleftharpoons Pa$	-1.95				5	8
		$Pa^{4+} + 4e^- \rightleftharpoons Pa$	-1.7				6	9
		$Pa^{4+} + e^- \rightleftharpoons Pa^{3+}$	-1.0				5	10
		$PaO_2^+ + 4H^+ + 5e^- \rightleftharpoons Pa + 2H_2O$	-1.0				1	11
		$PaO_2 + 4H^+ + e^- \rightleftharpoons Pa^{3+} + 2H_2O$	-0.5				5	12
		$U^{3+} + 3e^- \rightleftharpoons U$	-1.798		-70		2,1	13
		$UO + 2H^+ + 2e^- \rightleftharpoons U + H_2O$	-1.438				2	14
		$UO_2 + 4H^+ + 4e^- \rightleftharpoons U + 2H_2O$	-1.444				2	15
		$UO_2 + 2H_2O + 4e^- \rightleftharpoons U + 4OH^-$	-2.39		-1220		1	16
		$U_2O_3 + 6H^+ + 6e^- \rightleftharpoons 2U + 3H_2O$	-1.346				2	17
		$U(OH)_3 + 3e^- \rightleftharpoons U + 3OH^-$	-2.17				1	18
		$U(OH)_4 + 4H^+ + 4e^- \rightleftharpoons U + 4H_2O$	-1.353				2	19
		$U^{4+} + e^- \rightleftharpoons U^{3+}$	-0.607		-50		2,1	20
		$UO_2 + 4H^+ + e^- \rightleftharpoons U^{3+} + 2H_2O$	-0.382				2	21
		$U(OH)^{3+} + H^+ + e^- \rightleftharpoons U^{3+} + H_2O$	-0.538				2	22
		$U(OH)_4 + 4H^+ + e^- \rightleftharpoons U^{3+} + 4H_2O$	-0.019				2	23
		$UO_2^+ + 4H^+ + e^- \rightleftharpoons U^{4+} + 2H_2O$	0.612		-3130		2,1	24
1		$UO_2^{2+} + 4H^+ + 2e^- \rightleftharpoons U^{4+} + 2H_2O$	0.327	±1	-1270		1,7	25
		$U + 3H^+ + 3e^- \rightleftharpoons UH_3$	0.256				2	26
		$U^{3+} + 3H^+ + 6e^- \rightleftharpoons UH_3$	-0.772				2	27
		$UO + 5H^+ + 5e^- \rightleftharpoons UH_3 + H_2O$	-0.422				2	28
		$UO_2 + 7H^+ + 7e^- \rightleftharpoons UH_3 + 2H_2O$	-0.716				2	29

Electronically conducting phase	Intermediate species	Composition of the solution	Solvent	Temperature °C	Pressure Torr	Measuring method
a	b	c	d		e	d
Ind.	UH_3, U_2O_3		1	25	760	IV
Ind.	UH_3, $U(OH)_4$		1	25	760	IV
Ind.	UO, U_2O_3		1	25	760	IV
Ind.	UO_2		1	25	760	IV
Ind.	UO_2, UO_3		1	25	760	IV
Ind.	UO_2, $UO_3 \cdot H_2O$		1	25	760	IV
Ind.	UO_2, $UO_3 \cdot 2H_2O$		1	25	760	IV
Ind.	UO_2, U_3O_8		1	25	760	IV
Hg(DME)		$HClO_4$: 0.01 up to 0.5 m; $NaClO_4$: UO_2NO_3	1	25 ± 0.05		II
Ind.	U_2O_3, UO_2		1	25	760	IV
Ind.	U_2O_3, $U(OH)_4$		1	25	760	IV
Ind.			1	25	760	IV
Ind.			1	25	760	IV
Ind.	$U(OH)_3$, $U(OH)_4$		1	25	760	IV
Ind.	$U(OH)_4$		1	25	760	IV
Ind.	$U(OH)_4$, UO_3		1	25	760	IV
Ind.	$U(OH)_4$, $UO_3 \cdot H_2O$		1	25	760	IV
Ind.	$U(OH)_4$, $UO_3 \cdot 2H_2O$		1	25	760	IV
Ind.	$U(OH)_4$		1	25	760	IV
Ind.	$U(OH)_4$, U_3O_8		1	25	760	IV
Ind.	U_3O_8		1	25	760	IV
Ind.	U_3O_8, UO_3		1	25	760	IV
Ind.	U_3O_8, $UO_3 \cdot H_2O$		1	25	760	IV
Ind.	U_3O_8, $UO_3 \cdot 2H_2O$		1	25	760	IV
Np			1	25	760	IV
Ind.			1	25	760	IV
Ind.			1	25	760	IV
Ind.			1	25	760	IV
Np	$Np(OH)_3$		1	25	760	IV
		$NpO_2(NO_3)_2$: $6H_2O$	1	25		I
Ind.	NpO_2, NpO_2OH		1	25	760	IV
Ind.	$Np(OH)_3$, NpO_2		1	25	760	IV

Comparison electrode	Liquid junction	Electrode reaction	Standard value	Uncertainty	Temperature Coefficient	Notes	Reference	Electrode Reference Number
			V	mV	μV/°C			
d			f		g	h		
		$U_2O_3 + 12H^+ + 12e^- \rightleftharpoons 2UH_3 + 3H_2O$	-0.545				2	30
		$U(OH)_4 + 7H^+ + 7e^- \rightleftharpoons UH_3 + 4H_2O$	-0.664				2	31
		$U_2O_3 + 2H^+ + 2e^- \rightleftharpoons 2UO + H_2O$	-1.163				2	32
		$UO_2^{2+} + 2e^- \rightleftharpoons UO_2$	0.221				2	33
		$UO_3 + 2H^+ + 2e^- \rightleftharpoons UO_2 + H_2O$	0.657				2	34
		$UO_3 \cdot H_2O + 2H^+ + 2e^- \rightleftharpoons UO_2 + 2H_2O$	0.368				2	35
		$UO_3 \cdot 2H_2O + 2H^+ + 2e^- \rightleftharpoons UO_2 + 3H_2O$	0.387				2	36
		$U_3O_8 + 4H^+ + 4e^- \rightleftharpoons 3UO_2 + 2H_2O$	0.533				2	37
1		$UO_2^{2+} + e^- \rightleftharpoons UO_2^+$	0.062		580		1,8	38
		$2UO_2 + 2H^+ + 2e^- \rightleftharpoons U_2O_3 + H_2O$	-1.738				2	39
		$2U(OH)_4 + 2H^+ + 2e^- \rightleftharpoons U_2O_3 + 5H_2O$	-1.375				2	40
		$UO_2^+ + 3H^+ + e^- \rightleftharpoons U(OH)^{3+} + H_2O$	0.546				2	41
		$UO_2^{2+} + 3H^+ + 2e^- \rightleftharpoons U(OH)^{3+} + H_2O$	0.299				2	42
		$U(OH)_4 + e^- \rightleftharpoons U(OH)_3 + OH^-$	-2.20				1	43
		$UO_2^{2+} + 2H_2O + 2e^- \rightleftharpoons U(OH)_4$	0.040				2	44
		$UO_3 + H_2O + 2H^+ + 2e^- \rightleftharpoons U(OH)_4$	0.475				2	45
		$UO_3 \cdot H_2O + 2H^+ + 2e^- \rightleftharpoons U(OH)_4$	0.186				2	46
		$UO_3 \cdot 2H_2O + 2H^+ + 2e^- \rightleftharpoons U(OH)_4 + H_2O$	0.204				2	47
		$UO_4^{2-} + 4H_2O + 2e^- \rightleftharpoons U(OH)_4 + 4OH^-$	-1.618				1	48
		$U_3O_8 + 4H_2O + 4H^+ + 4e^- \rightleftharpoons 3U(OH)_4$	0.260				2	49
		$3UO_2^{2+} + 2H_2O + 2e^- \rightleftharpoons U_3O_8 + 4H^+$	-0.403				2	50
		$3UO_3 + 2H^+ + 2e^- \rightleftharpoons U_3O_8 + H_2O$	0.904				2	51
		$3UO_3 \cdot H_2O + 2H^+ + 2e^- \rightleftharpoons U_3O_8 + 4H_2O$	0.038				2	52
		$3UO_3 \cdot 2H_2O + 2H^+ + 2e^- \rightleftharpoons U_3O_8 + 7H_2O$	0.093				2	53
		$Np^{3+} + 3e^- \rightleftharpoons Np$	-1.856		-54		1,5	54
		$Np^{4+} + e^- \rightleftharpoons Np^{3+}$	0.147		1360		1	55
		$NpO_2^+ + 4H^+ + 2e^- \rightleftharpoons Np^{3+} + 2H_2O$	0.451				2	56
		$NpO_2^+ + 4H^+ + e^- \rightleftharpoons Np^{4+} + 2H_2O$	0.749		-3130		2,1	57
		$Np(OH)_3 + 3H^+ + 3e^- \rightleftharpoons Np + 3H_2O$	-1.420				2	58
		$NpO_2^{2+} + e^- \rightleftharpoons NpO_2^+$	1.130	±10	580		9,1	59
		$NpO_2OH + H^+ + e^- \rightleftharpoons NpO_2 + H_2O$	1.253				2	60
		$NpO_2 + H_2O + H^+ + e^- \rightleftharpoons Np(OH)_3$	-0.962				2	61

Electronically conducting phase	Intermediate species	Composition of the solution	Solvent	Temperature °C	Pressure Torr	Measuring method
a	b	c	d		e	d
Ind.	$Np(OH)_3$, $Np(OH)_4$		1	25	760	IV
Ind.	$Np(OH)_4$, NpO_2OH		1	25	760	IV
Ind.	NpO_2OH, $NpO_2(OH)_2$		1	25	760	IV
Pu			1	25	760	IV
Au		$HClO_4$: 0.1 N; Pu^{4+}: (6.37 up to 28.71) · 10^{-5} m; Pu^{3+} : (3.859 up to 6.093) · 10^{-4} m	1	25		I
Ind.	PuO_2		1	25	760	IV
Ind.			1	25	760	IV
Ind.			1	25	760	IV
Ind.			1	25	760	IV
Pu	$Pu(OH)_3$		1	25	760	IV
Pu	$Pu(OH)_3$		1	25	760	IV
Ind.	$Pu(OH)_4$		1	25	760	IV
Ind.	PuO_2		1	25	760	IV
Ind.			1	25	760	IV
Ind.	PuO_2 , PuO_2OH		1	25	760	IV
Ind.	$PuO_2(OH)_2$, PuO_2		1	25	760	IV
Ind.	$Pu(OH)_3$, PuO_2		1	25	760	IV
Ind.	$Pu(OH)_4$		1	25	760	IV
Ind.	$Pu(OH)_3$, $Pu(OH)_4$		1	25	760	IV
Ind.	$Pu(OH)_3$, $Pu(OH)_4$		1	25	760	IV
Ind.	$Pu(OH)_4$, PuO_2OH		1	25	760	IV
Ind.	$Pu(OH)_4$, $PuO_2(OH)_2$		1	25	760	IV
Ind.	PuO_2OH, $PuO_2(OH)_2$		1	25	760	IV
Ind.	PuO_2OH, $PuO_2(OH)_2$		1	25	760	IV
Au		$HClO_4$: 0.1 N; Na acetate : 0.1 N; Pu^{3+}: (3.859 up to 6.093) · 10^{-4} m	1	25		I
Am			1	25	760	IV
Ind.			1	25	760	IV
Am	Am_2O_3		1	25	760	IV
Ind.	AmO_2		1	25	760	IV
Ind.			1	25	760	IV

Comparison electrode	Liquid junction	Electrode reaction	Standard value	Uncertainty	Temperature Coefficient	Notes	Reference	Electrode Reference Number
			V	mV	µV/°C			
d			f		g	h		
		$Np(OH)_4 + H^+ + e^- \rightleftharpoons Np(OH)_3 + H_2O$	-0.928				2	62
		$NpO_2OH + H_2O + H^+ + e^- \rightleftharpoons Np(OH)_4$	1.219				2	63
		$NpO_2(OH)_2 + H^+ + e^- \rightleftharpoons NpO_2OH + H_2O$	1.310				2	64
		$Pu^{3+} + 3e^- \rightleftharpoons Pu$	-2.031		+60		2,5,1	65
3	HCl 0.1N	$Pu^{4+} + e^- \rightleftharpoons Pu^{3+}$	1.006	± 3	+1400		10,1	66
		$PuO_2 + 4H^+ + e^- \rightleftharpoons Pu^{3+} + 2H_2O$	0.862				2	67
		$PuO_2^{2+} + 4H^+ + 3e^- \rightleftharpoons Pu^{3+} + 2H_2O$	1.017				2	68
		$PuO_2^+ + 4H^+ + e^- \rightleftharpoons Pu^{4+} + 2H_2O$	1.157		-3820		2,1	69
		$PuO_2^{2+} + 4H^+ + 2e^- \rightleftharpoons Pu^{4+} + 2H_2O$	1.042		-1560		2,1	70
		$Pu(OH)_3 + 3H^+ + 3e^- \rightleftharpoons Pu + 3H_2O$	-1.592				2	71
		$Pu(OH)_3 + 3e^- \rightleftharpoons Pu + 3OH^-$	-2.42				1	72
		$Pu(OH)_4 + 4H^+ + e^- \rightleftharpoons Pu^{3+} + 4H_2O$	1.182				2	73
		$PuO_2^{2+} + 2e^- \rightleftharpoons PuO_2$	1.095				2	74
		$PuO_2^{2+} + e^- \rightleftharpoons PuO_2^+$	0.928		+710		2,1	75
		$PuO_2OH + H^+ + e^- \rightleftharpoons PuO_2 + H_2O$	1.908				2	76
		$PuO_2(OH)_2 + 2H^+ + 2e^- \rightleftharpoons PuO_2 + 2H_2O$	1.485				2	77
		$PuO_2 + H_2O + H^+ + e^- \rightleftharpoons Pu(OH)_3$	-0.455				2	78
		$PuO_2^{2+} + 2H_2O + 2e^- \rightleftharpoons Pu(OH)_4$	0.935				2	79
		$Pu(OH)_4 + H^+ + e^- \rightleftharpoons Pu(OH)_3 + H_2O$	-0.135				2	80
		$Pu(OH)_4 + e^- \rightleftharpoons Pu(OH)_3 + OH^-$	-0.963				1	81
		$PuO_2OH + H_2O + H^+ + e^- \rightleftharpoons Pu(OH)_4$	1.588				2	82
		$PuO_2(OH)_2 + 2H^+ + 2e^- \rightleftharpoons Pu(OH)_4$	1.325				2	83
		$PuO_2(OH)_2 + H^+ + e^- \rightleftharpoons PuO_2OH + H_2O$	1.062				2	84
		$PuO_2(OH)_2 + e^- \rightleftharpoons PuO_2OH + OH^-$	0.234				1	85
3	HCl 0.1N	$[Pu(acetate)_5]^- + e^- \rightleftharpoons Pu^{3+} + 5\,(acetate)^-$	-0.33				10	86
		$Am^{3+} + 3e^- \rightleftharpoons Am$	-2.38		+89		5,11,12,1	87
		$Am^{4+} + e^- \rightleftharpoons Am^{3+}$	2.181		+1530		2,1	88
		$Am_2O_3 + 6H^+ + 6e^- \rightleftharpoons 2Am + 3H_2O$	-1.676				2	89
		$AmO_2 + 4H^+ + e^- \rightleftharpoons Am^{3+} + 2H_2O$	1.856				2	90
		$AmO_2^+ + 4H^+ + 2e^- \rightleftharpoons Am^{3+} + 2H_2O$	1.721				2	91

Electronically conducting phase	Intermediate species	Composition of the solution	Solvent	Temperature °C	Pressure Torr	Measuring method
a	b	c	d		e	d
Ind.			1	25	760	IV
Ind.			1	25	760	IV
Am	Am $(OH)_3$		1	25	760	IV
Ind.	Am $(OH)_4$		1	25	760	IV
Ind.	AmO_2OH		1	25	760	IV
Ind.	Am_2O_3, AmO_2		1	25	760	IV
Ind.	AmO_2, AmO_2OH		1	25	760	IV
Ind.			1	25	760	IV
Ind.	Am_2O_3, $Am(OH)_4$		1	25	760	IV
Ind.	Am $(OH)_3$, AmO_2		1	25	760	IV
Ind.	Am $(OH)_3$, Am $(OH)_4$		1	25	760	IV
Ind.	Am $(OH)_4$, AmO_2OH		1	25	760	IV
Ind.	AmO_2OH		1	25	760	IV
Ind.	AmO_2OH, $AmO_2(OH)_2$		1	25	760	IV
Cm			1	25	760	IV
Bk			1	25	760	IV
Ind.			1	25	760	IV
Cf			1	25	760	IV
Fm			1	25	760	IV
No			1	25	760	IV

Comparison electrode	Liquid junction	Electrode reaction	Standard value V	Uncertainty mV	Temperature Coefficient μV/°C	Notes	Reference	Electrode Reference Number
d			f		g	h		
		$AmO_2^{2+} + 4H^+ + 3e^- \rightleftharpoons Am^{3+} + 2H_2O$	1.694				2	92
		$AmO_2^+ + 4H^+ + e^- \rightleftharpoons Am^{4+} + 2H_2O$	1.261				2	93
		$Am(OH)_3 + 3H^+ + 3e^- \rightleftharpoons Am + 3H_2O$	-1.878				2	94
		$Am(OH)_4 + 4H^+ + e^- \rightleftharpoons Am^{3+} + 4H_2O$	1.746				2	95
		$AmO_2OH + 5H^+ + 2e^- \rightleftharpoons Am^{3+} + 3H_2O$	1.639				2	96
		$2AmO_2 + 2H^+ + 2e^- \rightleftharpoons Am_2O_3 + H_2O$	-0.072				2	97
		$AmO_2OH + H^+ + e^- \rightleftharpoons AmO_2 + H_2O$	1.418				2	98
		$AmO_2^{2+} + e^- \rightleftharpoons AmO_2^+$	1.639				2	99
		$2Am(OH)_4 + 2H^+ + 2e^- \rightleftharpoons Am_2O_3 + 5H_2O$	-0.185				2	100
		$AmO_2 + H_2O + H^+ + e^- \rightleftharpoons Am(OH)_3$	0.533				2	101
		$Am(OH)_4 + H^+ + e^- \rightleftharpoons Am(OH)_3 + H_2O$	0.420				2	102
		$AmO_2OH + H_2O + H^+ + e^- \rightleftharpoons Am(OH)_4$	1.530				2	103
		$AmO_2^{2+} + H_2O + e^- \rightleftharpoons AmO_2OH + H^+$	1.804				2	104
		$Am_2(OH)_2 + H^+ + e^- \rightleftharpoons AmO_2OH + H_2O$	1.930				2	105
		$Cm^{3+} + 3e^- \rightleftharpoons Cm$	-2.70				6	106
		$Bk^{3+} + 3e^- \rightleftharpoons Bk$	-2.4				6	107
		$Bk^{4+} + e^- \rightleftharpoons Bk^{3+}$	1.6				1	108
		$Cf^{3+} + 3e^- \rightleftharpoons Cf$	-2.1				6	109
		$Fm^{3+} + 3e^- \rightleftharpoons Fm$	-2.1				6	110
		$No^{3+} + 3e^- \rightleftharpoons No$	-2.5				6	111

Actinides 1

Standard value V	Solvent	Electrode Reference Number
	Ac	
-2.6	1	1
	Th	
-2.48	1	7
-2.028	1	4
-1.920	1	3
-1.899	1	2
-1.789	1	5
-1.650	1	6
	Pa	
-1.95	1	8
-1.7	1	9
-1.0	1	10
-1.0	1	11
-0.5	1	12
	U	
-2.39	1	16
-2.20	1	43
-2.17	1	18
-1.798	1	13
-1.738	1	39
-1.618	1	48
-1.444	1	15
-1.438	1	14
-1.375	1	40
-1.353	1	19
-1.346	1	17
-1.163	1	32
-0.772	1	27
-0.716	1	29
-0.664	1	31
-0.607	1	20
-0.545	1	30
-0.538	1	22
-0.422	1	28
-0.403	1	50
-0.382	1	21
-0.019	1	23
0.038	1	52
0.040	1	44
0.062	1	38
0.093	1	53
0.186	1	46
0.204	1	47
0.221	1	33
0.256	1	26
0.260	1	49
0.299	1	42
0.327	1	25
0.368	1	35
0.387	1	36
0.475	1	45
0.533	1	37
0.546	1	41
0.612	1	24
0.657	1	34
0.904	1	51
	Np	
-1.856	1	54
-1.420	1	58
-0.962	1	61
-0.928	1	62
0.147	1	55
0.451	1	56
0.749	1	57
1.130	1	59
1.219	1	63
1.253	1	60
1.310	1	64
	Pu	
-2.42	1	72
-2.031	1	65
-1.592	1	71
-0.963	1	81
-0.455	1	78
-0.33	1	86
-0.135	1	80
0.234	1	85
0.862	1	67
0.928	1	75
0.935	1	79
1.006	1	66
1.017	1	68
1.042	1	70
1.062	1	84
1.095	1	74
1.157	1	69
1.182	1	73
1.325	1	83
1.485	1	77
1.588	1	82
1.908	1	76
	Am	
-2.38	1	87
-1.878	1	94
-1.676	1	89
-0.185	1	100
-0.072	1	97
0.420	1	102
0.533	1	101
1.261	1	93
1.418	1	98
1.530	1	103
1.639	1	96
1.639	1	99
1.694	1	92

Actinides 2

Standard value V	Solvent	Electrode Reference Number	Standard value V	Solvent	Electrode Reference Number	Standard value V	Solvent	Electrode Reference Number
1.721	1	91						
1.746	1	95						
1.804	1	104						
1.856	1	90						
1.930	1	105						
2.181	1	88						
	Cm							
-2.70	1	106						
	Bk							
-2.4	1	107						
1.6	1	108						
	Cf							
-2.1	1	109						
	Fm							
-2.1	1	110						
	No							
-2.5	1	111						

NOTES: Actinides

1) Distribution between two phases

2) Solubility data

BIBLIOGRAPHY: Actinides

1) Data reported by A.J. De Bethune, N.A. Swendeman-Loud, "Standard Aqueous Electrode Potentials and Temperature Coefficients at 25 ^{o}C" (Clifford A. Hampel, Ill.)

2) Data reported by M. Pourbaix, "Atlas d'Equilibres Electrochimiques à 25 ^{o}C" (Gauthier-Villars Editeur, Paris, 1963), pp. 198 - 212

3) W.C. Waggener, R.W. Stoughton, J. Phys. Chem., 56 (1952), 1

4) H.W. Dodgen, G.K. Rollefson, J. Am. Chem. Soc., 71 (1949), 2600

5) G.A. Krestov, Radiokhim., 7 (1965), 68 - 78

6) G.A. Krestov, Radiokhim., 5 (1963), 258

7) J. Sobkovski, S.J. Minc, J. Inorg. Nucl. Chem., 19 (1961), 101 - 106

8) D.M.H. Kern, E.F. Orlemann, J. Am. Chem. Soc., 71 (1949), 2102 - 2106

9) J.R. Brand, Diss. Abstr., 28 B (1968), 4475

10) K. Schwabe, D. Nebel, Z. Physik. Chem. (Leipzig), 220 (1962), 339 - 354

11) S.R. Gunn, B.B. Cunningham, J. Am. Chem. Soc., 79 (1957), 1563 - 1565

12) R.A. Penneman, L.B. Asprey, Proc. Intern. Conf. Peaceful Uses At. Energy, Geneva, 1955, 838, Vol. VII, 355

Appendix

Electronically conducting phase	Intermediate species	Composition of the solution	Solvent	Temperature °C	Pressure Torr	Measuring method
a	b	c	d		e	d
Li(Hg)		LiBr : 0.0199 up to 0.1118 M	13	25 ± 0.05		I
Na(Hg)		NaBr : 0.0200 up to 0.0894 M	13	25± 0.05		I
Na-coated Pt		$NaClO_4$: $1.97 \cdot 10^{-2}$ up to $1.15 \cdot 10^{-1}$ M	29	25		I
K(Hg)		KBr : 0.0206 up to 0.0952 M	13	25± 0.05		I
Rb(Hg)		$RbClO_4$: $2.30 \cdot 10^{-2}$ M	29	25		I
Rb			38	25	760	IV
Rb			39	25	760	IV
Rb			40	25	760	IV
Rb			41	25	760	IV
Rb			15	25	760	IV
Rb			42	25	760	IV
Rb			43	25	760	IV
Rb			44	25	760	IV
Rb			45	25	760	IV
Rb			20	25	760	IV
Rb			22	25	760	IV
Rb			24	25	760	IV
Rb			27	25	760	IV
Rb			46	25	760	IV
Cs(Hg)		$CsClO_4$: $(0.71 \text{ up to } 1.01) \cdot 10^{-2}$ M	29	25		I
Cu		$CuCl_2$: $4.8 \cdot 10^{-4}$ up to $2.5 \cdot 10^{-3}$ M	1	25		I
Ag		$AgNO_3$: 0.01 and 0.1 m; NH_4NO_3 : 1.0 m	2	-30		I
Ag	AgBr	DBr : 0.005 up to 0.1 m	35	25 ± 0.1		I
Ag	AgBr	CH_3COOH, CH_3COOK, KBr	18	25		I
Ag	$AgBrO_3$	$NaClO_4$: $(0.759 \text{ up to } 11.952) \cdot 10^{-2}$ M; $AgBrO_3$: $(1.862 \text{ up to } 2.220) \cdot 10^{-2}$ M	18	25		V
Ag	AgCl	HCl : 0.005 m; KCl: up to 3.0 m	1	25.00± 0.02		I
Ag	AgCl	HCl : 0.00452 up to 0.2873 m	7	25		I
Ag	AgCl	HCl : 0.001 up to 0.1 m	13	25		I
Ag	AgCl	HCl : 0.0138 up to 0.0345 M; $LiClO_4$: 0.1 M	17	25		I
Ag	AgI	CH_3COOH, CH_3COOK, KI	18	25		I
Ag	$AgMnO_4$	$NaClO_4$: $(0.250 \text{ up to } 2.500) \cdot 10^{-2}$ M; $AgMnO_4$: $(1.209 \text{ up to } 6.068) \cdot 10^{-5}$ M	1	25		V

Comparison electrode	Liquid junction	Electrode reaction	Standard value	Uncertainty	Temperature Coefficient	Notes	Reference	Electrode Reference Number
			V	mV	μV/°C			
d			f		g	h		
5		$Li^+ + e^- \rightleftharpoons Li$	-3.097	±1.5			1	1
5		$Na^+ + e^- \rightleftharpoons Na$	-2.787	±1.5			1	2
34		$Na^+ + e^- \rightleftharpoons Na$	-2.653	±2			2	3
5		$K^+ + e^- \rightleftharpoons K^+$	-2.998	±1.5			1	4
34		$Rb^+ + e^- \rightleftharpoons Rb$	-2.982	±3			2	5
		$Rb^+ + e^- \rightleftharpoons Rb$	-3.55				3	6
		$Rb^+ + e^- \rightleftharpoons Rb$	-3.32				3	7
		$Rb^+ + e^- \rightleftharpoons Rb$	-3.15				3	8
		$Rb^+ + e^- \rightleftharpoons Rb$	-3.46				3	9
		$Rb^+ + e^- \rightleftharpoons Rb$	-3.12				3	10
		$Rb^+ + e^- \rightleftharpoons Rb$	-3.50				3	11
		$Rb^+ + e^- \rightleftharpoons Rb$	-3.06				3	12
		$Rb^+ + e^- \rightleftharpoons Rb$	-3.70				3	13
		$Rb^+ + e^- \rightleftharpoons Rb$	-3.06				3	14
		$Rb^+ + e^- \rightleftharpoons Rb$	-2.96				3	15
		$Rb^+ + e^- \rightleftharpoons Rb$	-2.95				3	16
		$Rb^+ + e^- \rightleftharpoons Rb$	-2.92				3	17
		$Rb^+ + e^- \rightleftharpoons Rb$	-3.13				3	18
		$Rb^+ + e^- \rightleftharpoons Rb$	-2.95				3	19
34		$Cs^+ + e^- \rightleftharpoons Cs$	-3.012	±3			2	20
4		$Cu^{2+} + 2e^- \rightleftharpoons Cu$	0.3352	±0.4			4	21
29		$Ag^+ + e^- \rightleftharpoons Ag$	0.49				5	22
38		$AgBr + e^- \rightleftharpoons Ag + Br^-$	0.06014				6	23
1		$AgBr + e^- \rightleftharpoons Ag + Br^-$	0.0965		-810		7	24
		$AgBrO_3 + e^- \rightleftharpoons Ag + BrO_3^-$	0.4997		-1000	1	8	25
1		$AgCl + e^- \rightleftharpoons Ag + Cl^-$	0.22238	±40			9	26
1		$AgCl + e^- \rightleftharpoons Ag + Cl^-$	-0.0103				10	27
1		$AgCl + e^- \rightleftharpoons Ag + Cl^-$	0.0233		-1200		11	28
1		$AgCl + e^- \rightleftharpoons Ag + Cl^-$	-0.610	±5			12	29
1		$AgI + e^- \rightleftharpoons Ag + I^-$	-0.0996		-700		7	30
		$AgMnO_4 + e^- \rightleftharpoons Ag + MnO_4^-$	0.2149		-2000	1	13	31

Electronically conducting phase	Intermediate species	Composition of the solution	Solvent	Temperature °C	Pressure Torr	Measuring method
a	b	c	d		e	d
Ag	$AgMnO_4$	$NaClO_4$: (0.785 up to 14.680) · 10^{-2} M; $AgMnO_4$: (6.436 up to 10.840) · 10^{-5} M	18	25		V
Ag	AgSCN	$HClO_4$: (0.491 up to 15.183) · 10^{-3} M; KSCN : (0.486 up to 46.174) · 10^{-3} M	1	25		I
Ag	AgSeCN	KSeCN + KCO_3 : up to 0.07 m	1	25		I
Ag	Ag acetate	Acetic acid : 0.03872 up to 0.14470 m	18	25± 0.05		I
Au		HCl : 0.1 N, 1 N, 2 N	1	25± 0.02		I
Au-coated Pt	NH_3	NH_4NO_3 : 10 M; NH_3 : 0.195 up to 2.17 M; $[Au(NH_3)_4](NO_3)_3$: (1.48 up to 2.09) · 10^{-3} M	1	25± 0.1		I
Au-coated Pt	NH_3	NH_4NO_3: 10 M; NH_3: 0.195 up to 2.17 M; $[Au(NH_3)_4](NO_3)_3$: (1.48 up to 2.09) · 10^{-3} M	1	25± 0.1		I
Au-coated Pt	NH_3	NH_4NO_3 : 10 M; NH_3 : 0.195 up to 2.17 M; $[Au(NH_3)_4](NO_3)_3$: (1.48 up to 2.09) · 10^{-3} M	1	25± 0.1		I
Au		$Na_3[Au(S_2O_3)_2]$: 8.5 · 10^{-5} and 3.4 · 10^{-3} M; KCl : 0.60 up to 1.16 N	1	25		II
Ba			1	25	760	IV
Ba			13	25	760	IV
Ba			11	25	760	IV
Ba			12	25	760	IV
Ba			15	25	760	IV
Zn(Hg)		$ZnCl_2$: 10^{-4} up to 1.0 m	1	25		I
Zn(Hg)		$K_2Zn(OH)_4$: 0.126 up to 0.6400 M; KOH : 7.03	1	25± 0.1		I
Cd(Hg)		$CdCl_2$: up to 0.02 m	18	25		I
Hg		$Hg_2(ClO_4)_2$: (0.2364 up to 13.320) · 10^{-3} m; $HClO_4$: (0.2908 up to 10) · 10^{-3} m; $NaClO_4$: 0.001 up to 0.050 m	1	25 ± 0.01		I
Hg		$Hg_2(ClO_4)_2$: 3 · 10^{-4} up to 1.4 · 10^{-3} M; $HClO_4$: 5 · 10^{-3} M	23	25		I
Hg		$HClO_4$: 0.1 N	36	25		II
Hg		$HClO_4$:0.1 N	36	25		II
Hg		$Hg_2(ClO_4)_2$: 3 · 10^{-4} up to 1.4 · 10^{-3} M; $HClO_4$: 5 · 10^{-3} M	23	25		I
Hg		$HClO_4$: 0.1 N	36	25		II
Hg	Hg_2Br_2	KBr: 0.01 up to 0.05 m; KCl: 0.01 up to 0.05 m; KNO_3 : 0.05 down to 0.01 m	1	25± 0.01		I
Hg	Hg_2Cl_2	HCl: (0.456 up to 10.230) · 10^{-2} m	18	25± 0.03		I

Comparison electrode	Liquid junction	Electrode reaction	Standard value	Uncertainty	Temperature Coefficient	Notes	Reference	Electrode Reference Number
			V	mV	µV/°C			
d			f		g	h		
		$AgMnO_4 + e^- \rightleftharpoons Ag + MnO_4^-$	0.2055		-1800	1	14	32
1		$AgSCN + e^- \rightleftharpoons Ag + SCN^-$	0.08944				15	33
4		$AgSeCN + c^- \rightleftharpoons Ag + SeCN^-$	-0.0288	± 0.2	-94	2	16	34
1		Ag acetate + $e^- \rightleftharpoons$ Ag + acetate$^-$	0.4266	± 0.6			17	35
4		$Au(Cl)_2^- + e^- \rightleftharpoons Au + 2Cl^-$	1.1575	± 0.5			18	36
39		$[Au(NH_3)_2]^+ + e^- \rightleftharpoons Au + 2NH_3$	0.563	± 6		3,4	19	37
39		$[Au(NH_3)_4]^{3+} + 3e^- \rightleftharpoons Au + 4NH_3$	0.325	± 3		3,4	19	38
39		$[Au(NH_3)_4]^{3+} + 2e^- \rightleftharpoons [Au(NH_3)_2]^+ + 2NH_3$	0.206	± 3		3,4	19	39
3		$[Au(S_2O_3)_2]^{3-} + e^- \rightleftharpoons Au + 2S_2O_3^{2-}$	0.153				20	40
		$Ba^+ + e^- \rightleftharpoons Ba$	-4.20				21	41
		$Ba^{2+} + 2e^- \rightleftharpoons Ba$	-3.040				22	42
		$Ba^{2+} + 2e^- \rightleftharpoons Ba$	-2.930				22	43
		$Ba^{2+} + 2e^- \rightleftharpoons Ba$	-2.912				22	44
		$Ba^{2+} + 2e^- \rightleftharpoons Ba$	-2.870				22	45
4		$Zn^{2+} + 2e^- \rightleftharpoons Zn$	-0.76194	± 0.15			23	46
22		$Zn(OH)_4^{2-} + 2e^- \rightleftharpoons Zn + 4OH^-$	-1.205			4	24	47
4		$Cd^{2+} + 2e^- \rightleftharpoons Cd$	-0.412				25	48
1		$Hg_2^{2+} + 2e^- \rightleftharpoons 2Hg$	0.7889	± 0.2	-514		26	49
7		$Hg_2^{2+} + 2e^- \rightleftharpoons 2Hg$	0.460	± 5		5	27	50
40		$Hg_2^{2+} + 2e^- \rightleftharpoons 2Hg$	0.069	± 1			28	51
40		$Hg^{2+} + 2e^- \rightleftharpoons Hg$	0.065	± 2			29	52
7		$2Hg^{2+} + 2e^- \rightleftharpoons Hg_2^{2+}$	0.614	± 5			30	53
40		$2Hg^{2+} + 2e^- \rightleftharpoons Hg_2^{2+}$	0.061	± 5			28	54
3		$Hg_2Br_2 + 2e^- \rightleftharpoons 2Hg + 2Br^-$	0.13921	±0.04	-0.156		29	55
1		$Hg_2Cl_2 + 2e^- \rightleftharpoons 2Hg + 2Cl^-$	0.2453	± 0.2			30	56

Electronically conducting phase	Intermediate species	Composition of the solution	Solvent	Temperature °C	Pressure Torr	Measuring method
a	b	c	d		e	d
Hg	Hg_2SO_4	H_2SO_4 : 0.008 up to 0.032 m	1	25 ± 0.05		I
Hg	$Hg_2(acetate)_2$	Na acetate : (12.023up to 29.666) · 10^{-3} m; acetic acid : (120.49 up to 297.30) · 10^{-3} m	1	25± 0.05		I
Hg	$Hg_2(benzoate)_2$	Benzoic acid	18	25		I
Hg	$Hg_2(formate)_2$	Formic acid	1	25± 0.1		I
Hg	$Hg_2(picrate)_2$	Picric acid	1	25		I
Tl(Hg)		$TlClO_4$: 0.0040 up to 0.0075	29	25± 0.2		I
Tl(Hg)	TlCl	HCl : 0.004220 up to 0.01546 m	37	24		I
Tl	TlCl	HCl : 0.1374 up to 0.7220 M	20	30		I
Pb(Hg)		$Pb(ClO_4)_2$; $HClO_4$	1	25		I
Pb	$PbCl_2$	HCl: 0.1101 up to 0.8926 M	20	30		I
Pt	HNO_3 , HNO_2	HNO_3 : 2.5 up to 11.4 M; $NaNO_2$	1	25		I
Sb-coated Pt		$SbOClO_4$: (6.244 up to 27.873) · 10^{-4} M; $HClO_4$: 0.3520 up to 2.4643 M	1	25		I
Sb-coated Pt	Sb_4O_6	$SbOClO_4$: (6.244 up to 27.873) · 10^{-4} M; $HClO_4$: 0.3520 up to 2.4643 M	1	25		I
Ind.		$K_2S_4O_6$, $K_2S_2O_3$	1	25		V
Se			1	25	760	IV
Ind.			1	25	760	IV
Ind.	SeO_2		1	25	760	IV
Ind.			1	25	760	IV
Pt	Br_2 (in sol.)	KBr: 0.005 up to 0.1 M; Br_2 : 2 · 10^{-4} up to 6 · 10^{-4} M; KCl : 0.1 M; acetic acid : up to 90 %	1	25		I
Pt	I_2 (s)	HIO_3 : 10^{-3} up to 5 · 10^{-2} M; $HClO_4$: 0.1 M	1	25 ± 0.1		I
Ind.		$Na_3[Fe(CN)_5NH_3]$ + $Na_2[Fe(CN)_5NH_3]$: 0.005 up to 0.016 M; H_2SO_4 : 1.0 up to 5.0 N	1	25 ± 0.1		I
Pt		HCl : 10^{-3} M; $[Co(phenanthroline)_3]^{2+}$ + $[Co(phenanthroline)_3]^{3+}$: 10^{-4} M	1	25± 0.1		I
Ind.			1	25		V
Ind.			1	25		V
Ind.			1	25		V
Ind.			1	25		V
Ind.			1	25		V
Ind.			1	25		V
Ind.			1	25		V

Comparison electrode	Liquid junction	Electrode reaction	Standard value V	Uncertainty mV	Temperature Coefficient μV/°C	Notes	Reference	Electrode Reference Number
d			f		g	h		
33		$Hg_2SO_4 + 2e^- \rightleftharpoons 2Hg + SO_4^{2-}$	0.6135		-8500	6	31	57
33		$Hg_2(acetate)_2 + 2e^- \rightleftharpoons 2Hg + 2\ acetate^-$	0.5111	± 0.1	-600		32	58
33		$Hg_2(benzoate)_2 + 2e^- \rightleftharpoons 2Hg + 2\ benzoate^-$	0.2535	± 0.5			33	59
3, 27, 33		$Hg_2(formate)_2 + 2e^- \rightleftharpoons 2Hg + 2\ formate^-$	0.566	± 3	167		34	60
35		$Hg_2(picrate)_2 + 2e^- \rightleftharpoons 2Hg + 2\ picrate^-$	0.4924				35	61
34		$Tl^+ + e^- \rightleftharpoons Tl$	-0.3817		191		36	62
1		$TlCl + e^- \rightleftharpoons Tl + Cl^-$	-0.973	± 10			37	63
1		$TlCl + e^- \rightleftharpoons Tl + Cl^-$	-0.5994	± 11			38	64
1		$Pb^{2+} + 2e^- \rightleftharpoons Pb$	-0.1237	± 0.6	-438		39	65
1		$PbCl_2 + 2e^- \rightleftharpoons Pb + 2Cl^-$	-0.3368	± 20			40	66
3		$HNO_3 + 2H^+ + 2e^- \rightleftharpoons HNO_2 + H_2O$	0.945				41	67
1		$SbO^+ + 2H^+ + 3e^- \rightleftharpoons Sb + H_2O$	0.2040	± 0.3	-360		42	68
1		$Sb_4O_6 + 12H^+ + 12e^- \rightleftharpoons 4Sb$	0.1504	± 0.5	-360		42	69
		$S_4O_6^{2-} + 2e^- \rightleftharpoons 2S_2O_3^{2-}$	0.169	± 10		7	43	70
		$Se_8^{2+} + 2e^- \rightleftharpoons 8Se$	-0.8				44	71
		$2Se_4^{2+} + 2e^- \rightleftharpoons Se_8^{2+}$	-1.0				44	72
		$4SeO_2 + 16H^+ + 14e^- \rightleftharpoons Se_4^{2+} + 8H_2O$	-0.72				44	73
		$ClO_4^{\cdot} + e^- \rightleftharpoons ClO_4^-$	2.75	± 30			45	74
35		$Br_2 + 2e^- \rightleftharpoons 2Br^-$	1.078	± 1			46	75
1		$2IO_3^- + 12H^+ + 10e^- \rightleftharpoons I_2 + 6H_2O$	1.1942	± 0.2	-300	8	47	76
		$[Fe(CN)_5NH_3]^{2-} + e^- \rightleftharpoons [Fe(CN)_5NH_3]^{3-}$	-0.378			9	48	77
3		$[Co(phenanthroline)_3]^{3+} + e^- \rightleftharpoons [Co(phenanthroline)_3]^{2+}$	0.419	± 1			49	78
		$Pr^{3+} + e^- \rightleftharpoons Pr^{2+}$	1.8	± 200		10	50	79
		$Nd^{3+} + e^- \rightleftharpoons Nd^{2+}$	1.8	± 200		10	50	80
		$Tb^{3+} + e^- \rightleftharpoons Tb^{2+}$	1.6	± 200		10	50	81
		$Dy^{3+} + e^- \rightleftharpoons Dy^{2+}$	1.7	± 200		10	50	82
		$Ho^{3+} + e^- \rightleftharpoons Ho^{2+}$	1.7	± 200		10	50	83
		$Er^{3+} + e^- \rightleftharpoons Er^{2+}$	1.6	± 200		10	50	84
		$Tm^{3+} + e^- \rightleftharpoons Tm^{2+}$	2.0	± 200		10	50	85

Standard value V	Solvent	Electrode Reference Number
	Li	
-3.097	13	1
	Na	
-2.787	13	2
-2.653	29	3
	K	
-2.998	13	4
	Rb	
-3.70	44	13
-3.55	38	6
-3.50	42	11
-3.46	41	9
-3.32	39	7
-3.15	40	8
-3.13	27	18
-3.12	15	10
-3.06	43	12
-3.06	45	14
-2.982	29	5
-2.96	20	15
-2.95	22	16
-2.95	46	19
-2.92	24	17
	Cs	
-3.012	29	20
	Cu	
0.3352	1	21
	Ag	
-0.0288	1	34
0.08944	1	33
0.2149	1	31
0.22238	1	26
0.49	2	22
-0.0103	7	27
0.0233	13	28
-0.610	17	29
-0.0996	18	30
0.0965	18	24
0.2055	18	32
0.4266	18	35
0.4997	18	25
0.06014	35	23
	Au	
0.153	1	40
0.206	1	39
0.325	1	38
0.563	1	37
1.1575	1	36
	Ba	
-4.20	1	41
-2.930	11	13
-2.912	12	44
-3.040	13	42
-2.870	15	45
	Zn	
-1.205	1	47
-0.76194	1	46
	Cd	
-0.412	18	48
	Hg	
0.13921	1	55
0.4924	1	61
0.5111	1	58
0.566	1	60
0.6135	1	57
0.7889	1	49
0.2453	18	56
0.2535	18	59
0.460	23	50
0.614	23	53
0.061	36	54
0.065	36	52
0.069	36	51
	Tl	
-0.973	37	63
-0.5994	20	64
-0.3817	29	62
	Pb	
-0.1237	1	65
-0.3368	20	66
	N	
0.945	1	67
	Sb	
0.1504	1	69
0.2040	1	68
	S	
0.169	1	70
-1.0	1	72
-0.8	1	71
-0.72	1	73
	Cl	
2.75	1	74
	Br	
1.078	1	75
	I	
1.1942	1	76
	Fe	
-0.378	1	77
	Co	
0.419	1	78
	Pr	
1.8	1	79
	Nd	
1.8	1	80

Standard value V	Solvent	Electrode Reference Number	Standard value V	Solvent	Electrode Reference Number	Standard value V	Solvent	Electrode Reference Number
	Tb							
1.6	1	81						
	Dy							
1.7	1	82						
	Ho							
1.7	1	83						
	Er							
1.6	1	84						
	Tm							
2.0	1	85						

NOTES: Appendix

1) Solubility data
2) Liquid junction electric tension eliminated by the method of Owen and King (J. Am. Chem. Soc., 63 (1941), 1711
3) Potentiometric titrations with hydrazine
4) Liquid junction electric tension negligible
5) 0.1 M Et_4NClO_4 liquid junction
6) K_2SO_4 liquid junction
7) Heat capacity data
8) Correction for liquid junction electric tension applied
9) Potentiometric titrations using $KMnO_4$, $Ce(SO_4)_2$ or $K_2Cr_2O_7$ as the titrants
10) Pulse radiolysis measurements

BIBLIOGRAPHY: Appendix

1) K.K. Kundu, A.K. Rakshit, M.N. Das, Electrochim. Acta, 17 (1972), 1921 - 1937
2) M. L'Her, J. Courtot-Coupez, Bull. Soc. Chim. France, 1972, 3645 - 3653
3) N. Tanaka, T. Ogata, Inorg. Nucl. Chem. Lett., 10 (1974), 511 - 515
4) F. Grønlund, S. Noer, J. Electrochem. Soc., 121 (1974), 25 - 29
5) O.R. Brown, S.A. Thorton, J. Chem. Soc., Faraday Trans. I, 1974, 1009 - 1017
6) M.H. Lietzke, T.J. Lemmonds, J. Inorg. Nucl. Chem., 36 (1974), 2299 - 2301
7) U.N. Dash, B. Nayak, Austr. J. Chem., 28 (1975), 1657 - 1661
8) U.N. Dash, B. Nayak, Thermochim. Acta, 11 (1975), 17 - 24
9) N.P. Komar, A.Z. Kaftanov, B.A. Dunai, Elektrokhimiya, 8 (1972), 1177 - 1179
10) K.K. Kundu, D. Jana, M.N. Das, J. Chem. Eng. Data, 19 (1974), 329 - 333
11) V.V. Aleksandrov, B.N. Bezpalyi, Zh. Fiz. Khim., 49 (1975), 1314 - 1316
12) Yu. A. Nevskaya, P.I. Korotkova, T.N. Sumarova, Elektrokhimiya, 11 (1975), 1660 - 1664
13) U.N. Dash, J. Mohanty, Thermochim. Acta, 12 (1975), 189 - 196
14) U.N. Dash, Thermochim. Acta, 12 (1975), 197 - 202
15) S.C. Lal, B. Prasad, Indian J. Chem., 13 (1975), 372 - 374
16) R.C. Das, G. Sahu, D. Satynarayana, S.N. Misra, Electrochim. Acta, 19 (1974), 887 - 890

BIBLIOGRAPHY: Appendix

17) B. Nayak, U.N. Dash, J. Electroanal. Chem., 41 (1973), 323 - 328

18) I.G. Murgulescu, I. Vartires, Rev. Roum. Chim., 13 (1968), 1397 - 1407

19) L.H. Skibsted, J. Bjerrum, Acta Chem. Scand. A, 28(1974), 764 - 770

20) J. Pouradier, M.C. Gadet, J. Chim. Phys. Physico-Chim. Biol., 66 (1969), 109 - 112

21) S.N. Pobedinskii, G.A. Krestov, L.L. Kuzmin, Izv. Vyss. Uchebn. Zaved. Khim. Khim. Tekhnol., 6 (1963), 768 - 773

22) N.E. Komutov, Zh. Fiz. Khim., 42 (1968), 2223 - 2229

23) H.S. Dunsmore, R. Paterson, J. Chem. Soc. Faraday Trans. I, 72 (1970), 495 - 503

24) D.P. Boden, R.B. Wylie, V.J. Spera, J. Electrochem. Soc., 118 (1971), 1298 - 1301

25) A.W. Andrews, D.A. Armitage, R.W. C. Broadbank, K.W. Morcom, B.L. Muju, Trans. Faraday Soc., 67 (1971), 128 - 131

26) Kuan Pan, Jin-Jyi Chang, Shu-Ching Hsin, J. Chin. Chem. Soc., 18 (1971), 1 - 4

27) J.F. Coetzee, J.J. Campion, D.R. Liberman, Anal. Chem., 45 (1973), 343 - 347

28) M. Bréant, C. Buisson, J. Electroanal. Chem., 24 (1970), 145 - 154

29) W. Leuschke, K. Schwabe, J. Electroanal. Chem., 25 (1970), 219 - 226

30) B. Nayak, D.K. Saku, Electrochim. Acta, 16 (1971), 1757 - 1761

31) L. Sharma, B. Prasad, J. Indian Chem. Soc., 47 (1970), 379 - 383

32) B.K. Choudhary, B. Prasad, Indian J. Chem., 11 (1973), 931 - 933

33) U.N. Dash, Austr. J. Chem., 28 (1975), 1653 - 1656

34) N.I. Ostannii, L.A. Zharkova, B.V. Erofeev, G.D. Kazakova, Zh. Fiz. Khim., 48(1974), 358 - 361

35) A.K. Covington, K.V. Srinivasan, J. Chem. Thermodyn., 3 (1971), 795 - 800

36) D.P. Boden, L.M. Mukherjee, Electrochim. Acta, 18(1973), 781 - 787

37) C.H. Contreras-Ortega, P.A. Rock, J. Electrochem. Soc., 121(1974), 1048 - 1050

38) F.L. Bates, Y.T. Nee, J. Electrochem. Soc., 121 (1974), 79 - 82

39) V.P. Vasil'ev, S.R. Glavin, Elektrokhimiya, 7 (1971), 1395

40) F.L. Bates, Y.T. Nee, J. Electrochem. Soc., 121 (1974), 79 - 82

41) N.I. Alekseeva, Ya. D. Zymmer, V.A. Nikol'skii, Elektrokhimiya, 7 (1971), 873 - 879

42) V.P. Vasil'ev, V.I. Shorokhova, Elektrokhimiya, 8 (1972), 185 - 190

43) J.W. Cobble, H.P. Stephens, I.R. McKinnon, E.F. Westrum,Jr., Inorg. Chem., 11 (1976), 1669-1674

44) A.H. Stiller, Dissert. Abstr. B, 34 (1973), 1418 - 1419

45) I.V. Shimonis, Elektrokhimiya, 9 (1973), 1787 - 1789

46) L.G. Lavrenova, T.V. Zegzhda, V.M. Shul'man, Elektrokhimiya, 7 (1971), 83 - 86

47) R.D. Spitz, H.A. Liefbafsky, J. Electrochem. Soc., 122 (1975), 363 - 367

48) W.U. Malik, R. Bembi, J.K. Dwivedi, Indian J. Chem., 13 (1975), 589 - 591

49) L.G. Lavrenova, T.V. Zegzhda, L.M. Shul'man, Elektrokhimiya, 8 (1972), 1153 - 1155

50) Y. Tendler, M. Faraggi, J. Chem. Phys., 57 (1972), 1358 - 1359

Index to Tables

Actinides, 398–408
Actinium, 398
Aluminum, 120–123
Americium, 402
Antimony, 206–209, 414
Arsenic, 202–205
Astatine, 298–301

Barium, 84–89, 412
Berkelium, 404
Beryllium, 62–65
Bismuth, 210–213
Boron, 114–119
Bromine, 284–289

Cadmium, 100–105, 412
Calcium, 70–78
Californium, 404
Carbon, 146–151
Cerium, 386
Cesium, 18–21, 410
Chlorine, 278–283
Chromium, 258–263
Cobalt, 336–343
Copper, 22–35, 410
Curium, 404

Dysprosium, 390

Erbium, 392
Europium, 390

Fermium, 404
Fluorine, 274–277

Gadolinium, 390
Gallium, 124–127
Germanium, 156–159
Gold, 54–59, 412
Group I, 2–59, 410–412
Group II, 62–112, 412–414
Group III, 118–144, 414
Group IV, 146–186, 414
Group V, 188–226, 414
Group VI, 228–271, 414
Group VII, 274–317
Group VIII, 320–383, 414

Hafnium, 184–186
Holmium, 392

Indium, 128–131
Iodine, 290–297
Iridium, 372–375
Iron, 320–335

Lanthanides, 386–396
Lanthanum, 386
Lead, 164–173, 414
Lithium, 2–5, 410
Lutetium, 392

Magnesium, 66–69
Manganese, 302–309
Mercury, 106–112, 412–414
Molybdenum, 264–267

Neodymium, 388
Neptunium, 400
Nickel, 344–349
Niobium, 220–223
Nitrogen, 188–193
Nobelium, 404

Osmium, 366–371
Oxygen, 228–231

Palladium, 360–365
Phosphorus, 194–201
Platinum, 376–383, 414
Plutonium, 402
Polonium, 254–257
Potassium, 10–13, 410
Praseodymium, 388
Promethium, 390
Protoactinium, 398

Radium, 90–93
Rhenium, 314–317
Rhodium, 356–359
Rubidium, 14–17, 410
Ruthenium, 350–355

Samarium, 390
Scandium, 138–141
Selenium, 244–247, 414
Silicon, 152–155
Silver, 36–53, 410–412
Sodium, 6–9, 410
Strontium, 80–83
Sulfur, 232–243

Tantalum, 224–226
Technetium, 310–313
Tellurium, 248–253
Terbium, 390
Thallium, 132–137, 414
Thorium, 398
Thulium, 392
Tin, 160–163
Titanium, 174–179
Tungsten, 268–271

Uranium, 398

Vanadium, 214–219

Ytterbium, 392
Yttrium, 142–144

Zinc, 94–99, 412
Zirconium, 180–183